ANALYTICAL NEBULIZERS

ANALYTICAL NEBULIZERS
Fundamentals and Applications

Edited by

ANTONIO CANALS
Department of Analytical Chemistry and Food Science and University Institute of Materials, Faculty of Science, University of Alicante, Alicante, Spain

MIGUEL ÁNGEL AGUIRRE
Department of Analytical Chemistry and Food Science and University Institute of Materials, Faculty of Science, University of Alicante, Alicante, Spain

MAZAHER AHMADI
Department of Analytical Chemistry, Faculty of Chemistry, Bu-Ali Sina University, Hamedan, Iran

ELSEVIER

Elsevier
Radarweg 29, PO Box 211, 1000 AE Amsterdam, Netherlands
The Boulevard, Langford Lane, Kidlington, Oxford OX5 1GB, United Kingdom
50 Hampshire Street, 5th Floor, Cambridge, MA 02139, United States

ISBN: 978-0-323-91181-8

For information on all Elsevier publications
visit our website at https://www.elsevier.com/books-and-journals

Publisher: Susan Dennis
Acquisitions Editor: Charlotte Rowley
Editorial Project Manager: Catherine Costello
Production Project Manager: Surya Narayanan Jayachandran
Cover Designer: Matthew Limbert

Typeset by STRAIVE, India

Contents

8. Electrospray

Alfonso M. Gañán-Calvo, José M. López-Herrera, and Luis B. Modesto-López

9. Ultrasonic nebulizers

Krzysztof Jankowski

3

Applications

10. Applications in single-particle inductively coupled plasma-mass spectrometry

Deyana Georgieva

11. Application of online sample introduction systems for single-cell analysis with ICP-MS

Roberto Álvarez-Fernández García, Mario Corte-Rodríguez, Paula García-Cancela, Jörg Bettmer, and María Montes-Bayón

12. Nonconventional applications of nebulizers: Nanomaterials synthesis

Zahra Amouzegar, Fariba Mollarasouli, Sepideh Asadi, Ali Fathi, Mazaher Ahmadi, Abbas Afkhami, and Tayyebeh Madrakian

13. Nebulizer-based detectors for liquid chromatography

Tadesse Haile Fereja and Guobao Xu

Contributors

Abbas Afkhami Department of Analytical Chemistry, Faculty of Chemistry, Bu-Ali Sina University; D-8 International University, Hamedan, Iran

Miguel Ángel Aguirre Department of Analytical Chemistry and Food Science and University Institute of Materials, Faculty of Science, University of Alicante, Alicante, Spain

Mazaher Ahmadi Department of Analytical Chemistry, Faculty of Chemistry, Bu-Ali Sina University, Hamedan; Autophagy Research Center, Shiraz University of Medical Sciences, Shiraz, Iran

Roberto Álvarez-Fernández García Department of Physical and Analytical Chemistry, Faculty of Chemistry and Health Research Institute of Asturias (ISPA), University of Oviedo, Oviedo, Spain

Zahra Amouzegar Department of Analytical Chemistry, Faculty of Chemistry, Bu-Ali Sina University, Hamedan, Iran

Sepideh Asadi Department of Analytical Chemistry, Faculty of Chemistry, Bu-Ali Sina University, Hamedan, Iran

Sonia Bahrani Nanotechnology Department, Borjobaru Fars Company, Fars Science and Technology Park; Health Policy Research Center, Health Institute, Shiraz University of Medical Sciences, Shiraz, Iran

Jörg Bettmer Department of Physical and Analytical Chemistry, Faculty of Chemistry and Health Research Institute of Asturias (ISPA), University of Oviedo, Oviedo, Spain

Antonio Canals Department of Analytical Chemistry and Food Science and University Institute of Materials, Faculty of Science, University of Alicante, Alicante, Spain

Mario Corte-Rodríguez Department of Physical and Analytical Chemistry, Faculty of Chemistry and Health Research Institute of Asturias (ISPA), University of Oviedo, Oviedo, Spain

Marta Costas-Rodriguez Department of Green Chemistry and Technology, Ghent University; Department of Chemistry, Ghent University, Campus Sterre, Ghent, Belgium

Jefferson Santos de Gois Department of Analytical Chemistry, Rio de Janeiro State University, Rio de Janeiro, Brazil

Ali Fathi Department of Analytical Chemistry, Faculty of Chemistry, Bu-Ali Sina University, Hamedan, Iran

Tadesse Haile Fereja State Key Laboratory of Electroanalytical Chemistry, Changchun Institute of Applied Chemistry, Chinese Academy of Sciences, Changchun, Jilin, People's Republic of China; Department of Pharmacy, College of Medicine and Health Sciences, Ambo University, Ambo, Ethiopia

Alfonso M. Gañán-Calvo Departamento de Ingeniería Aeroespacial y Mecánica de Fluidos, ETSI; Laboratory of Engineering for Energy and Environmental Sustainability, Universidad de Sevilla, Sevilla, Spain

Paula García-Cancela Department of Physical and Analytical Chemistry, Faculty of Chemistry and Health Research Institute of Asturias (ISPA), University of Oviedo, Oviedo, Spain

Deyana Georgieva Department of Analytical Chemistry and Computer Chemistry, University of Plovdiv "Paisii Hilendarski", Plovdiv, Bulgaria

Krzysztof Jankowski Faculty of Chemistry, Department of Analytical Chemistry, Warsaw University of Technology, Warsaw, Poland

José M. López-Herrera Departamento de Ingeniería Aeroespacial y Mecánica de Fluidos, ETSI, Universidad de Sevilla, Sevilla, Spain

Tayyebeh Madrakian Department of Analytical Chemistry, Faculty of Chemistry, Bu-Ali Sina University, Hamedan; Autophagy Research Center, Shiraz University of Medical Sciences, Shiraz, Iran

Luis B. Modesto-López Departamento de Ingeniería Aeroespacial y Mecánica de Fluidos, ETSI, Universidad de Sevilla, Sevilla, Spain

Fariba Mollarasouli Department of Chemistry, Yasouj University, Yasouj, Iran

María Montes-Bayón Department of Physical and Analytical Chemistry, Faculty of Chemistry and Health Research Institute of Asturias (ISPA), University of Oviedo, Oviedo, Spain

Mariusz Ślachciński Poznan University of Technology, Faculty of Chemical Technology, Poznan, Poland

Guobao Xu State Key Laboratory of Electroanalytical Chemistry, Changchun Institute of Applied Chemistry, Chinese Academy of Sciences, Changchun, Jilin; University of Science and Technology of China, Hefei, Anhui, People's Republic of China

Acknowledgments

We acknowledge all the people who helped with the cover images for this book:

- Electrospray in Taylor cone-jet mode (courtesy of Alfonso M. Gañán-Calvo)
- Ultrasonic nebulizer (courtesy of Mariusz Ślachciński)
- MultiNeb by Ingeniatrics Tecnologías S.L., Universidad de Alicante, Universidad de Sevilla (courtesy of Alfonso M. Gañán-Calvo)

SECTION 1

Fundamentals

CHAPTER

1

An introduction to nebulizers

Mazaher Ahmadi[a], *Miguel Ángel Aguirre*[b], *and Antonio Canals*[b]

[a]Department of Analytical Chemistry, Faculty of Chemistry, Bu-Ali Sina University, Hamedan, Iran, [b]Department of Analytical Chemistry and Food Science and University Institute of Materials, Faculty of Science, University of Alicante, Alicante, Spain

1.1 Analytical nebulizers

In analytical plasma-based techniques, sample introduction is one of the most important steps. There are a variety of sample introduction options available, including nebulizers, electrothermal vaporization, chemical vapor generation, and laser ablation devices. Because of their readiness for use, ease of coupling with autosamplers, and the ability to provide accurate results when used appropriately, nebulizers connected to a spray chamber are the most extensively utilized among other sample introduction systems.

When looking for nebulizers on the internet, the reader must be careful because the aerosol generation can be referred to as air blast atomization in the medical literature, and the nebulizers can be referred to as atomizers, so proper terminology must be used when the subject is for analytical purposes, where the term atomization is a well-known phenomenon in atomic spectrometry (i.e., the conversion of volatilized analyte into free atoms). In analytical chemistry, the word nebulizer refers to a device that may transform the bulk liquid into an aerosol. The aerosol is the most popular way to present the sample to the detector. Particularly, in plasma-based techniques, the primary purpose of breaking down the liquid sample into small droplets is to increase the surface area of the sample so that both sample transported is increased and the processes of desolvation, vaporization, atomization, excitation, and/or ionization can run smoothly into the plasma.

Analytical Nebulizers
https://doi.org/10.1016/B978-0-323-91181-8.00001-1

A large number of commercial analytical techniques are based on nebulizers for sample introduction systems, necessitating a better grasp of the fundamental ideas and applications of this critical phenomenon. However, despite its ease of implementation, the nebulizer is considered as a black box, and it may not be touched for long periods of time for cleaning or replacement. As a result, in most circumstances, the selection and use of a correct nebulizer for a specific application are crucial, and it is critical to comprehend all ideas and challenges linked to nebulizer selection. The right nebulizer may improve sensitivity, precision, minimize interferences, shorten washout times, eliminate the need for preceding sample preparation stages, and even allow for the analysis of micro-quantities of materials without the need for special equipment.

1.2 Types of analytical nebulizers

Aerosol generation requires the contribution of a certain amount of energy to the liquid mass in order to overcome the cohesive forces of the liquid (e.g., surface tension and viscosity), dividing it into droplets. Depending on the energy source, the most common nebulizers used in chemical analysis can be classified as (i) pneumatic nebulizer, (ii) hydraulic nebulizer, (iii) ultrasonic nebulizer, and (iv) electrostatic nebulizer. Additionally, pneumatic nebulizers can be categorized as either concentric or nonconcentric nebulizers (Fig. 1.1). Other criteria such as

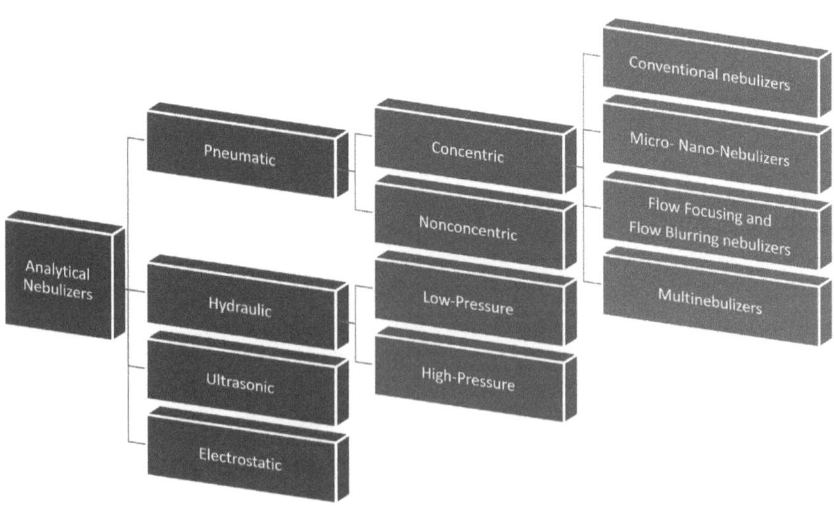

FIG. 1.1 Classification of analytical nebulizers based on the energy source. *No permission required.*

solutions containing high amounts of total dissolved solid (TDS), hydrofluoric (HF) acid, organic solvents, particles (i.e., slurry solutions), when the amount of sample is limited, may be used to create more categories (Fig. 1.2).

Pneumatic nebulizers generate aerosols with a high-velocity gas stream. This kind of nebulizer combines a variety of nebulizer geometries and sizes, as well as different liquid-gas interaction efficiencies, and therefore, different aerosol characteristics. The distinction of pneumatic concentric and nonconcentric nebulizers is the initial categorization; after that, all varieties are associated with size, materials, and geometry. These distinctions allow each nebulizer type to have its own set of uses and features. For example, when the sample is limited, it is notoriously difficult to analyze; in these circumstances, micro- and nanonebulizers are the best alternatives. Another example is the cross-flow nebulizer for the analysis of slurry samples. In cross-flow nebulizers, the interaction between liquid and gas streams is perpendicular and the liquid inlet is positioned vertically and the gas inlet is positioned horizontally (i.e., at an angle of 90°). This configuration is ideal to analyze samples containing high amount of solid particles (i.e., salty solutions and suspensions) and to prevent the possible tip blockage.

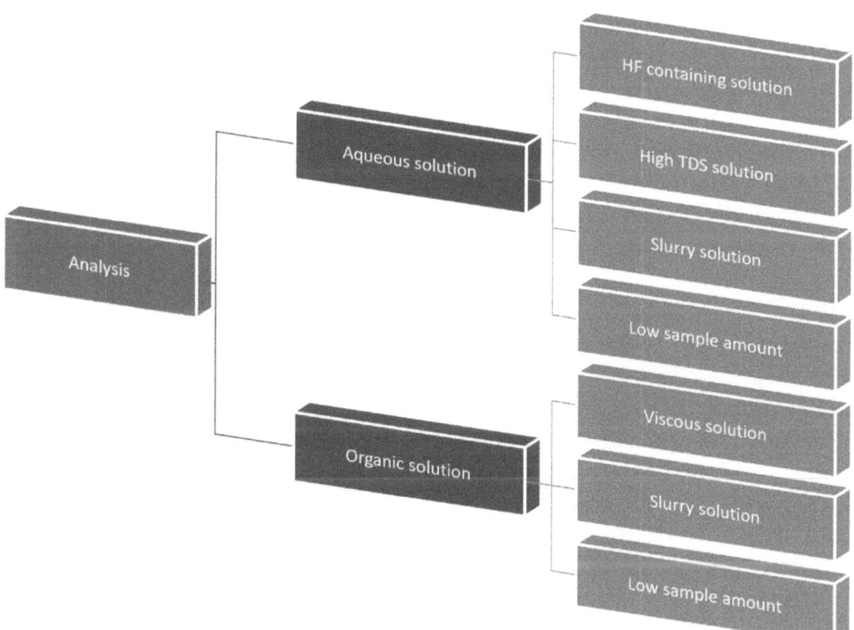

FIG. 1.2 Different kind of analysis of liquid samples to be performed by analytical nebulizers. *No permission required.*

1.3 Analytical nebulizers for sample introduction

Nebulizers have a wide range of applications in science and technology; for instance, converting a liquid solution into an aerosol is a critical step in a variety of applications, including combustion systems, industrial operations, pharmacy, and agriculture, among others. In chemical analysis, the main aim of nebulizers is to create an aerosol to prepare the liquid sample for analysis. In both spectrometry and chromatography, nebulizers are employed. In spectrometry, the liquid sample is introduced as an aerosol in the atomization/excitation/ionization source using the analytical nebulizer. On the other hand, they are also commonly used as a soft ionization source. The hyphenation of chromatography techniques to mass spectrometers has recently been extensively investigated, and novel soft ionization nebulizers have attracted the interest of the miniaturization community. Thermospray ionization, sonic spray ionization, electrospray ionization, and the hybrid electrosonic spray ionization approach are among the spray ionization techniques developed for usage in liquid chromatography/capillary electrophoresis coupled mass spectrometry interfaces.

The growing importance of nanoparticles and nanostructured materials in human life and industry has prompted the creation of a new branch of analytical science called Analytical Nanoscience and Nanotechnology, in which nanoparticles and nanostructured materials are used as analytical tools or as the subject of analysis (Valcárcel, Simonet, & Cárdenas, 2008). The purpose of the study as the subject of analysis is to obtain information on several properties such as chemical composition, morphology, shape, size and size distribution, structure, and concentration. With the increased usage of plasma and laser-based techniques in the previous decade, several analytical methods have been created to address the demands for a range of information linked to nanoscale materials. Among the latter, single-particle inductively coupled plasma mass spectrometry (SP-ICP-MS) is cited as a potent and adaptable nanoparticle characterization approach (Laborda, Bolea, & Jiménez-Lamana, 2014). In SP-ICP-MS, a continuous nebulization of a dilute suspension of nanoparticles into a typical ICP-MS technique is used in the procedure. The primary goal of the sample introduction system is to guarantee that a single nanoparticle transported by aerosol droplets enters constantly and individually into the plasma. In this case, the nebulization mechanism and the quality of the resulting aerosol are critical.

Individual cell heterogeneity is seen in tissues and organisms, but it may also be seen in groups of isogenic cells cultivated under similar circumstances. In reality, variations in genes, transcripts, and proteins between cells can cause a variety of illnesses, including cancer, neurological problems, and developmental abnormalities. Similarly, because of their importance in activities such as cell respiration, growth, enzymatic reactions, cell proliferation, and immunological function, metal control is closely regulated by cells. Changes in this homeostasis have been linked to the onset of serious pathological disorders. As a result, measuring cell-to-cell transitory variations in metal ion concentrations might lead to the creation of "signatures" that describe how cells manage metal ion levels and how they affect cellular activity (Yu, He, Chen, & Hu, 2020). The extensive use of metallodrugs to treat a variety of ailments, on the other hand, implies that these metallic structures are incorporated into cells. The intercellular variation in the absorption of these structures can be crucial in determining the onset of drug resistance processes, for example. The development of analytical/biotechnological tools that enable the assessment of heterogeneity between cells and within populations that were previously considered to be homogenous is an ongoing field of inquiry to investigate all of these possibilities. As a result, the development of single-cell ICP-MS (SC-ICP-MS) has opened a new area of research which allows the rapid detection and analysis of cells in a variety of matrices and applications.

1.4 Analytical nebulizers beyond sample introduction

1.4.1 Aerosol-assisted nanomaterial synthesis procedures

Nanomaterials can be synthesized using liquid- or gas-phase aerosol-assisted nanomaterial synthesis procedures (Lu et al., 1999). By nebulizing solvents of a particular chemical formula, different methods such as pneumatic nebulization, ultrasonic nebulization, electrospray nebulization, and spray pyrolysis can be used to produce aerosol particles. The role of nebulization in determining particle size is critical. In the nebulization process, the original precursor is fed into a special nebulizer to generate many microsized droplets. Several types of driving forces, such as electrostatic, gravity, centrifugal, and ultrasonic forces can be used to aid the nebulization operation, depending on the droplet size requirements. In this technique, the liquid feed is converted into small droplets by nebulization. Various nebulizers can be used and fine droplets of different sizes can be acquired. The fundamental theory behind nanoparticle synthesis is that the solvent in the nebulized droplets evaporates causing the droplets to freeze and create nanomaterials. As an aerosol-assisted approach in nanomaterial synthesis, the gas-phase route is based on the generation of the precursor vapor. In this route, the precursor vapor can be a flow of the input gas, evaporated liquid, or sublimated solid. As a result of the chemical reactions in the reactor, supersaturation occurs. It is an important driving force for the nucleation step of nanoparticles. The formed nuclei are the centers of condensation and coagulation. Generally, the process conditions have a significant effect on the size and morphology of the final nanoparticles. Commonly, the developed approach is named based on the reactor used. Therefore, they are called furnace generation, glowing-wire aerosol generation, spark-discharge generation, plasma, and laser ablation. For material synthesis, there is not any monopolization of these reactors, and it is possible to combine them with other nebulizers, probes, or aerosol sources. To produce high-quality nanomaterials, gas-phase approaches need supernumerary considerations relating to desired nanomaterials and reactor type.

1.4.2 Nebulizer-based detectors

Nebulizer-based detectors overcome the limitations presented over conventional detectors used in high-performance liquid chromatography (HPLC). For instance, ultraviolet detection is limited to compounds that possess a chromophore, mass spectrometry to those compounds that ionize, and refractive index detectors are chromatographically inferior as they are extremely sensitive to temperature (Bailey, Plante, Crafts, & Acworth, 2010). In nebulizer-based detectors, the mobile phase is converted into an aerosol using a nebulizer. Aerosol formation necessitates the addition of energy to the liquid mass to overcome the liquid's cohesive forces and divide it into droplets. Once the aerosol is formed, the aerosol characteristics are modified by many processes, including the droplet impact on surfaces within the experimental setup. These losses reduce the number of particles and mass of analyte that reaches the detector, reducing signal levels. However, increasing the volatility of the mobile phase of the HPLC will decrease the level of these losses since the droplets decrease in size due to evaporation, leading to increased signal. Another important characteristic of the nebulizer-based detectors is that the liquid collected during these droplet impact processes can be drained smoothly to prevent the re-nebulization of this waste and also to avoid pulsation of the flow system.

1.5 What this book provides for the readers

This book is presented in three sections discussing fundamentals, designs, and application of analytical nebulizers.

In the first section, fundamental aspects of analytical nebulizers are discussed. This chapter summarizes the role of nebulizers in analytical chemistry, offering a wide range of principles of nebulization and their main and most recent applications in analytical chemistry. This chapter also provides an introduction to nebulizer-assisted nanomaterial synthesis. Chapter 2 presents a general classification of nebulizers, including new designs and recent applications of each type of nebulizer in spectrochemical analysis using plasma-based techniques. This chapter also discusses spray chambers and describes how to evaluate performance by applying simple experiments, and troubleshooting. Flow charts about the selection of nebulizers, performance diagnosis, and troubleshooting are included in this chapter. Chapter 3 presents fundamental and theoretical studies of the aerosol generation of the most common nebulization mechanism used in analytical chemistry. Aerosol generation is of high importance since many applications of these nebulizers extremely depend on the characteristics of their aerosol. Chapter 4 provides the most common techniques employed for aerosol characterization and the most important parameters for nebulizer comparison. This chapter also presents typical examples of diameter distributions of droplets generated by a given nebulizer.

In the second section, nebulizer designs are discussed. Nanonebulizers play a key role in interfacing different instruments. For instance, nanonebulizers are used for interfacing nano-HPLC with ICP-MS, among others. In addition, the advance in nanonebulizer design also allows the development of portable systems and miniaturized mass spectrometers. Chapter 5 shows the evolution of design and the recent applications of nanonebulizers. Multinebulizers as a new kind of nebulizers that recently appeared in the field of spectrochemical analysis are discussed in Chapter 6. This chapter shows the advantages of using the simultaneous nebulization of two or more liquids (both pneumatic and ultrasonic nebulization), the evolution in design (from modified nebulizer/spray chamber to multicapillary nebulizers), and the latest applications (chemical/photochemical vapor generation and online isotope dilution). Discrete droplet generation commonly consists of a piezoelectrically micropump that can produce isolated droplets and they can be varied by utilizing an appropriate pulse generator. Chapter 7 shows the evolution of design and the newest applications. Electrospray nebulizer takes advantage of the large electric field applied since it causes charge separation that generates an aerosol when the repulsion force overcomes surface tension. Chapter 8 shows the physics, operation principles, and scaling laws of electrosprays in a comprehensive review with a special emphasis on the so-called steady Taylor cone-jet mode. For an ultrasonic nebulizer, a liquid sample is pumped onto the surface of a rapidly vibrating solid surface (i.e., piezoelectric transducer) and a checkerboard-like wave pattern appears in the film that forms as the liquid spreads over the surface. Chapter 9 shows the evolution of design and recent applications.

In the last section, the applications of analytical nebulizers are discussed. Chapter 10 discusses the applications of analytical nebulizers in SP-ICP-MS. SP-ICP-MS is becoming an important tool for the characterization of nanoparticles. The method allows determining the size, size distribution, and particle number concentrations of nanoparticles in suspensions after a mere few minutes of measurement. This chapter discusses the instrumental developments of recent years, focusing on the nebulizer point of view. Developments in SC-ICP-MS have increased significantly with essential efforts in the introduction of cell suspensions into the spectrometer by developing nebulizer devices that allow maximum transport efficiencies. Chapter 11 discusses the instrumental developments of recent years, focusing on the application of on-line sample introduction systems for single-cell analysis with ICP-MS. Chapter 12 reviews nonconventional applications of nebulizers such as aerosol-assisted synthesis nanoparticles, ultrasonic nebulization extraction, and nebulizer-based thin films for optoelectronic devices. Finally, Chapter 13 reviews the operating principles, nebulizer technology, and recent applications of the most common detectors for liquid chromatography.

Finally, with the contents of this book, the authors have been tried to give a wide, current, and sufficiently deep vision of both the fundamentals and applications of aerosols in science and technology.

References

Bailey, B., Plante, M., Crafts, C., & Acworth, I. (2010). *Advances in universal detection* (pp. 1–4). A Dionex Company. Application Note.

Laborda, F., Bolea, E., & Jiménez-Lamana, J. (2014). Single particle inductively coupled plasma mass spectrometry: A powerful tool for nanoanalysis. *Analytical Chemistry, 86*(5), 2270–2278. https://doi.org/10.1021/ac402980q.

Lu, Y., Fan, H., Stump, A., Ward, T. L., Rieker, T., & Brinker, C. J. (1999). Aerosol-assisted self-assembly of mesostructured spherical nanoparticles. *Nature, 398*(6724), 223–226. https://doi.org/10.1038/18410.

Valcárcel, M., Simonet, B. M., & Cárdenas, S. (2008). Analytical nanoscience and nanotechnology today and tomorrow. *Analytical and Bioanalytical Chemistry, 391*(5), 1881–1887. https://doi.org/10.1007/s00216-008-2130-9.

Yu, X., He, M., Chen, B., & Hu, B. (2020). Recent advances in single-cell analysis by inductively coupled plasma-mass spectrometry: A review. *Analytica Chimica Acta, 1137*, 191–207. https://doi.org/10.1016/j.aca.2020.07.041.

CHAPTER

2

Types of nebulizers in plasma-based techniques: How to choose an appropriate nebulizer

Jefferson Santos de Gois[a] *and Marta Costas-Rodriguez*[b,c]

[a]Department of Analytical Chemistry, Rio de Janeiro State University, Rio de Janeiro, Brazil, [b]Department of Green Chemistry and Technology, Ghent University, Ghent, Belgium, [c]Department of Chemistry, Ghent University, Campus Sterre, Ghent, Belgium

2.1 Introduction

Sample introduction systems are known as the Achilles' heel for plasma-based techniques, where different configurations are available such as those based on nebulizers, electrothermal vaporization, and laser ablation systems. Among those, nebulizers coupled to a spray chamber are the most widely used due to the readiness for its use; it is easy to couple with autosamplers and provides good and accurate results when applied properly. The conversion of the liquid sample into an aerosol is an important process for different applications, not only analytical purposes, such as combustion systems, medicine, industrial processes, and as we will discuss in this chapter, for the introduction of samples in plasma-based techniques. In this chapter, we will present one overview for the selection of nebulizers for plasma-based techniques.

When searching for nebulizers, the reader must be careful; it is possible to find the aerosol generation in the medical literature as "airblast atomization" and the nebulizers called "atomizers"; therefore, attention must be given to the proper terms when the subject is for analytical purposes, where the term atomization is a well-known phenomenon in atomic spectrometry.

Analytical Nebulizers
https://doi.org/10.1016/B978-0-323-91181-8.00004-7

The term nebulizer, in plasma-based techniques, relies on an apparatus that is able to turn liquid into an aerosol, which is later carried to a spray chamber so that only a homogenous and very thin aerosol reaches the plasma; in some cases, it can be connected directly to the torch for direct injection nebulization (B'Hymer & Caruso, 2000; Meyer, 2013). The main goal of breaking the liquid sample into little droplets is to increase the surface area of the sample so that the processes of solvent vaporization, analyte atomization, excitation, and ionization are efficient (Borkowska-Burnecka, Leśniewicz, & Zyrnicki, 2006; Twyman, 2005).

The commercial instrumentation for inductively coupled plasma techniques is almost exclusively based on nebulizers for sample introduction systems, leading to the need for understanding what are the main concepts and application of this phenomenon that is so important for analysis (Meyer, 2013).

Due to its simplicity in design, sometimes, one can easily make mistakes, and problems that can jeopardize an analysis may arise; therefore, the selection and application of a proper nebulizer for a given application are important for most cases, and it is crucial to understand all concepts and related problems regarding the selection of a nebulizer.

The selection of a proper nebulizer may increase the sensitivity, reduce interferences, reduce washout times, reduce the need for previous sample preparation steps, or even enable the possibility of analyzing microvolumes of samples without special apparatus; also, there are specific nebulizers for applications that involve the coupling with other techniques such as liquid chromatography (Burgener & Makonnen, 2020).

Regarding the materials, nebulizers are available from different materials, such as borosilicate glass, quartz, polytetrafluoroethylene (PTFE), perfluoroalkoxy (PFA), or polyetheretherketone (PEEK); therefore, the analyst can select the friendliest material for each sample matrix (e.g., matrices rich in hydrofluoric acid are not compatible with glass nebulizers) (Burgener & Makonnen, 2020).

Therefore, this chapter aims to provide basic concepts on nebulization as well as an overview of the available configurations at the moment so that the reader can be able to select the proper nebulizer for its application.

2.2 Types of nebulizers for plasma-based techniques

Numerous configurations from different brands can be found, and it is easy to confuse which one is better for the application. So, firstly, we can classify the mechanism by which the aerosol is formed; the first classification may be done regarding ultrasonic and pneumatic nebulizers; secondly, we can classify between concentric and cross-flow nebulizers, and finally according to regular or micro (or nano) flow nebulizers. Obviously, this is only a suggestion, and more classifications can be made taking into account other characteristics such as direct injection nebulizers, high efficiency, nebulizers for high dissolved solid samples, samples containing organic solvents, samples containing hydrofluoric acid, among others. These classifications may be helpful to understand and select the appropriate device for the application.

2.2.1 Ultrasonic nebulizer

An ultrasonic nebulizer (USN) relies on the use of a piezoelectric crystal vibrating at high frequency to create the aerosol. This is possible due to the converse piezoelectric effect, which makes the piezoelectric crystal vibrate when an electric field is applied to it (opposite to the direct piezoelectric effect, which generates electricity when stress is applied) (de Gois et al., 2012; Flament, Leterme, & Gayot, 2010; Jankowski, Karmasz, Ramsza, & Reszke, 1997).

The piezoelectric vibration breaks the solution into a very thin and dense aerosol with uniform particle size. This is one of the most efficient ways to generate an aerosol for analytical purposes, leading to an increase in sensitivity and therefore a lower limit of detections when compared to pneumatic nebulizers (de Gois et al., 2012). However, due to the very dense aerosol produced, USN normally needs to be coupled with a desolvation system so that the plasma is not overloaded or even extinguished, and this system is more susceptible to nonspectral interferences if samples and analytical standards present significant differences in composition. On the other hand, the cost of a USN with a desolvation system is significantly higher than a pneumatic nebulizer; also, the piezoelectric must be replaced from time to time due to loss in its efficiency, leading to a high cost to acquire and maintain this type of nebulizer (Burgener & Makonnen, 2020).

2.2.2 Pneumatic nebulizers

A pneumatic nebulizer (PN) uses a jet of compressed gas that breaks the liquid into an aerosol. This type of nebulizer compromises different geometries and sizes of nebulizers, with different sample introduction efficiency, applications, and differences in droplet size distributions (Jankowski et al., 1997).

Numerous versions of pneumatic nebulizers are available; therefore, we will focus only on the main characteristics of these devices. The first classification among them is the separation of cross-flow nebulizers and concentric nebulizers; after this classification, all variants are related to dimensions, materials, geometry, among others. These differences allow unique applications and characteristics for each nebulizer type. It is also important to pay attention to the material of the nebulizer; most devices are made of glass, and therefore they are not appropriate for solutions containing hydrofluoric acid; in these cases, polytetrafluoroethylene (PTFE), perfluoroalkoxy (PFA), and polyetheretherketone (PEEK) are more compatible. Also, in the case of organic solvents, pump tubing must be compatible with the solutions; most materials may be partially attacked by organic solvents. In cases where there is no pump tubing compatible with the sample matrix, a self-aspirating nebulizer may be used with PTFE tubing for sample aspirating (these tubes are mostly nonflexible). Self-aspirating nebulizers use the Bernoulli Effect, where there is a negative pressure at the tip of the nebulizer promoting the fluid to flow through the sample line without any external pumping action (Burgener & Makonnen, 2020).

Microsamples are generally a challenge to analyze; in these cases, micronebulizers, direct injection nebulizers, high-efficiency nebulizers, etc. are the best options to choose; however, they are prone to clocking with high dissolved solid solutions (de Gois et al., 2012). Thus, it is possible to notice the vast possibilities for nebulizer selection and we aim, in the next items, to introduce these features so that one can wisely select the proper nebulizer.

2.2.3 Cross-flow nebulizers

A cross-flow nebulizer generally is not self-aspirating; in this system, the aerosol is generated by the contact of the pumped solution with high-pressure gas stream in an angle of, normally 90 degrees, thus it is named as "cross-flow."

This type of nebulizer is mostly connected with double-pass spray chambers, resulting in droplets with low diameters that reach the plasma and thus high precision measurements. The use of double-pass spray chambers decreases the sample introduction efficiency; thus, this type of nebulizer is mostly found with most sensitive techniques such as ICP-MS.

Some advantages of this type of nebulizer are that different materials may be commercially found, thus fitting almost any application; it may handle high dissolved solid solutions, it is easy to unclog without damaging the nebulizer, and it is relatively easy to use. Fig. 2.1 presents the scheme of the nebulization using cross-flow nebulizers.

2.2.4 Concentric nebulizers

Probably the most common nebulizer found in commercial instruments, this type of nebulizer promotes the generation of aerosol from the contact of the gas in high pressure with the liquid solution that flows in parallel with the gas. A representative figure of this type of nebulizer

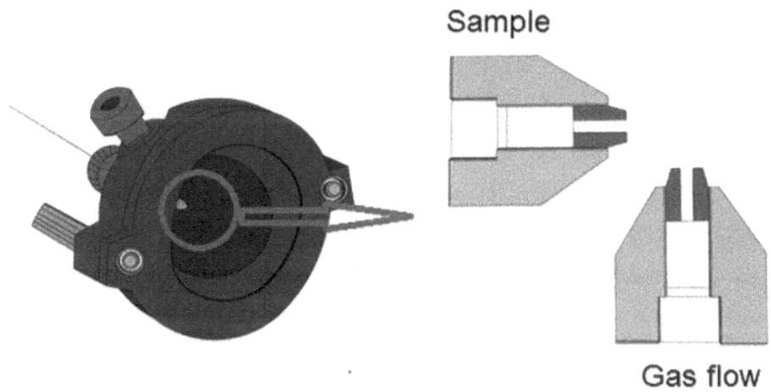

FIG. 2.1 Scheme of a cross-flow nebulizer for plasma-based analytical techniques. *No permission required.*

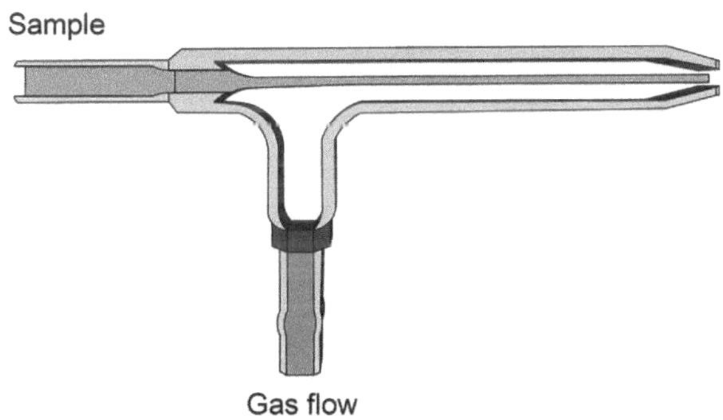

FIG. 2.2 Representation of a concentric nebulizer for plasma-based analytical techniques. *No permission required.*

may be found in Fig. 2.2. This type of nebulizer may be found in configurations that can handle the self-aspirating mode; however, it is important to notice that self-aspirating nebulizers may be extremely dependent on the physical-chemical characteristics of the solutions and therefore are prone to nonspectral interferences. These nebulizers may be found also for regular sample flows or microsample flows.

2.2.4.1 Routine or unchallenging samples

This application would require less from the nebulizer since the sample volume is not a problem and the samples do not contain high dissolved solid content. In this case, regular concentric nebulizers would fit the purpose; these types of nebulizers may be found for sample flow around a few milliliters per minute, made of different materials fitting the application and also capable of dealing with some amount of dissolved solids (even though this is not the best option for solutions with dissolved solids) (de Gois et al., 2012; Glass Expansion Newsletter, 2019).

Most of the variations of this type of nebulizer are regarding the position and dimensions of the inner capillary, the materials that it is built of, sample flow, and technical parameters that it will operate within. Obviously, these characteristics may deliver unique characteristics for each nebulizer and different results in the limit of detection, precision, and mainly in sample consumption; however, these differences may not be high in orders of magnitude. Many configurations are available and it is not our purpose to sell the brands or present specific characteristics of commercially available nebulizers, but to understand the application of nebulizers so one has its own judgment for selecting it.

This type of nebulizer is in general suited for regular and routine applications, without much challenge in the sample matrix, where the volume of the sample is not a problem and it does not contain dissolved solids so that the nebulizers can clog (de Gois et al., 2012; Glass Expansion Newsletter, 2019).

2.2.4.2 *Samples containing high amounts of solids*

Examples of nebulizers for samples containing solids are V-groove type, Babington Nebulizers, among others. This type of nebulizer may be found to be made of different materials fitting the application, and it is also the best choice for samples coining dissolved solids. They are also designed to facilitate unclogging without damaging the nebulizer, and therefore are very useful and robust nebulizers that can be found operating in sample uptakes of a few milliliters per minute or even microflows (Burgener & Makonnen, 2020).

This nebulizer is therefore the choice for challenging samples regarding solid content and it would fit for direct analysis of liquid samples that may contain some dissolved solids, for example, samples that overcome extraction procedures where the sample matrix is not completely decomposed (ultrasonic-assisted extraction) or samples prepared in alkaline media (de Gois et al., 2018; Frena et al., 2014; Tavares et al., 2020).

2.2.4.3 *Small volume of samples*

The small volume of samples may disable the analysis if a proper samples introduction system is not available; diluting the sample sometimes is impossible because it would decrease analyte concentration and sometimes make it below the limit of detection. Microsamples are therefore challenging and different approaches can be used for the analysis, e.g., using a graphite furnace, flow injection analysis, among others (Kurfürst, 1998).

As a very straightforward approach, the use of a micronebulizer is a very attractive option. These nebulizers have their dimensions reduced so that the aerosol generation is stable; moreover, the sample introduction efficiency is higher than that of regular size nebulizers making the sensitivity, even at a low sample flow, sometimes higher than for regular flow nebulizers.

Different configurations of micronebulizers can be found in the literature; due to the very low sample volume, sometimes it can be connected to the torch directly without a spray chamber, which can represent almost 100% sample introduction efficiency (regular nebulizers may range from 2% to 10%). Moreover, studies with this type of nebulizer have demonstrated the potential to reduce nonspectral interferences (de Gois et al., 2012).

Despite the outstanding capabilities, the use of micronebulizers in routine analysis or samples containing high dissolved solids is challenging; since they are prone to clogging, they are very sensitive (easy to break) and it is not easy to unclog due to the reduced dimensions.

Thus, this nebulizer is the choice for microvolumes of samples that do not contain dissolved solids; also, it can be applied for challenging matrices (without dissolved solids) to reduce the matrix effects in the analysis.

2.2.5 How to choose?

A scheme for selecting a nebulizer is presented in Fig. 2.3; this can be used as a general way to select a nebulizer that will be compatible with the samples; other effects such as reducing interferences, improving detection limits, removing organic solvents are not considered for this selection. Please note that the scheme starts with a sample in aqueous media; therefore, in the case of samples in organic medium or slurries, they must undergo sample preparation to make the analyte available in aqueous media before the analysis.

It is possible to observe, following the scheme presented in Fig. 2.3, that there is a nebulizer type that fits several applications; on the other hand, nebulizers are very restricted. Thus, one can easily find on sellers' websites the nebulizer that can be used for all the applications needed.

2.3 Fundamentals

The main goal of a nebulizer is to deliver the sample in a suitable way to the plasma so that all processes of desolvation, dissociation, atomization, and excitation or ionization occur properly.

One question to be raised is how tiny the droplets of the aerosol must be to be inserted into the plasma ideally. The answer would be disappointing; ideally, no droplets must be inserted into the plasma, because it would require energy to remove the solvent, which is a loss of energy; therefore, the best way to insert the analyte into the plasma would be

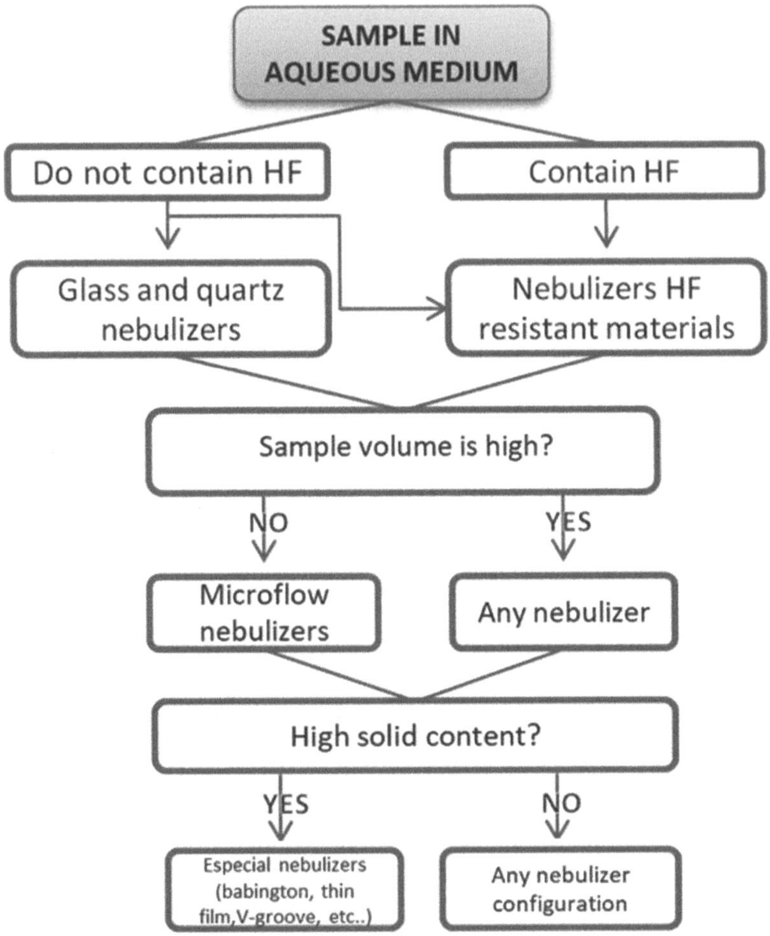

FIG. 2.3 Simplified scheme for selecting a nebulizer for analysis using plasma-based analytical techniques. *No permission required.*

as a gas, and for that, many sample introduction systems are available, such as chemical and photochemical vapor generation, electrothermal vaporization, among others. However, they are relatively difficult to use and are not able to convert any element of the periodic table into vapor (Sturgeon, 2017). Therefore, to date, nebulization is the most general manner and trustable technique to insert the analyte into the plasma and to deal with that, we must fit some requirements in order to perform one analysis successfully.

Returning to the question of how tiny the droplets must be to be inserted into the plasma, an acceptable answer is as tiny as possible, as large droplets may cool the plasma or may not be vaporized; sometimes the outer part of the droplet is vaporized while the inside of the droplet remains cooled. Also, if a high load of the sample with large droplets reaches the plasma, it can be extinguished. This is the main goal of the spray chamber that is connected to the nebulizer, avoiding the large droplets generated by the nebulizer to reach the plasma (Burgener & Makonnen, 2020). The size of the droplets generated by a nebulizer may be calculated by the Nukiyama and Tanasawa equation, introduced in 1938 and revised by Ashgriz in 2011 (Ashgriz, 2011; Nukiyama, 1938).

We can separate the phenomena of the droplet size selection according to the primary and tertiary aerosols. The primary aerosol is the aerosol generated by the nebulizer with droplets of different sizes; it is a nonhomogenous aerosol (for regular nebulizers) that cannot be carried directly to the plasma. The tertiary aerosol is the aerosol that reaches the plasma; this must be as homogenous as possible and present droplets ideally lower than 8 μm. Therefore, nebulizers that can produce a high number of droplets lower than 8 μm are considered good choices and may increase the detection capabilities of the analytical technique (Burgener & Makonnen, 2020).

The sample introduction efficiency is another important parameter; this measures the amount of the sample that reaches the plasma in regard to the aspirated solution. Regular nebulizers present about 2%–5% of the sample introduction efficiency; ultrasonic nebulizers may deliver up to 20% of the sample into the plasma, while microflora nebulizers also present a high sample introduction efficiency (higher than 20%). Obviously, these features are related not only to the nebulizer but to the whole sample introduction system (nebulizer and spray chamber); if a heated spray chamber is used, the sample introduction efficiency may be higher. Also, there are direct injection nebulizers that may deliver up to 100% of the sample into the plasma at a very low flow so that the plasma is not extinguished.

Sample flow, as already mentioned, is also an important point to be considered. There are nebulizers for regular sample flows (milliliters per minute) and nebulizers for microflows (microliters per minute); it is important to notice that regular nebulizers have their sample and gas flow, as well as backpressure, relatively standardized and they cannot operate using parameters other than those indicated by the manual; otherwise, the aerosol formation is not stable and one will not be able to obtain good standard deviations and limit of detection from the analysis. Therefore, microflows will necessarily require microflow nebulizers, while regular flows will require standard nebulizers (Burgener & Makonnen, 2020).

And finally, pumping the sample is almost a mandatory requirement for analysis in plasma-based techniques; this comes due to the instability of self-aspiration regarding the sample matrix and also due to problems that may arise such as salting the orifice of the nebulizer changing the sample flow. Thus, the use of a peristaltic pump may arise as a key factor to provide stable sample flow at all times. The disadvantage of using a peristaltic pump may come with the pulses that it may deliver; to reduce pulsation, low diameter tubes may be used and replaced constantly.

2.4 Applications and future perspectives

Although a wide variety of nebulizers and/or models are commercially available to date, studies on new designs and evaluation of their performance are of increasing interest. The nebulizer must be carefully selected for a given analysis and sample. Sample preparation and analytical characteristics (sensitivity, detection limits, and precision) must be considered. Detection limits are affected by the droplet size distribution as larger droplets entering into the plasma cause irregularities. Precision is affected by the droplet size distribution and variable droplet sizes degrade stability and precision (Burgener & Makonnen, 2020). Other aspects to be considered are robustness, 'ease of use,' versatility for coupling to commercial spray chambers, lifetime, and cost (Todoli & Mermet, 2008).

Pneumatic nebulizers are the most commonly used for liquid sample introduction in plasma-based techniques. In general, pneumatic nebulizers can exhibit 0.5%–5% transport efficiency with a sample flow rate of approximately $1\,mL\,min^{-1}$. However, there is an increasing analytical need for reducing the sample flow rate. The reasons are (i) to increase analyte transport efficiency; (ii) to reduce the amount of sample; (iii) to reduce nonspectral interferences, and (iv) to reduce the waste amount (Canals & Aguirre, 2015). The use of a micronebulizer can be advantageous for low volume samples, simple matrices, interface with separation devices or to improve detection limits (LODs). Different approaches and applications of micronebulizers coupled to a spray chamber or a desolvation system and direct injection nebulizers have been reviewed (Canals & Aguirre, 2015; Todoli & Mermet, 2008). Recent applications of micronebulizers include nanoparticle characterization and single-cell analysis (Corte-Rodríguez, Álvarez-Fernández, García-Cancela, & Montes-Bayón, 2020; Mozhayeva & Engelhard, 2020). The use of a micronebulizer, e.g., one operating at approximately $10\,\mu L\,min^{-1}$, can improve the transport efficiency to 60% or up to 80% (Olesik & Gray, 2012).

Direct injection nebulizers (DIN), which introduce the primary aerosol directly into the plasma, exhibit a transport efficiency of 100% (Canals & Aguirre, 2015). The use of a high-efficiency nebulizer (DIHEN), which is a modified DIN, has been shown to be an interesting approach for metal determination and isotope ratio measurements in biological, petroleum-related samples and radioactive wastes (Giusti et al., 2007; Todoli & Mermet, 2008). A demountable DIHEN (d-DIHEN) has been used as an interface for coupling FIA (Wang & Hansen, 2001, 2002) and CE (Bendahl, Gammelgaard, Jøns, Farver, & Hansen, 2001) with ICP-MS (Canals & Aguirre, 2015). Matusiewicz and Ślachciński (2011) used the d-DIHEN as an interface between a microchip-based CE and a microwave-induced plasma optical emission spectrometry (MIP-OES) for copper speciation. Louvat et al. used a μ-dDIHEN nebulizer composed of a demountable direct injection high-efficiency nebulizer (d-DIHEN), a flow injection analysis (FIA) valve, and a gas displacement pump (GDP) connected to a μ-mass-flow meter for three applications: multielement analyses of natural water samples by sector field (SF)-ICP-MS, boron isotope ratio measurement of geological samples by multicollector (MC)-ICP-MS, and gold nanoparticle characterization by single particle (sp)-ICP-MS. The limits of quantification (LOQs) of the majority of the trace elements were lower (up to 15 times) with the μ-dDIHEN as a lower sample volume was used. A transport efficiency of 85% was obtained for gold NPs of 40 nm certified size (Louvat, Tharaud, Buisson, Rollion-Barda, & Benedetti, 2019).

An increasing number of studies on multimode nebulizer configurations have been reported. By using dual or triple nebulizer arrangements, sensitivity was improved by factors ranging from 20 to 100 compared to that with a conventional nebulizer (Asfaw & Wibetoe, 2006; Mulugeta, Wibetoe, Engelsen, & Asfaw, 2009). Multimode approaches are especially interesting for the determination of volatile species and/or online matrix matching calibration (Bings, von Niessen, & Schaper, 2014; Ślachciński, 2019).

Also, nebulization devices based on Flow Focusing (FF) and Flow Blurring (FB) are gaining interest for sample introduction in ICP-based techniques. FB technology is an evolution of the FF system, both developed by Gañán-Calvo (1998, 2005) and currently available in the market (Agilent Technologies®, 2022; OneNeb®, Ingeniatrics Tecnologías S. L, 2022). In the FF systems, the liquid stream is aligned with the outlet of the orifice by the gas stream. The FB system produces a turbulent mixing between the liquid and the gas in the liquid capillary to break up the liquid before entering into the gas stream (Rosell-Llompart & Gañán-Calvo, 2008). FB nebulizers produce finer aerosols with similar energy consumption in comparison with other pneumatic nebulizers. The droplet sizes generated are determined by the speed of the gas interacting with the liquid, by the shape of the gas orifice, the relative sizes of the gas and liquid orifices, and the turbulence around the gas and liquid orifices (Burgener & Makonnen, 2020). As the gas and liquid streams are premixed, the evaporation of the liquid in the gas orifice and clogging risks are minimized; thus, its use is very convenient for samples with high salt content.

Novel designs based on multiple nebulizers and FF and FB technologies have been proposed for analysis of complex (organic) matrices by ICP-based techniques and for on-line calibration (Aguirre, Kovachev, Hidalgo, &

Canals, 2012; Pereira, Aguirre, Nóbrega, Hidalgo, & Canals, 2012; Pinheiro, Aguirre, Nóbrega, & Canals, 2021; MultiNeb®, Ingeniatrics Tecnologías S. L, 2022). The use of these systems was advantageous to mitigate spectroscopic and matrix-based interferences.

The state of the art of nebulizers, such as nanonebulizers, multinebulizers, discrete droplet generation, and electrospray nebulizers, showing a promising perspective for sample introduction in plasma-based techniques, will be discussed further in this book.

References

Agilent Technologies®. (2022). https://www.agilent.com/Agilent404?s=https://www.agilent.com/en-US/products-services/Parts-Supplies/Spectroscopy/ICP-OES/ICP-OES-Nebulizers-Accessories/Pages/ (last accessed 06 July 2022).

Aguirre, M. A., Kovachev, N., Hidalgo, M., & Canals, A. (2012). Analysis of biodiesel and oil samples by on-line calibration using a flow blurring® multinebulizer in ICP-OES without oxygen addition. *Journal of Analytical Atomic Spectrometry, 27*, 2102–2110.

Asfaw, A., & Wibetoe, G. (2006). Dual mode sample introduction for multi-element determination by ICP-MS: The optimization and use of a method based on simultaneous introduction of vapor formed by NaBH4 reaction and aerosol from the nebulizer. *Journal of Analytical Atomic Spectrometry, 21*, 1027–1035.

Ashgriz, N. (Ed.). (2011). *Handbook of atomization and sprays.*

Bendahl, L., Gammelgaard, B., Jøns, O., Farver, O., & Hansen, S. H. (2001). Interfacing capillary electrophoresis with inductively coupled plasma mass spectrometry by direct injection nebulization for selenium speciation. *Journal of Analytical Atomic Spectrometry, 16*, 38–42.

B'Hymer, C., & Caruso, J. A. (2000). Chapter 7 Nebulizer sample introduction for elemental speciation. In *Comprehensive analytical chemistry: Vol. 33* (pp. 213–226).

Bings, N. H., von Niessen, J. O. O., & Schaper, J. N. (2014). Liquid sample introduction in inductively coupled plasma atomic emission and mass spectrometry—Critical review. *Spectrochimica Acta Part B, 100*, 14–37. https://doi.org/10.1016/j.sab.2014.08.011.

Borkowska-Burnecka, J., Leśniewicz, A., & Zyrnicki, W. (2006). Comparison of pneumatic and ultrasonic nebulizations in inductively coupled plasma atomic emission spectrometry—Matrix effects and plasma parameters. *Spectrochimica Acta Part B: Atomic Spectroscopy, 61*(5), 579–587.

Burgener, J. A., & Makonnen, Y. (2020). Nebulization systems. In *Sample introduction systems in ICPMS and ICPOES* (pp. 57–142).

Canals, A., & Aguirre, M. A. (2015). *Roles of nebulizers in analytical chemistry. Encyclopedia of analytical chemistry.* John Wiley & Sons Ltd, doi:10.1002/9780470027318.a9391.

Corte-Rodríguez, M., Álvarez-Fernández, R., García-Cancela, P., & Montes-Bayón, M. (2020). Single cell ICP-MS using online sample introduction systems: Current developments and remaining challenges. *TrAC Trends in Analytical Chemistry, 132*, 116042. https://doi.org/10.1016/j.trac.2020.116042.

de Gois, J. S., Lucena, I., de Oliveira Cezario, P. S., da Silva, A. P., Lima, I. C. A., & Luna, A. S. (2018). A high-throughput method for multi-element determination in green coffee beans using diluted nitric acid and ultrasound energy. *Analytical Methods, 10*(14), 1656–1661.

de Gois, J. S., Maranhão, T.d. A., Oliveira, F. J. S., Frescura, V. L. A., Curtius, A. J., & Borges, D. L. G. (2012). Analytical evaluation of nebulizers for the introduction of acetic acid extracts aiming at the determination of trace elements by inductively coupled plasma mass spectrometry. *Spectrochimica Acta Part B: Atomic Spectroscopy, 77*, 35–43.

Flament, M. P., Leterme, P., & Gayot, A. (2010). Study of the technological parameters of ultrasonic nebulization. *Drug Development and Industrial Pharmacy, 27*(7), 643–649. https://doi.org/10.1081/DDC-100107320.

Frena, M., Quadros, D. P. C., Castilho, I. N. B., de Gois, J. S., Borges, D. L. G., Welz, B., & Madureira, L. A. S. (2014). A novel extraction-based procedure for the determination of trace elements in estuarine sediment samples by ICP-MS. *Microchemical Journal, 117*, 1–6.

Gañán-Calvo, A. M. (1998). Generation of steady liquid microthreads and micron-sized monodisperse sprays in gas stream. *Physical Review Letters, 80*, 285–288.

Gañán-Calvo, A. M. (2005). Enhanced liquid atomization: From flow focusing to flow blurring. *Applied Physics Letters, 86*, 214101.

Giusti, P., Ordóñez, Y. N., Lienemann, C. P., Schaumlöffel, D., Bouyssiere, B., & Lobiński, R. (2007). μFlow-injection–ICP collision cell MS determination of molybdenum, nickel and vanadium in petroleum samples using a modified total consumption micronebulizer. *Journal of Analytical Atomic Spectrometry, 22*, 88–92. https://doi.org/10.1039/b611542j.

Glass Expansion Newsletter. (2019). A guide to choosing nebulizers and spray chambers for your ICP-OES (47). www.geicp.com/careers. (Accessed 22 November 2021).

Jankowski, K., Karmasz, D., Ramsza, A., & Reszke, E. (1997). Characteristics of nebulizers for microwave induced plasma atomic emission spectrometry. II. Ultrasonic nebulizers. *Spectrochimica Acta Part B: Atomic Spectroscopy, 52*(12), 1813–1823.

Kurfürst, U. (1998). Solid sample analysis: Edition anglaise. https://books.google.com/books?hl=pt-BR&lr=&id=z3evkGNamysC&pgis=1. (Accessed 21 November 2021).

Louvat, P., Tharaud, M., Buisson, M., Rollion-Barda, C., & Benedetti, M. F. (2019). μ-dDIHEN: A new micro-flow liquid sample introduction system for direct injection nebulization in ICP-MS. *Journal of Analytical Atomic Spectrometry, 34*, 1553–1563.

Matusiewicz, H., & Ślachciński, M. (2011). Interfacing a microchip-based capillary electrophoresis system with a microwave induced plasma spectrometry for copper speciation. *Central European Journal of Chemistry, 9*, 896–903.

Meyer, G. (2013). In *Nebulizers for inductively coupled plasma spectroscopy*. https://www.spectroscopyonline.com/view/nebulizers-inductively-coupled-plasma-spectroscopy. (Accessed 21 November 2021).

Mozhayeva, D., & Engelhard, C. (2020). A critical review of single particle inductively coupled plasma mass spectrometry—A step towards an ideal method for nanomaterial characterization. *Journal of Analytical Atomic Spectrometry, 35*, 1740. https://doi.org/10.1039/c9ja00206e.

MultiNeb®, Ingeniatrics Tecnologías S. L. (2022). https://www.ingeniatrics.com/producto/multineb/ (last accessed 06 July 2022).

Mulugeta, M., Wibetoe, G., Engelsen, C. J., & Asfaw, A. (2009). Multivariate optimization and simultaneous determination of hydride and non-hydride-forming elements in samples of a wide pH range using dual-mode sample introduction with plasma techniques: Application on leachates from cement mortar material. *Analytical and Bioanalytical Chemistry, 393*, 1015–1024.

Nukiyama, S. (1938). An experiment on the atomization of liquid by means of an air stream (2). *Journal of the Society of Materials Science, Japan, 4*, 138–143. https://ci.nii.ac.jp/naid/10011183802. (Accessed 21 November 2021).

Olesik, J. W., & Gray, P. J. (2012). Considerations for measurement of individual nanoparticles or microparticles by ICP-MS: Determination of the number of particles and the analyte mass in each particle. *Journal of Analytical Atomic Spectrometry, 27*, 1143–1155.

OneNeb®, Ingeniatrics Tecnologías S. L. (2022). https://www.ingeniatrics.com/producto/oneneb/ (last accessed 06 July 2022).

Pereira, C. D., Aguirre, M. A., Nóbrega, J. A., Hidalgo, M., & Canals, A. (2012). Correction of matrix effects for As and Se in ICP OES using a flow blurring® multiple nebulizer. *Journal of Analytical Atomic Spectrometry, 27*, 2132–2137.

Pinheiro, F. C., Aguirre, M. A., Nóbrega, J. A., & Canals, A. (2021). Dispersive liquid–liquid microextraction of Cd, Hg and Pb from medicines prior to ICP OES determination according to the United States Pharmacopeia. *Analytical Methods, 13*, 5670–5678.

Rosell-Llompart, J., & Gañán-Calvo, A. M. (2008). Turbulence in pneumatic flow focusing and flow blurring regimes. *Physical Review E, 77*, 036321.

Ślachciński, M. (2019). Modern chemical and photochemical vapour generators for use in optical emission and mass spectrometry. *Journal of Analytical Atomic Spectrometry, 34*, 257. https://doi.org/10.1039/c8ja00383a.

Sturgeon, R. E. (2017). Photochemical vapor generation: A radical approach to analyte introduction for atomic spectrometry. *Journal of Analytical Atomic Spectrometry, 32*(12), 2319–2340.

Tavares, K.d. N., de Lucena, I. O., Toaldo, I. M., da Silva Haas, I. C., Luna, A. S., & de Gois, J. S. (2020). Optimized sample preparation for sulfur determination in animal feed by inductively coupled plasma – optical emission spectrometry (ICP-OES) with correlation to the total protein content. *Analytical Letters, 53*(14), 2252–2265. https://doi.org/10.1080/00032719.2020.1736090.

Todoli, J.-L., & Mermet, J.-M. (2008). *Liquid sample introduction in ICP spectrometry: A practical guide* (1st ed.). Elsevier.

Twyman, R. M. (2005). ATOMIC EMISSION SPECTROMETRY|Principles and instrumentation. In *Encyclopedia of analytical science* (2nd ed., pp. 190–198).

Wang, J., & Hansen, E. H. (2001). Interfacing sequential injection on-line preconcentration using a renewable micro-column incorporated in a "Lab-on-Valve" system with direct injection nebulization inductively coupled plasma mass spectrometry. *Journal of Analytical Atomic Spectrometry, 16*, 1349–1355.

Wang, J., & Hansen, E. H. (2002). Coupling sequential injection on-line preconcentration using a PTFE beads column to direct injection nebulization inductively coupled plasma mass spectrometry. *Journal of Analytical Atomic Spectrometry, 17*, 1278–1283.

CHAPTER

3

Aerosol generation

Fariba Mollarasouli[a], *Sonia Bahrani*[b,c], *and Mazaher Ahmadi*[d]

[a]Department of Chemistry, Yasouj University, Yasouj, Iran, [b]Nanotechnology Department, Borjobaru Fars Company, Fars Science and Technology Park, Shiraz, Iran, [c]Health Policy Research Center, Health Institute, Shiraz University of Medical Sciences, Shiraz, Iran, [d]Department of Analytical Chemistry, Faculty of Chemistry, Bu-Ali Sina University, Hamedan, Iran

3.1 Introduction

The suspension of liquid (droplets) or solid particles in a gaseous environment is referred to as an aerosol. The term, which refers to particle suspension in a liquid, was created to represent the gas phase parallel to hydrosol (Greek for "water particle"). Aerosols are two-phase systems made up of an enclosing gas phase and a solid or suspended liquid phase. An "aerosol" is a dispersed phase of a liquid in the form of dust or spray, or a solid such as dust or luminous gas that ignites in the air, according to the Gibbs definition. Aerosol diameters can vary from 0.01 to 100 μm, while the exact limit is unknown. Aerosols are created by either converting gases into particles or dissolving liquids or solids into smaller parts (Hidy and Brock, 2016).

Analytical chemistry is only one of the numerous scientific and technical fields where nebulizers have a wide range of applications. Nebulizers are mostly used in chemical analysis to create an aerosol that will be used to prepare the sample liquid for analysis. For analytical applications, the ideal nebulizer would have the following qualities (Lefebvre and McDonell, 2017). Ability to produce an aerosol that is as repeatable, fine, and monodisperse as feasible; being able to operate under a variety of liquid and (where necessary) gas flow rates; chemically inert, able to operate with slurries and organic and aqueous solutions—even those with a high salt content—without experiencing clogging issues; aerosol properties are

Analytical Nebulizers
https://doi.org/10.1016/B978-0-323-91181-8.00008-4

independent of matrix solutions; ease of handling and maintenance; and low potential for harm when operating and installing. Indeed, there is no perfect nebulizer that meets each of these requirements at once. However, there is a large range of them with various working principles, forms, and sizes, which has led to the development of several applications in chemical analysis. Both atomic and molecular spectrometry employ nebulizers. The liquid sample is introduced into the atomization, excitation, and ionization source using nebulizers in the first method. Nebulizers have been developed in this field to broaden their range of application in the examination of micro- and nanoscale materials (Canals and Aguirre, 2006). Nebulizers, on the other hand, are mostly used in molecular spectrometry as a gentle ionization approach. Recent research has focused on the hyphenation of microfluidic systems to mass spectrometers, and the miniaturization community has taken an interest in novel nebulizers for soft ionization. Thermospray ionization, sonic spray ionization, electrospray ionization, and the hybrid electrosonic spray ionization approach are among the spray ionization methods that have been created for use in CE- and LC-MS interfaces.

In order to overcome the cohesive forces of the liquid and separate it into droplets, the mass of the liquid must be given a particular amount of energy during aerosol production. The most popular nebulizers used in chemical analysis may be divided into four categories based on the energy source: pneumatic, hydraulic, ultrasonic, and electrostatic. This chapter summarizes aerosol generation using the most commonly used analytical nebulizers.

3.2 Fundamentals

3.2.1 Most commonly used analytical nebulizers

The three types of nebulizers most frequently employed for producing aerosols in analytical chemistry are pneumatic, ultrasonic, and hydraulic nebulizers. Concentric, cross-flow, V-groove, thin film, parallel path, improved parallel route, Hildebrand grid, flow blurring, etc., are examples of pneumatic nebulizers. Vibrating transducers and vibrating mesh are components of ultrasonic nebulizers. Both low-pressure and high-pressure nebulizers are hydraulic devices.

A high-velocity gas stream and a relatively low-velocity liquid stream collide in a pneumatic nebulizer, where the compressed gas's kinetic energy is employed to create aerosols. The most popular method for bringing liquid samples into atomic spectrometry is a pneumatic nebulizer since it is affordable, simple to use, and sturdy enough. Sample introduction can be done with or without pumping. In concentric pneumatic nebulizers, an input connects the pumped liquid capillary to the compressed gas capillary. Increased liquid/gas interactions are caused by the gas stream's high tangential velocity against the liquid. The capillary's location in this type of nebulizer is movable, allowing the user to alter the liquid flow rate and consequently the degree of liquid/gas interaction. However, ICP technologies mostly employ nonadjustable ones. In cross-flow nebulizers, the contact between the liquid and gas streams occurs perpendicularly in cross-flow nebulizers. The gas intake is positioned horizontally, whereas the liquid inlet is vertical. Babington received a patent for the nebulizer in 1969 (Babington et al., 1969). An orifice-equipped hollow glass sphere makes up the original Babington nebulizer. The different orifices correspond to distinct gas outputs. A capillary is used to carry the liquid sample, which is then placed perpendicularly on the sphere's surface. When the liquid comes into contact with the gas outlet orifices, an aerosol is produced. The Babington operating concept is the foundation of the V-groove nebulizer. It is made up of a capillary and a V-shaped groove with one opening. The orifice is for the gas outflow, whereas the capillary is for the liquid output. The solution leaves the capillary and falls through the bottom hole by sliding down the sloped groove. The aerosol is created when the liquid sheet crosses over the gas exit and the two streams interact. A thin film nebulizer is either a spherical in shape with liquid poured over it and a gas hole in the side, or it is an inclined sheet with the liquid running down it in a thin film. The sheet or sphere's gas hole was made to blast gas through the liquid, creating a mist. Although most of the liquid would miss the gas hole, the approach proved beneficial for nonplugging. The liquid is delivered to a location close to the gas orifice in the parallel route nebulizer. The liquid is drawn into the gas stream by the low pressure created by the gas flow exiting the aperture. Induction is the method used by cross-flow and V-groove nebulizers to create a mist. The extended parallel route approach makes use of the capillary gas streams' parabolic flow patterns. The Hildebrand nebulizer is made up of two parallel and back-to-back circular platinum grids that are spaced apart. By interacting with the gas stream passing through the top of the first grid, the sample, which is deposited there by external pumping, produces the aerosol. The second grid serves as a surface for the impact that reduces the droplet size. Flow focusing and flow blurring nebulizers are two relatively recent technologies for aerosol formation. An outline of the hydrodynamic principles of this novel nebulization technology is provided by Gañán-Calvo (2005).

In an ultrasonic nebulizer, a liquid sample is pushed onto a piezoelectric transducer's surface, which is quickly vibrating. As the liquid spreads across the surface, a wave pattern that resembles a checkerboard develops in the film that forms. The wave crest height in the film grows together with the surface vibration's amplitude. When the vibrating surface's amplitude rises to a certain level, the wave crests in the film become unstable and collapse, which results in the formation of aerosols. The ability of ultrasonic nebulizers to produce very fine, low-velocity aerosols is one of its intriguing features (Canals and Aguirre, 2006). The current generation of ultrasonic nebulizers employs vibrating transducers, which convey the sample to the front of the transducer before an Ar sweep gas blows away the resultant mist. The sample is transferred to the rear of the vibrating mesh in the other type of ultrasonic nebulizer and is forced through holes in the mesh (Burgener and Makonnen, 2020). This creates a consistent mist of droplets of the same size, sends the mist to the torch, and prevents the spray chamber walls from getting wet.

In hydraulic nebulizers, when a high-velocity liquid jet that originates from a small aperture strikes a solid surface positioned in front of the opening, it produces an aerosol (Ashgriz, 2011).

In the following sections, the aerosol generation mechanism for each nebulizer type will be discussed.

3.2.2 Aerosol generation mechanism

3.2.2.1 Pneumatic nebulizers

A high-velocity gas stream and liquid solution contact results in pneumatically created aerosols. Droplets are produced when a portion of the kinetic energy of the gas stream is transferred to the liquid bulk. It is challenging to conduct a full analysis of the pneumatic aerosol production mechanism used in ICP nebulizers because it happens so quickly—on the scale of several microseconds. Based on information from engineering applications, Sharp presented an outstanding assessment of the pneumatic aerosol production concept in 1988. This study highlighted the intricacy of the processes that take place when the aerosol is formed by a liquid-gas interaction (Sharp, 1988). It may be concluded that at least for the instruments and operational settings employed in plasma spectrometry, the precise control of pneumatic aerosol formation is subpar. This is due to the fact that the majority of research into the mechanics behind aerosol formation has taken place in branches like engineering and aeronautics. Compared to what is often employed in ICP-MS and ICP-AES, the nebulizers used in these regions, as well as the liquid and gas flows operated, are many orders of magnitude greater. It should also be remembered that there are several processes involved in the formation of droplets. The fact that developing models and further interpreting the findings of measuring aerosol properties are the sole ways to analyze the aerosol formation process adds to the challenge. The advent of additional phenomena that happen just after the nebulization event, that is, close to the nebulizer tip, causes the aerosol characteristics to change quickly after the droplets are created. The experimental assessment of the aerosol properties is consequently particularly challenging due to this rapid shift.

Two distinct processes may be used to create pneumatic aerosols: creating waves on the solution surface, and developing and breaking up these waves to create droplets.

(i) *Wave generation*: The liquid surface of the column at the nebulizer sample capillary's outflow can be taken to be flat. There is little energy transferred when it comes into contact with a high-velocity gas stream because only tangential forces are at play. In actuality, the gas and liquid stream's velocity at the liquid surface is less than 3% of that of the gas stream (Sharp, 1988). The minimal amount of energy that is transmitted from the gas stream in these conditions causes waves to emerge on the liquid surface. Surface tension and gravity are two factors that work against the forces that cause instability. As a result, the gas stream needs a certain amount of extra energy to defeat these forces. Under the typical operating conditions seen in ICP pneumatic nebulizers, the perturbations produced on liquid surfaces have wavelengths on the order of many tens of micrometers.

(ii) *Wave growing and breakup*: The degree of gas-liquid contact rises after waves are created on the liquid surface. As a result, the gas interacts with each of these waves perpendicularly. The amount of the force acting on a single wave is directly related to the relative velocities of the gas and liquid, the size of the wave, and the drag factor. The flow regime is another key consideration. A turbulent flow encourages the production and growth of the waves by allowing the gas to penetrate the liquid bulk. This cycle is repeated until the waves become unstable and the aerosol droplets form. Three separate processes can result in the destruction of waves: (i) direct surface stripping; (ii) the formation of thin sheets or ligaments from the wave crests; and (iii) wave disintegration. When the surface tension of the sample is overcome by the gas energy, process (i) results in producing an aerosol with very small droplets. The generation of tiny droplets is also encouraged by procedure (ii). When the wave acceleration created by the gas is sufficient, process (iii) is visible in the interim. According to this process, the gas can permeate the liquid bulk and produce open-ended gas columns and chaotic instabilities. Large liquid masses may thus get separated, resulting in the formation of coarse droplets.

The difference in velocities between the liquid and gas streams affects aerosol generation. The measurement of the quantity of energy or the gas velocity at the exit of pneumatic nebulizers is therefore an intriguing point. According to reports (Sharp, 1988), the critical gas velocity needed for aerosol production is around $4.9\,m\,s^{-1}$. So, to produce tiny droplets, supersonic gas streams are ideal. The sonic velocity, however, has been hypothesized to represent the maximum gas velocity at the tip of a converging nozzle. This is due to the gas's ability to be compressed. Thus, while the density reduces faster than the velocity grows after the sonic velocity is attained, the velocity increases slower than the density declines in the acceleration area. To get the argon stream to the converging nozzle throat's sonic velocity, around 2 bar of pressure must be supplied. To achieve conventional nebulizer gas flow rates, several pneumatic nebulizers need greater pressures. There is no additional change in gas velocity seen at pressures higher than the critical limit. The mass of the gas should grow as the pressure applied increases the gas's energy. In actuality, a rise in gas density results from a pressure increase over the critical value. Since the nebulizer nozzle serves as a mass flow controller, measuring the back pressure using a manometer inserted from the instrument mass flow controller to the nebulizer is a good way to identify any nebulizer obstruction. The pressure needed to maintain the gas flow rate will rise if a partial tip blockage is created.

In flow focusing and flow blurring technologies the aerosol generation mechanism is different (Gañán-Calvo, 2005). The liquid capillary and outflow aperture are exactly aligned during flow focusing nebulization. The liquid vein is concentrated in the center of the output orifice by the gas stream within a certain range of physical conditions and geometrical arrangements. The hydrodynamic flow focusing method creates aerosols that are pneumatically fine and essentially monodisperse. Additionally, the basic equations regulating the physical principle that defines the flow focusing allow for the prediction of aerosol properties. Additionally, there is a significant reduction in the likelihood of clogging since the liquid jet does not contact the output orifice edges. The technology of "flow blurring" is an advancement of "flow focusing." The company Ingeniatrics Tecnologías S.L. manufactures the OneNeb nebulizer (Ingeniatrics, 2022), implementing the Flow Blurring technology. Despite its simplicity in the nebulization mechanism and nebulizer design, the flow blurring technique is simple, repeatable, and reliable and produces aerosols with a high degree of efficiency. With the same amount of energy, Flow Blurring technology produces significantly finer particles than conventional pneumatic nebulization techniques. This technique is therefore seen to be quite effective, even when compared to Flow Focusing technology. In order to achieve a good turbulent mixing between the liquid and gas streams, which produces very thin primary aerosols, flow blurring necessitates a specialized yet straightforward geometry. At the nebulizer tip, the gas flow pattern abruptly splits in two, according to the Flow Blurring mechanism. As a result, some of the gas moves upstream into the liquid capillary and agitates the incoming liquid.

3.2.2.2 *Ultrasonic nebulizers*

An alternating electrical field used to generate the vibrations of a piezoelectric crystal (transducer) is the basis of an ultrasonic nebulizer's operation. Sound waves having a frequency greater than 20,000 Hz are known as ultrasounds (Flament et al., 2001). In a periodic disturbance known as a sound wave, molecules in certain parts of a material medium are briefly moved out of their equilibrium locations and feel a counteracting force as a result of the medium's elasticity. The molecules oscillate about their mean location as a result of this force, which propagates the disturbance wave. The size of this force affects the wave's propagation speed. Alternating positive and negative deviations from the mean values of density, pressure, temperature, particle velocity, and particle acceleration are involved in the propagation of sound waves across the medium. Cavitation, which is the creation and collapse of tiny bubbles in the liquid, occurs when the pressure amplitude is sufficiently high and produces large variations in pressure. The formation is due to the sound waves' negative pressure component, which causes part of the liquid's vapor to escape the solution as tiny bubbles. These bubbles then serve as weak points for the liquid to further separate, creating bigger voids. The cavities then collapse violently with a hammering motion in the opposite half of the sound wave cycle when the pressure turns positive, creating high local instantaneous pressures and temperatures. A water geyser appears at the surface as a result of the instantaneous particle velocities reaching supersonic speed during the collapse of the bubbles. Periodic hydraulic shocks send the liquid's surface into a rapid oscillatory motion, causing standing capillary waves of limited amplitude to develop on its surface and the standing capillary wave to spontaneously excite. The capillary waves' amplitude will increase if the vibration intensity, or the mean power transmitted per surface unit, is high enough. However, nonlinearities will cause the capillary waves' form to gradually depart from a sine wave. Finally, because of their instability at high amplitudes, droplets will be ejected from the crests of the wave, which causes the liquid to atomize (Hess, 2000; Kuttruff, 2012).

3.2.2.3 *Hydraulic nebulizers*

The jet impact nebulizer developed by Doherty and Hieftje was the first hydraulic nebulizer utilized in atomic spectrometry (Doherty and Hieftje, 1984). The liquid pressure ranged from 3 to 6 bar, while the orifice diameter ranged from 20 to 100 m. The jet impact nebulizer produces a very stable aerosol and achieves transport efficiency and limits of detection values that are comparable to those of cross-flow and concentric pneumatic nebulizers. However, it requires a very constant liquid flow rate and is highly susceptible to tip blockage with high salt content solutions (Doherty and Hieftje, 1984). Berndt created the high-pressure hydraulic nebulizer at a later time (Berndt, 1988). Since the orifice diameter of the high-pressure hydraulic nebulizer is smaller than that of the jet impact nebulizer, higher liquid pressure is needed. In order to inject the liquid in a steady and continuous way, the high-pressure hydraulic nebulizer thus requires an HPLC pump. The main aerosols produced have greater analytical performance because they are substantially finer than those produced by traditional concentric pneumatic nebulizers. Additionally, nebulization of acid solutions and solutions with a high salt content has minimal impact on memory (Canals and Aguirre, 2006).

3.2.2.4 *Thermospray*

By first heating the liquid stream and then creating an adiabatic expansion of the solvent vapor, the thermospray creates aerosols. The electrical current heats the metal capillary, which causes a certain vaporized combination (vaporized and residual liquid) to develop. The leftover liquid is transformed into a hot aerosol by the expanding vapor inside the tube. Due to preheating action and the production of finer aerosol, these nebulizers operate better and are more sensitive to analyte transfer than typical pneumatic nebulizers. This technique does have certain problems, though, including the necessity for a high-pressure pump to enter the liquid, high solvent loading, the possibility of clogging when using high salt content, high cost, and most critically, breakdown of thermally labile molecules (Yang et al., 1998).

3.2.2.5 *Electrospray*

In electrospray, the reciprocal repulsion of similar charges that have collected on the surface generates the energy that disrupts the liquid surface. The shape of the liquid begins to change when a modest flow rate of electrically conductive liquid is pushed through a capillary (such as a hypodermic needle or stainless steel capillary) and subjected to a very high voltage (2–6 kV). Assuming a positive potential, positive ions will gather at the surface of the solution, creating an electrical pressure that tends to increase the surface area. Surface tension forces, which have a tendency to constrict or reduce the surface area, work against this pressure to cause the liquid to assume a cone-like form (Taylor, 1964). A liquid jet is blasted through the surface's apex when the electrical pressure exceeds the surface tension forces. The liquid jet becomes unstable and the charged droplet production starts at a certain distance from the nozzle. With an increase in the electric field, the length of the intact liquid jet diminishes. Even more, the cone vanishes as the voltage rises, and droplets are expelled from the capillary tip (Canals and Aguirre, 2006). Nebulization is continuous if the electrical pressure is kept above the critical level that is consistent with the liquid flow rate. Better nebulization is achieved by a coaxial sheath gas (dry nitrogen gas) flow surrounding the capillary.

Through solvent evaporation, the charged droplets get smaller with the help of nitrogen flow (Banerjee and Mazumdar, 2012).

According to the findings of numerous experimental studies on electrostatic nebulization, the size of the droplets produced is generally influenced by the applied voltage, surface tension, electrode size and configuration, liquid flow rate, and electrical properties of the liquid, such as dielectric constant and electrical conductivity. The maximum voltage that can be applied without experiencing too many corona losses produces the tiniest droplets (Ashgriz, 2011). The process by which the charged droplets produce solute ions is still up for debate (Canals and Aguirre, 2006). According to the charge residue concept, ions come from tiny droplets that contain a single molecule of the analyte. The Rayleigh limit is achieved as the newly created droplet travels along a pressure gradient toward the analyzer and the solvent evaporates, causing the droplet's diameter to decrease and its surface field to grow. The size of the charge is sufficient to dissipate the surface tension keeping the droplet together, resulting in a Coulomb explosion. As a result of the instability, the droplet fragments into a number of smaller droplets, each of which continues to evaporate until it reaches the Rayleigh limit and breaks apart. One possible outcome of this process is the production of an ion that contains just one analyte molecule. As the remainder of the solvent evaporates, the molecule keeps part of the charge from its droplet to become a free ion. The ion desorption model, the second method of ion production, postulates that before a droplet reaches the final stage, its surface electric field will be strong enough to lift an analyte ion at the droplet's surface over the energy barrier preventing its departure.

3.3 Conclusions

The nebulizer's function is to transform a purportedly homogenous sample solution into an aerosol for effective and repeatable conveyance to the plasma, where it will be quickly and consistently desolvated, evaporated, and dissociated, then stimulated and/or ionized. Each of those procedures affects the one after it. The analytical signal cannot be highly reproducible if the processes are not highly repeatable in time, location, and completeness. Since the surface area to volume ratio affects how quickly a droplet's solvent evaporates, the smaller the droplet, the more quickly and thoroughly it will desolvate, vaporize, dissociate, and go through excitation and/or ionization. Nevertheless, other factors limit the use of the nebulizers that create the tiniest droplet sizes. The truth is that no nebulizer performs all tasks well; each has strengths and weaknesses, and each is better suited to a specific set of conditions, whether it be in terms of cost, practicality, robustness, stability, usability, reproducibility, size of the aerosol it produces, or other factors that may be relevant.

References

Ashgriz, N. (2011). *Handbook of atomization and sprays: Theory and applications*. Springer Science & Business Media.

Babington, R. S., Yetman, A. A., & Slivka, W. R. (1969). *Method of atomizing liquids in a monodispersed spray*. Google Patents.

Banerjee, S., & Mazumdar, S. (2012). Electrospray ionization mass spectrometry: A technique to access the information beyond the molecular weight of the analyte. *International Journal of Analytical Chemistry, 2012*.

Berndt, H. (1988). High pressure nebulization: A new way of sample introduction for atomic spectroscopy. *Fresenius' Zeitschrift für Analytische Chemie, 331*(3), 321–323.

Burgener, J. A., & Makonnen, Y. (2020). Nebulization systems. In D. Beauchemin (Ed.), *Sample introduction systems in ICPMS and ICPOES* (pp. 57–142). Elsevier.

Canals, A., & Aguirre, M.Á. (2006). Roles of nebulizers in analytical chemistry. In *Encyclopedia of analytical chemistry: Applications, theory and instrumentation* (pp. 1–45). John Wiley & Sons.

Doherty, M. P., & Hieftje, G. (1984). Jet-impact nebulization for sample introduction in inductively coupled plasma spectrometry. *Applied Spectroscopy, 38*(3), 405–412.

Flament, M., Leterme, P., & Gayot, A. (2001). Study of the technological parameters of ultrasonic nebulization. *Drug Development and Industrial Pharmacy, 27*(7), 643–649.

Gañán-Calvo, A. M. (2005). Enhanced liquid atomization: From flow-focusing to flow-blurring. *Applied Physics Letters, 86*(21), 214101.

Hess, D. R. (2000). Nebulizers: Principles and performance. *Respiratory Care, 45*(6), 609.

Hidy, G. M., & Brock, J. R. (2016). *The dynamics of aerocolloidal systems: International reviews in aerosol physics and chemistry. Vol. 1*. Elsevier.

Ingeniatrics. (2022). *cited 2022 7/22/2022*. Available from: https://www.ingeniatrics.com/producto/oneneb.

Kuttruff, H. (2012). *Ultrasonics: Fundamentals and applications*. Springer Science & Business Media.

Lefebvre, A. H., & McDonell, V. G. (2017). *Atomization and sprays*. CRC Press.

Sharp, B. L. (1988). Pneumatic nebulisers and spray chambers for inductively coupled plasma spectrometry. A review. Part 1. Nebulisers. *Journal of Analytical Atomic Spectrometry, 3*(5), 613–652.

Taylor, G. I. (1964). Disintegration of water drops in an electric field. *Proceedings of the Royal Society of London. Series A: Mathematical and Physical Sciences, 280*(1382), 383–397.

Yang, C., et al. (1998). Thermospray nebulizer as sample introduction technique for microwave plasma torch atomic emission spectrometry. *Spectrochimica Acta Part B: Atomic Spectroscopy, 53*(10), 1427–1435.

CHAPTER

4

Aerosol characterization

Miguel Ángel Aguirre and Antonio Canals

Department of Analytical Chemistry and Food Science and University Institute of Materials, Faculty of Science, University of Alicante, Alicante, Spain

4.1 Introduction

A nebulizer is a device that converts a bulk liquid into a mist (i.e., aerosol) in a process called nebulization. The nebulization process can be considered as a disruption of surface tension by the action of diverse types of energy (e.g., pneumatic, ultrasonic, electric, among others) by means of the creation of an instability on the liquid surface. Surface tension is a cohesive force between molecules of the liquid, and it pulls the liquid into a sphere shape, since this has the minimum surface energy. In addition, liquid viscosity provides a stabilizing force by resisting any change in system geometry. However, when a disruptive force acts on a liquid surface, it can induce the liquid breakup if the magnitude of the disruptive force exceeds the surface tension.

Analytical nebulizers play a crucial role in atomic and molecular spectrometry, and liquid chromatography, among others, and the importance of understanding the fundamentals of aerosol generation and transport in order to improve analytical performance is absolutely clear in the different chapters of this book. For instance, it is well known that an ideal aerosol for ICP-based techniques should be very fine and monodispersive, and droplets should travel at uniform velocity, adequate to transport them into the instrument up to the aerosol transformation in the species to be detected. Unfortunately, aerosols produced by conventional nebulizers are far from ideal since they are polydisperse and are with different velocities. Droplet size and velocity are key parameters to control the quality of an aerosol for usefulness in analytical applications. For example, droplets smaller than a certain diameter and with certain velocity will result in complete desolvation, vaporization, atomization, excitation, and

Analytical Nebulizers
https://doi.org/10.1016/B978-0-323-91181-8.00003-5

ionization processes inside the plasma. Droplets with greater diameter and velocity will only partially undergo with those processes or, in the extreme case, will pass through the plasma without having any impact on the analytical signal. Therefore, these critical parameters are described in the following subsections.

4.2 Fundamentals

4.2.1 Droplet size and velocity distributions

Aerosol characterization for nebulizer diagnostics offers valuable information on the fundamental aspects of aerosol generation and transport (Benson, Gimelshein, Levin, & Montaser, 2001; Benson, Zhong, Gimelshein, Levin, & Montaser, 2003). Droplet size and velocity distributions are the most important factors that determine the quality of an aerosol. For example, in atomic spectrometry small and monodisperse droplets, with adequate and uniform velocity, must be introduced into the central channel of the plasma for efficient desolvation, vaporization, atomization, excitation, and ionization (Clifford, Ishii, Montaser, & Meyer, 1990; Olesik, 1997).

A typical droplet size distribution is shown in Fig. 4.1A. The droplet size distribution is conventionally obtained by plotting the number of droplets as a function of the droplet diameter. This is also referred as a number distribution curve. In addition, to representing the droplet size distribution, it is also helpful to use a cumulative distribution (Fig. 4.1B). This is a plot of the integral of the frequency curve, and it represents the percentage of the total number of droplets in the aerosol contained below a given size and its typical shape is shown in Fig. 4.1B.

In some cases, the droplet size distribution is obtained by plotting a histogram of droplet size, since a limited number of droplets counted. Therefore, it is useful to present a fit of the measured data with an appropriate function and they should have some important properties (Lefebvre &

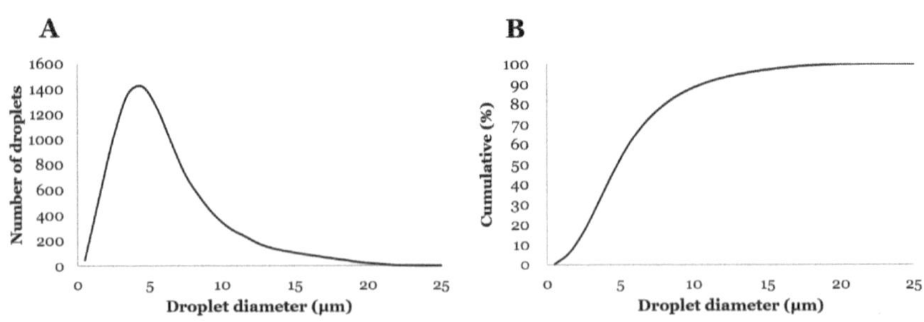

FIG. 4.1 Droplet size distribution based on: (A) number of droplets and (B) cumulative number of droplets in percentage. *No permission required.*

McDonell, 2017) including: (i) to provide a satisfactory fit to the droplet size data obtained in the experiment, (ii) to allow extrapolation to droplet sizes outside the range of measured values, (iii) to allow the calculation of different parameters of interest, and (iv) to provide some understanding about the basic mechanisms involved in the nebulization.

The distributions described in Fig. 4.1 represent the count distributions for a typical aerosol generated by a pneumatic nebulizer in analytical atomic spectrometry. The number distribution presents the fraction of the total number of droplets in any size range. In contrast, the graph of mass- or volume-based distribution displays how the total mass or volume of droplets is divided among the various droplet sizes. For analytical purposes, the mass or volume of the aerosol, instead of the number of droplets reaching the plasma, is more important. Thus, the mass or volume distribution is more appropriate for characterizing an analytical aerosol. Analogously, a cumulative mass- or volume-based distribution curve is more analytically relevant than a cumulative number distribution curve. For instance, the distributions depicted in Fig. 4.2 represent the differences between the count and volume distributions. The count distribution presents the total number of droplets in a given diameter. In contrast, the graph of volume distribution displays how the volume of droplets is divided over the diameter measured.

In order to characterize the droplet size distribution, several mean diameters are often calculated. As can be seen in Figs. 4.1 and 4.2, most droplet size distributions exhibit an asymmetrical shape with a long tail at large droplet diameter where the highest point of the droplet size distribution plot represents the number mode diameter (i.e., the most frequent droplet size found in a given aerosol). The mean diameter ($D_{1,0}$) is simply the mathematical mean of a number or volume distribution plot. The length mean diameter ($D_{2,1}$) is defined as the diameter of a droplet having the same surface/length ratio. Surface mean diameter, also known as Sauter mean diameter ($D_{3,2}$), is defined as the diameter of a droplet having the same volume/surface ratio. As a general expression of mean diameter:

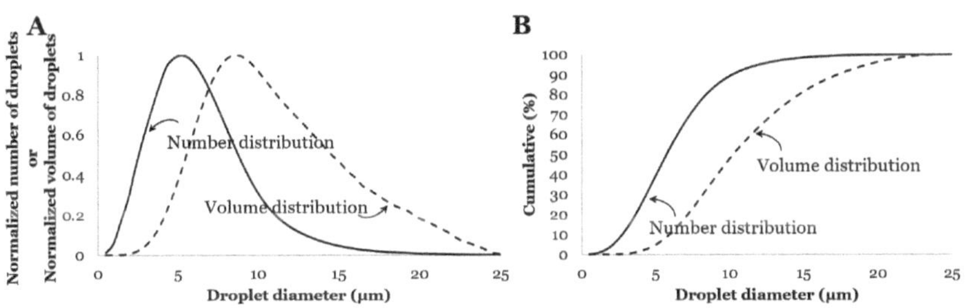

FIG. 4.2 Number and volume droplet size distributions: (A) the number and volume distributions and (B) cumulative distribution for count and volume distributions. *No permission required.*

$$D_{m,n} = \left(\frac{\sum n_i D_i^m}{\sum n_i D_i^n} \right)^{\frac{1}{m-n}} \tag{4.1}$$

where m and n values correspond to the effect investigated (e.g., 0, 1, 2, and 3 are related to count, length, surface, and volume, respectively); i denotes the size range considered; n_i is the number of droplets in size range i, and D_i is the diameter of size range i. Therefore, other mean diameters can be defined. For instance, $D_{2,0}$ is the surface average value of all the droplets in the aerosol, and $D_{3,0}$ is the diameter of a droplet whose volume is equal to the total number of droplets. In terms of cumulative number distribution, several characteristic diameters can also be defined. For instance, $D_{10\%}$, $D_{50\%}$, and $D_{90\%}$ are defined as the droplet diameter below which the cumulative numbers of droplets are 10%, 50%, and 90%, respectively. If the cumulative distribution is expressed in volume, the previous representative diameters should be written as $vD_{10\%}$, $vD_{50\%}$, and $vD_{90\%}$, respectively. For the sake of clarity, Fig. 4.3 shows the most important diameters mentioned above.

Among all the mean diameters calculated, the Sauter mean diameter ($D_{3,2}$) is the most important one to describe the quality of an aerosol since it is directly related to the analytical signals in ICP-based techniques (Liu & Montaser, 1994).

A number of mathematical functions have been proposed based on statistical distribution functions and empirical distributions to predict the evolution of the droplet size distribution. On the one hand, log-normal functions (i.e., statistical distribution) have been used for many years for aerosol characterization. One of the most significant characteristics of this function is that different log-normal distributions generated under similar circumstances (i.e., an aerosol generated by the same nebulizer under similar experimental conditions) have a nearly constant standard deviation (Gustavsson, 1984). This means that, for nebulizers with similar

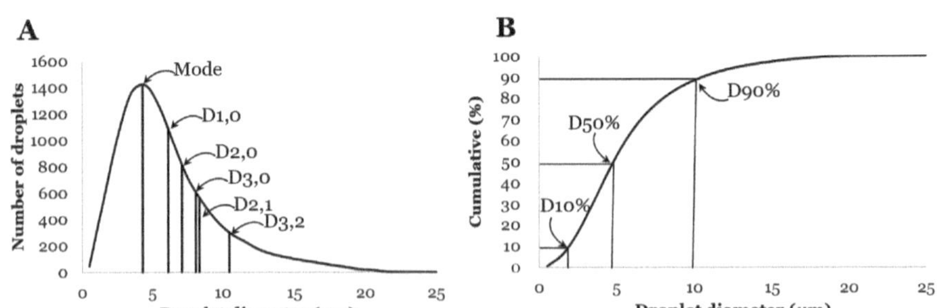

FIG. 4.3 Locations of various representative diameters in the droplet size distribution based on: (A) number of droplets and (B) cumulative number of droplets in percentage. *No permission required.*

relevant dimensions and specific geometries, the standard deviation of the droplet size distribution can be considered constant.

On the other hand, several empirical distributions have been proposed to characterize the distribution of droplet sizes in an aerosol. Some of the functions most frequently used in the analysis and correlation of droplet size data are given in the following sections. For many years, the model accepted in atomic spectrometry for the prediction of the Sauter mean diameter of pneumatically generated aerosols was that derived empirically by Nukiyama and Tanasawa (N-T) (Browner, Boom, & Smith, 1982). Nevertheless, Canals, Hernandis, and Browner (1990) checked the applicability of this model to concentric nebulizers, employed under normal nebulization conditions in atomic spectrometry, and they concluded that: (i) the $D_{3,2}$ values obtained using this model are higher than those measured using a laser diffraction technique, and (ii) the model does not reproduce the solvent nature effect, since trends in experimental $D_{3,2}$ values are different from those predicted by the model. An empirical method for $D_{3,2}$ prediction based on experimental data interpolation was proposed by Canals, Wagner, Browner, and Hernandis (1988). Using this method, the Sauter mean diameter of an aerosol, pneumatically generated under a given set of experimental conditions, was calculated by interpolation between two $D_{3,2}$ values measured in two similar nebulizers working under the same set of experimental conditions. Olesik and Bates (1995) checked the interpolation method and concluded that, under certain experimental conditions, the method is effective for predicting trends in $D_{3,2}$, when the liquid uptake rate and gas flow rate are modified, but is unable to predict the $D_{3,2}$ values accurately. Using this method, the calculated values were always of the same order of magnitude as the experimental ones (differences lower than 47%). Nevertheless, the method is scarcely used in practice, since it is valid only for concentric nebulizers of the same type (e.g., types A, C, and K) and with similar gas cross-sectional areas rather than the two used as a reference.

Years later, Gras, Alvarez, and Canals (2002) modified semiempirical models (i.e., the Rizk-Lefebvre model for concentric nebulizers and the El Shanawany-Lefebvre model for prefilming nebulizers) for $D_{3,2}$ prediction, which were not previously employed in atomic spectrometry, and the authors satisfactorily predicted the Sauter mean diameter of aerosols generated by three concentric pneumatic nebulizers with different dimensions and geometries (i.e., conventional A and C types, and a slurry nebulizer) and a single-bore, high-pressure pneumatic nebulizer. In addition, the authors evaluated different solvents (i.e., water, ethanol, methanol, 2-propanol, and 1-butanol) as a function of the liquid and gas flow rates. The authors concluded that by assuming a lognormal function, the shape of the primary droplet size distribution generated by the nebulizers tested can be predicted using the modified models. Three years later, Kahen,

Aeon, and Montaser (2005) proposed modified Nukijama-Tanasawa and Rizk-Lefebvre models to predict the $D_{3,2}$ values produced by a direct-injection, high-efficiency nebulizer (DIHEN) using several organic solvents (i.e., hexane, acetone, xylene, toluene, methanol, and ethanol).

As is abovementioned, the droplet velocity is another important factor affecting the quality of the aerosol. As the droplet size distribution, the velocity distribution is obtained by plotting the number of droplets as a function of their velocity. In this case, the mean velocity is the most used parameter. Taking into account that the droplet mass and velocity are the most important factors that determine the quality of an aerosol, several researchers agree that the momentum (i.e., momentum = mass · velocity) should be measured to evaluated the quality of the aerosol (Clifford et al., 1990; Matusiewicz, Ślachciński, Almagro, & Canals, 2009; Schaldach, Berndt, & Sharp, 2003).

4.2.2 Droplet size dispersion

Every droplet size distribution is characterized, at least, by two parameters that represent: (i) position (i.e., droplet mean diameter) and (ii) dispersion. The above section has been dedicated to calculate the droplet mean diameter. On the contrary, this section is dedicated to characterize the droplet size dispersion. The term dispersion expresses the range of droplet sizes in an aerosol and the relative span factor is the parameter most commonly used. This parameter indicates the uniformity of the droplet size distribution, and it is defined by

$$\text{Span} = \frac{D_{90\%} - D_{10\%}}{D_{50\%}} \tag{4.2}$$

where $D_{90\%}$, $D_{10\%}$, and $D_{50\%}$ refer to the droplet diameter below which a cumulative 90%, 10%, and 50% of a collected aerosol are found, respectively (Fig. 4.3B). The closer this number is to zero, the more uniform the spray will be.

4.2.3 Measurement of droplet size and velocity distributions

The problem of measuring the size and the velocity of small droplets is encountered in many fields of science and technology, and many different methods have been developed. However, special difficulties arise in the application of these methods to the measurement of droplet sizes and velocities in an aerosol, including: (i) the high number and varied size of droplets in an aerosol, (ii) the high and varied speed of the droplets, and (iii) the changes in droplet size due to the coalescence and evaporation. All these factors must be considered when choosing a droplet size measuring technique for any given application.

Aerosol diagnostic techniques can be categorized according to the nature of the analysis as: (i) electrical, (ii) mechanical, and (iii) optical methods. The ideal diagnostic technique must provide simultaneous information about the aerosol size and velocity distributions. A great number of new aerosol diagnostic techniques are now available to aid in the understanding of aerosol generation and transport processes, but many have not been applied to aerosols used in analytical applications. Only mechanical and optical methods are widely employed in this field. For this reason, those methods are described in the following sections.

4.2.3.1 Mechanical methods

Mechanical methods have the benefits of low cost and simplicity. Their main drawbacks are the extraction and collection of representative aerosol samples. Microscopic measurements are often made by collecting aqueous aerosol on glass slides coated with oil films or magnesium oxide (Browner, 1987). This technique generates a numerical count distribution, which can be converted into mass or volume distribution, being more analytically valuable. Nevertheless, this approach suffers from three main problems: (i) the smallest droplets in diameter are difficult to sample efficiently, the result being size distributions that differ from the original aerosol in favor of larger droplets; (ii) the aerosol is often dense enough to rapidly overload the glass slide; such overloading causes droplet coalescence and, consequently, a biased droplet distribution is obtained; and (iii) evaporation effects are important in measuring fine aerosol since the lifetime of small droplets is extremely short. Summarizing, the three problems described provide a droplet size distribution commonly biased toward larger droplet diameters.

Another mechanical method for droplet sizing involves a cascade impactor. Browner et al. have used this approach extensively in analytical atomic spectrometry (Browner et al., 1982; Novak & Browner, 1980).

In this device (Fig. 4.4), the aerosol is passed through different collection plates containing orifices that become progressively smaller in diameter, involving eight stages and a final absolute filter. The final filter collects droplets smaller than the impactor's minimum collection diameter. In general, the droplet size distribution is calculated by weighing the aerosol mass collected on each plate. Some examples of droplet size distributions of aerosol produced by concentric (Browner et al., 1982), cross-flow (Skogerboe & Freeland, 1985a, 1985b), and thermospray nebulizers (Koropchak, Aryamanya-Mugisha, & Winn, 1988) have been reported in the field of analytical atomic spectrometry.

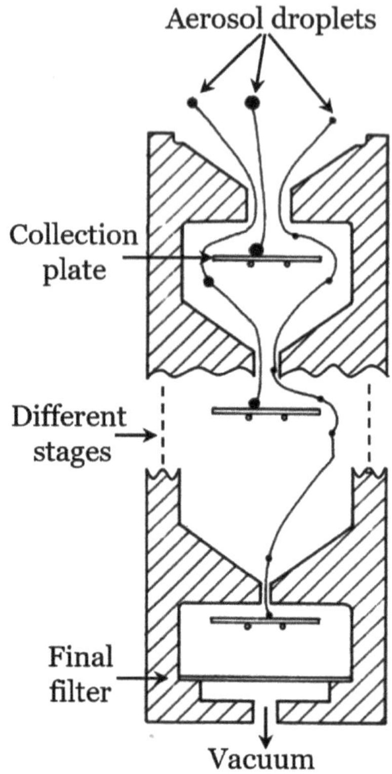

FIG. 4.4 Schematic representation of the principle of operation of an Andersen cascade impactor. *Adapted from Ali, M. (2010) Pulmonary drug delivery. In Kulkarni, V. S. (Ed.),* Handbook of noninvasive drug delivery systems *(pp. 209–246). Norwich: William Andrew Publishing. Copyright (2010), with permission from Elsevier.*

4.2.3.2 *Optical methods*

Optical methods consist of imaging and nonimaging techniques. Imaging techniques allow the aerosol droplets to be seen and include high-speed photography, particle imaging velocimetry (PIV) and holography, among others. Nonimaging techniques are based upon light-scattering measurements and include Fraunhofer light diffraction and light scattering interferometry (phase Doppler), among others. A variety of optical methods have been employed for aerosol characterization. The different modalities have their own advantages and limitations, but they all have the important attribute of allowing size/velocity measurements to be made without the insertion of a physical probe into the aerosol.

Undoubtedly, photography is currently one of the most accurate and least expensive techniques for measuring droplet size and velocity. The technique takes an aerosol image through the use of short-duration light pulses (microsecond to nanosecond duration) or high-speed cameras. This method can provide spatially resolved data on droplet size, shape, velocity, and number density. High-speed visualization is also a very effective tool for examining the mechanism of aerosol generation and droplet breakup (Sharp, 1988). Particle-imaging velocimetry (PIV) is an optical imaging technique for measuring droplet velocity at several points in a flow field simultaneously and instantaneously. The technique uses a thin sheet of light to illuminate a single plane in the flow field. PIV determines the velocity and the moving direction of droplets in the illuminated plane by evaluating the distance between two successive images of the same droplet and can be used from low velocities all the way up to supersonic speeds (Williams, Nguyen, Schreyer, & Smits, 2015). Aside from size, aerosol velocity is also an important parameter in determining the analytical performance. For instance, in ICP-based techniques, fast droplets do not spend enough time in the plasma for complete desolvation, vaporization, atomization, excitation, and/or ionization. This is particularly important for direct injection nebulizers where no spray chambers are used. Another important imaging method is the shadowgraphy and it consists of a pulsed backlight illumination (i.e., a pulsed laser or a flash lamp, depending on the velocity of the droplets), which illuminates the droplets passing the system between the light source and the camera with magnification lenses. The shadow picture consists of dark spots representing the droplet size and the images are treated with an image analysis software, which determines the diameters of the dark spots (Wang, Si, Tan, Zheng, & Yang, 2019).

Currently, nonimaging optical methods are used mostly by analytical chemists for measuring size and velocity distributions. In nonimaging optical methods, when a light beam is incident on a drop, scattering occurs in all directions. When droplet diameters are greater than the wavelength of the light beam, the intensity and the angular distribution of the scattered light can be described by the Mie theory. When droplets are much larger than the wavelength of light, theories of Fraunhofer diffraction, reflection, and refraction can be used (Kahen, Jorabchi, Gray, & Montaser, 2004). Related to the Fraunhofer diffraction, the form of the Malvern droplet size analyzer has been used most widely. Unfortunately, droplet velocity cannot be measured with this method. Nonimaging optical methods based on light-scatter interferometry, particularly the phase-Doppler particle analyzer (PDPA), provide size and velocity information simultaneously and therefore offer useful information for diagnostic studies of aerosol-related processes in nebulization devices and plasmas.

In the following sections, some of the most common nonimaging optical methods used by analytical chemists for aerosol characterization are discussed.

Fraunhofer laser diffraction

This technique is based on the diffraction of a collimated laser beam by the aerosol, resulting in a distribution of diffraction angles corresponding to the size distribution of the polydisperse aerosol (Mohamed, Fry, & Wetzel, 1981).

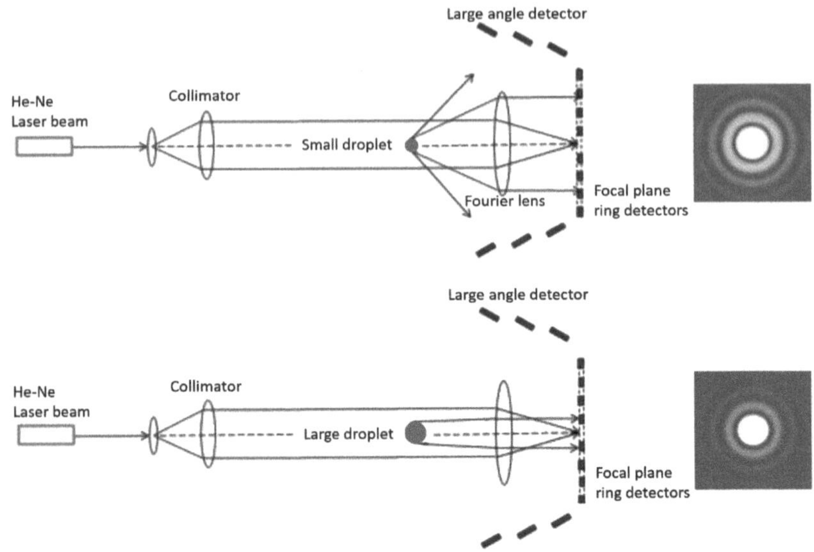

FIG. 4.5 A Fraunhofer laser-diffraction system. *Adapted from Gupta, A., & Yan, D. (2016). Particle size estimation and distributions. In A. Gupta, & D. Yan (Eds.),* Mineral processing design and operations *(pp. 33–69). Amsterdam: Elsevier. Copyright (2016), with permission from Elsevier.*

Fig. 4.5 shows the scheme of a Malvern system (Gupta & Yan, 2016). The light from a He-Ne laser is expanded into a beam of 1 cm diameter. If the parallel beam of light strikes a droplet, some of the light is diffracted by an angle depending on the droplet size, and a series of light and dark concentric rings is formed. This pattern is known as the Fraunhofer diffraction, and the space of the ring is a function of the droplet size. As droplets are made smaller, the diffraction pattern becomes wider. The pattern is transferred through a Fourier transform receiver lens to a multielement photodetector that registers the light energy distribution. Problems such as multiple scattering, positioning of the aerosol with respect to the lens, and optical alignments limit the accuracy of the measurements. In addition, the method does not provide any information regarding the velocity of the droplets.

Planar dropsizing

Planar dropsizing is a two-dimensional technique that can be used to study the structure of the aerosol. In this method, the flow field is illuminated with a laser sheet and the laser induced fluorescence (LIF) and the Mie scattering signals are recorded on two different coupled charged devices (CCD). The ratio of LIF/Mie signals is then calculated, providing a two-dimensional map of the Sauter mean diameter (D_{32}), defined as the volume-to-surface area ratio of aerosol in the illuminated field (Domann &

Hardalupas, 2003). The approach is based on the simple concept that, for a droplet which is excited, the fluorescence signal will give a measure of the droplet volume, whereas, the Mie scattering will represent the surface area. By dividing the LIF signal by the Mie signal, the Sauter mean diameter can be deduced.

Phase-Doppler particle analyzer

The phase-Doppler particle analyzer (PDPA) is a technique that provides simultaneous size and velocity information on individual droplets (Liu & Montaser, 1994). Therefore, size and velocity distributions for a polydisperse aerosol can be obtained using PDPA. Depending on the number of lasers used, different PDPA systems can be described. The simplest PDPA system is known as the one-component system because only one velocity component is measured simultaneously with the droplet size. In two- and three-component systems, droplet velocities are measured in two and three directions, respectively. Fig. 4.6 shows the setup of one-component PDPA system (Fulk, Beaudry, & Rochelle, 2017). As can be seen, the laser beam (Fig. 4.6, number 1) is divided, and one beam is phase shifted with a Bragg cell (Fig. 4.6, number 2). Then, the two beams intersect (i.e., measurement volume) in the aerosol with a given angle (Fig. 4.6, number 3). The intersection of the two beams creates an interference pattern, which means a series of light and dark fringes (Fig. 4.6, number 4).

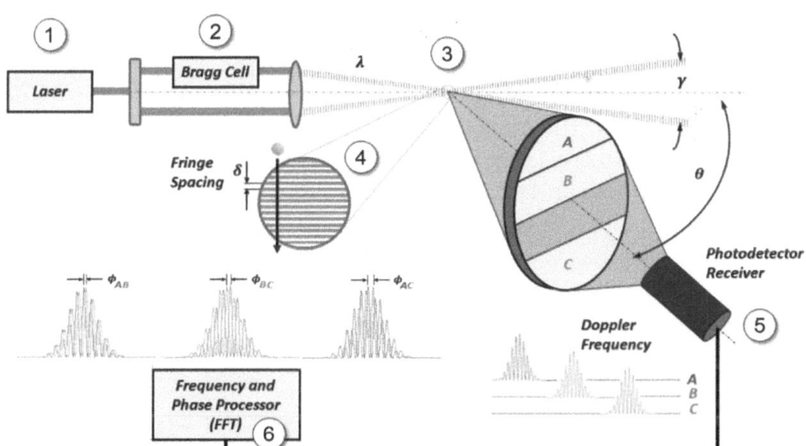

FIG. 4.6 One-component PDPA operation and data sequence. Each droplet that flows through the measurement volume scatters light from the interference pattern. The scattered light creates a signal by means of a photodetector. The phase difference of the signals is related to the particle diameter. *Adapted from Steven M. Fulk, Matthew R. Beaudry, Gary T. Rochelle (2017). Amine aerosol characterization by phase Doppler interferometry.* Energy Procedia, *114, 939–951. Copyright (2016), with permission from Elsevier.*

When a droplet crosses the fringes, it scatters the light, which is collected by a lens and focused onto a photo-detector, which converts the light into a signal. Commonly, three photomultiplier tubes (Fig. 4.6, number 5), placed at an off-axis angle, measure the scattered light. For velocity measurement, since the fringe distance is known and the time for the droplet to pass from one fringe to the next is measured by the detectors, the velocity can be calculated by dividing the fringe distance by the time. Besides, because of one laser beam being phase shifted with respect to the other one, the interference pattern is moving with a velocity depending on the phase difference of the Bragg cell. This is very useful since the velocity direction can be distinguished, depending on if the droplet is moving against or in the direction of the moving interference pattern. In order to measure the droplet diameter, three detectors are used located at different positions. Detectors receive the scattered light with a measurable phase difference since the optical path length of the scattered light is dependent on the position of the detectors, with the phase difference having a linear dependence on the droplet diameter. Therefore, three different frequency and phase differences are processed for each droplet: (i) A and B, ϕ_{AB}, (ii) B and C, ϕ_{BC}, and (iii) detectors A and C, ϕ_{AC} (Fig. 4.6, number 6). Due to the different elevation angles, the phase differences for the three detector pairs are different and the closely spaced pair yield a lower phase difference. Two of the three phase differences are linearly independent and can be used for two independent droplet diameter estimations. The third measurement can be used as a validation criterion.

Interferometric laser imaging for droplet sizing

Another planar-based sizing method involves imaging the interference patterns generated by the interaction of laser light with the glare points of droplets. Interferometric laser imaging for droplet sizing (ILIDS) provides the instantaneous size and spatial droplet distributions. Fig. 4.7 shows the

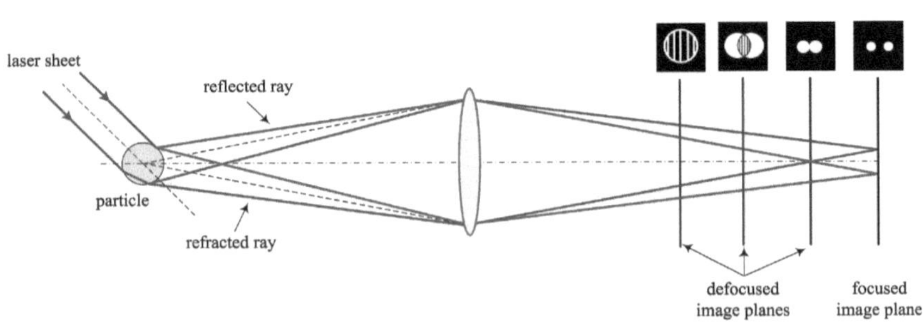

FIG. 4.7 Schematic illustration of interferometric laser imaging for droplet sizing (ILIDS). *Adapted from Qieni, L., Wenhua, J., Tong, L., Xiang, W., & Yimo, Z. (2014). High-accuracy particle sizing by interferometric particle imaging.* Optics Communications, 312, *312–318. Copyright (2014), with permission from Elsevier.*

basic principle of this optical method (Qieni, Wenhua, Tong, Xiang, & Yimo, 2014). A pulsed laser sheet is sent toward the aerosol and the reflected and first-order refracted rays are collected off-axis by a CCD camera. Two bright spots are observed in the focal plane, one glare point representing the reflected ray and the other the first-order refracted ray. The interference fringe pattern is produced in an out-of-focus image plane. For each droplet, the number of fringes is proportional to the droplet diameter, by a factor which depends on the receiving optics parameters (numerical aperture, focal length, and out-of-focus distance, among others) (Jorabchi, Brennan, Levine, & Montaser, 2006). One of the main advantages of this technique is its insensitivity to changes in refractive index, presenting an attractive optical method for dropsizing in high-temperature environments where refractive index is not well defined. Moreover, the CCD camera employed to capture interference images is also used for recording the moving droplet at two difference instances with a known time lag.

Experimental investigations using optical methods

Optical methods are commonly employed to compare droplet size/velocity distributions between different nebulizers. Although several researchers compared a couple of nebulizers, only few studies evaluate critically the aerosol characteristics of four or more nebulizers, being the works carried out by Almagro et al. (Almagro, Gañán-Calvo, Hidalgo, & Canals, 2006; Matusiewicz et al., 2009) and Inagaki et al. (Groombridge et al., 2012; Inagaki et al., 2014). In the first work of Almagro et al. (2006), the authors compared five commercial nebulizers [i.e., AriMist (AM), high-efficiency nebulizer (HEN), MicroMist (MM), MiraMist (MiM), and perfluoroalkyl nebulizer (PFA)] to the flow focusing pneumatic nebulizer (FFPN). In order to characterize the primary aerosols of different nebulizers, a Fraunhofer laser diffraction system was used to measure the volume median diameter ($vD_{50\%}$). Primary aerosols were sampled 1.5 mm from the nebulizer tip along the center line of the primary aerosols. The authors evaluated all nebulizers under different conditions (i.e., 0.6, 0.7, and 0.8 $L\,min^{-1}$ of gas flow rate and 100, 200, 400, and 1000 $\mu L\,min^{-1}$ of liquid flow rate), and all primary aerosols produced by FFPN were contained in droplets smaller than 15 μm. Besides, the FFPN produces a finer and narrower primary aerosol than the commercial pneumatic nebulizers evaluated. As an example, at a liquid flow rate of 200 $\mu L\,min^{-1}$ and a gas flow rate of 0.7 $L\,min^{-1}$, the median diameter ($vD_{50\%}$) values of the primary aerosols were 21.71, 31.75, 16.84, 13.13, 4.92, and 3.25 μm with the PFA, MM, AM, MiM, HEN, and FFPN, respectively. In the second study (Matusiewicz et al., 2009), the authors compared the flow blurring nebulizer (FBN) with six nebulizers [i.e., AM, demountable direct injection high efficiency nebulizer (DIHEN), HEN,

MiM, Meinhard nebulizer (MN), and ultrasonic nebulizer (NOVA-1)]. In this work, droplet size and velocity distributions of primary aerosols were determined using a two-dimensional phase Doppler particle analyzer. Primary aerosol was sampled at a distance of 5.0 mm from the nebulizer tip along the centerline of the aerosol. In this research, the authors characterized the aerosols of the different nebulizers using the optimized operating conditions of liquid and gas flow rate obtained by microwave plasma atomic emission spectrometry (MP-AES). The optimized conditions were different for every nebulizer and the cumulative number of droplets in percentage of primary aerosol contained in droplets having sizes smaller than 20 μm were 93%, 95%, and 85% for AM, MiM, and MN, respectively and near 100 % for DIHEN, FBN, and HEN. The authors demonstrated that the droplet size distribution of the aerosols from DIHEN, FBN, and HEN was much finer (the fraction of small droplets was higher and more monodisperse) than that produced by other nebulizers. In addition, the mean velocities of primary aerosols were 73, 45, 39, 36, 31, and 25 m s^{-1} for MiM, HEN, AM, FBN, MN, and DIHEN, respectively. The differences in velocity were a consequence of the different operation conditions. For example, the optimized gas flow rate for MiM had the highest value, and hence showed the highest mean velocity of primary aerosol. On the contrary, the gas flow rate for DIHEN was the lowest value, and hence showed the smallest mean velocity.

More recently, two research works published by Inagaki et al. showed the comparison between their proposed nebulizer (i.e., the modified high-performance triple-tube concentric nebulizer (HPCN) (Groombridge et al., 2012) and the concentric-type grid nebulizer (CGrid) (Inagaki et al., 2014) with different commercial nebulizers. In the first research (Groombridge et al., 2012), the authors compared the HPCN with four nebulizers [i.e., Conikal (Ck), MiM, MN, and SeaSpray (SS)]. The authors employed a Fraunhofer laser diffraction system as an optical method for measuring the droplet size distributions of the primary aerosol generated at 5 mm from the nebulizer tip through the centerline of the spray cone. In this study, the authors compared all the nebulizers under the same conditions (i.e., liquid flow rate of 800 μL min^{-1} and gas flow rate of 1 L min^{-1}). Under these conditions, the aerosol generated by HPCN was finer than those obtained with other nebulizers, obtaining a $D_{3,2}$, v$D_{50\%}$, and v$D_{90\%}$ of 3.4, 4.3, and 10.9 μm, respectively. In the second study (Inagaki et al., 2014), the same optical method was employed and the operating conditions for each nebulizer were optimized by inductively coupled plasma optical emission spectrometry (ICP-OES). Under optimized conditions, the developed nebulizers (i.e., CGrid and HPCN) were compared with two commercial nebulizers, MN and OneNeb (ON), and the values obtained (i.e., v$D_{50\%}$ and v$D_{90\%}$) were very similar between CGrid, HPCN, and ON, CGrid being slightly better.

Although optical methods are commonly employed to characterize the aerosol generated by different nebulizers, several studies used optical imaging methods to improve the analytical capabilities of flame- and ICP-based techniques. Hieftje and Malmstadt were the first authors to study the velocity of the droplets and their desolvatation in air-acetylene laminar flames using high-speed photography (Hieftje & Malmstadt, 1968, 1969). Following this initial work, Hieftje et al. employed a similar optical method to study the vaporization of different solutions and droplet acceleration in flame spectrometry (Bastiaans & Hieftje, 1974; Russo & Hieftje, 1980). Monnig and Koirtyohann (1985) continued the research, and the authors investigated the Mie scattering in order to monitor the desolvation and vaporization of analyte aerosol in ICP-based techniques. The Mie scattering profile showed that solvent vaporization is essentially complete at the analytical zone of the plasma. Moreover, the optical method revealed that the droplets formed are quickly vaporized for volatile matrices, but some refractory oxide particles persisted throughout the analytical zone in the plasma. Years later, Olesik et al. (Olesik, 1997; Olesik & Fister, 1991) published remarkable works researching the fate of droplet behavior in an ICP, using the laser light scattering. This optical method was employed to measure the number of incompletely desolvated droplets in the plasma. The authors demonstrated the estimated sizes of droplets incompletely desolvated in the plasma. For the particular conditions in the sample introduction used (Olesik, 1997), droplets that were originally about 13 μm in diameter or larger remained incompletely desolvated at 10 mm above the load coil. In another experiment where the nebulizer gas flow rate was increased, droplets originally 13 μm in diameter or larger were not completely desolvated until about 17 mm above the load coil.

Montaser and coworkers (Jorabchi et al., 2006; Jorabchi, Kahen, Gray, & Montaser, 2005; Kahen et al., 2004) continued the research of the fate of aerosol droplets in the ICP using different novel optical methods. In the first and second articles (Jorabchi et al., 2005; Kahen et al., 2004), particle image velocimetry was employed to study the in situ behavior of the aerosol before interaction with the plasma, while the individual surviving droplets are explored by particle tracking velocimetry. The aerosol produced by DIHEN and PFA were compared entering the ICP torch without the plasma ignited. The tertiary aerosol produced by the combination of PFA and spray chamber was highly confined to the center of the plasma. In contrast, DIHEN aerosol was not as homogeneous as the previous one and they could cause plasma instability. Afterward, particle tracking velocimetry was used to track droplets' surviving evaporation within the ICP itself and to produce qualitative distributions of the axial and radial velocity of surviving droplets. The tertiary aerosol (i.e., PFA and spray chamber) showed few and slower surviving droplets in comparison with DIHEN aerosol which were due to faster droplets were also more likely to survive the ICP as there was less time for their desolvation. The results highlighted the importance of the droplet velocity for efficient droplet evaporation. In the third work, size, velocity, and evaporation rate of droplets in an ICP were simultaneously measured using the ILIDS optical method. The results indicated that at normal operating conditions for DIHEN (i.e., gas flow rate of $0.2\,L\,min^{-1}$ and $85\,\mu L\,min^{-1}$ of liquid flow rate), surviving droplets in the analytical zone of the plasma possess diameters, velocities, and evaporation rates of 3–30 μm, $20–70\,m\,s^{-1}$, and $0.26–0.36\,mm^2\,s^{-1}$, respectively. The authors concluded that taking into account the diameter, velocity, and evaporation rate, the maximum estimated initial diameter is 18 μm for complete evaporation.

4.3 Conclusions

The study of the aerosol properties is an integral part of any development in the nebulization area toward the ultimate goal of minimum sample consumption, nearly monodisperse droplets and uniform velocity. A typical aerosol includes a wide range of droplet sizes and velocities. Some knowledge of droplet size and velocity distribution is helpful in the analytical applications of aerosols. Unfortunately, a robust correlation has not yet been developed to describe the majority of nebulization processes. Thus, only empirical correlations are available for predicting mean droplet sizes. However, in addition to mean droplet size, another parameter of importance in the definition of the droplet size distribution is the dispersion of droplet sizes (e.g., span) it contains. For analytical ICP spectrometry, the smaller the span (i.e., the greater the monodispersity of the aerosol), the better the quality of the aerosol. If an aerosol has a large span, one should expect a greater level of short-term fluctuation in desolvation, vaporization, atomization, excitation, and ionization processes in the plasma.

The use of optical methods has revolutionized the droplet size/velocity distribution measurements. Nevertheless, they have their own advantages and limitations, which mostly depend on the operating conditions and optical properties of the working liquid. On the basis of the review, several general conclusions can be made: (i) optical methods based on laser systems could provide fast measurements, which are suitable for online monitoring and for aerosol with a high rate of change; (ii) optical methods based on image analysis could provide a high number of images in a short time with the development of camera and lenses, but it requires a lot of work and time to extract data of droplets from a large number of images; (iii) the majority of optical methods depend on the optical properties of the droplets to be measured (i.e., refractive index); and (iv) planar methods are starting to make inroads owing to their highly time-efficient application coupled with ease of adding complementary measurements.

Acknowledgments

The authors are grateful to the Spanish Ministry of Science and Innovation (PID2021-126155OB-I00) and the Regional Government of Valencia (Spain) (CIPROM/2021/062) for the financial support. The authors extend their appreciation to Ministry of Science, Innovation and Universities for granting the Spanish Network of Excellence in Sample Preparation (RED2018-102522-T). This chapter is based upon work from the Sample Preparation Study Group and Network, supported by the Division of Analytical Chemistry of the European Chemical Society.

References

Almagro, B., Gañán-Calvo, A. M., Hidalgo, M., & Canals, A. (2006). Flow focusing pneumatic nebulizer in comparison with several micronebulizers in inductively coupled plasma atomic emission spectrometry. *Journal of Analytical Atomic Spectrometry, 21*(8), 770–777. https://doi.org/10.1039/b518282d.

Bastiaans, G. J., & Hieftje, G. M. (1974). Vaporization of individual solute particles in flame spectrometry. *Analytical Chemistry, 46*(7), 901–910. https://doi.org/10.1021/ac60343a009.

Benson, C. M., Gimelshein, S. F., Levin, D. A., & Montaser, A. (2001). Simulation of droplet heating and desolvation in an inductively coupled plasma—Part I. *Spectrochimica Acta - Part B Atomic Spectroscopy, 56*(7), 1097–1112. https://doi.org/10.1016/S0584-8547(01)00233-6.

Benson, C. M., Zhong, J., Gimelshein, S. F., Levin, D. A., & Montaser, A. (2003). Simulation of droplet heating and desolvation in inductively coupled plasma—Part II: Coalescence in the plasma. *Spectrochimica Acta - Part B Atomic Spectroscopy, 58*(8), 1453–1471. https://doi.org/10.1016/S0584-8547(03)00137-X.

Browner, R. F. (1987). Fundamental aspects of aerosol generation and transport. In *Inductively coupled plasma emission spectroscopy, Part II* Wiley & Sons.

Browner, R. F., Boom, A. W., & Smith, D. D. (1982). Aerosol transport model for atomic spectrometry. *Analytical Chemistry, 54*(8), 1411–1419. https://doi.org/10.1021/ac00245a034.

Canals, A., Hernandis, V., & Browner, R. F. (1990). Experimental evaluation of the Nukiyama-Tanasawa equation for pneumatic nebulisers used in plasma atomic emission spectrometry. *Journal of Analytical Atomic Spectrometry, 5*(1), 61–66. https://doi.org/10.1039/JA9900500061.

Canals, A., Wagner, J., Browner, R. F., & Hernandis, V. (1988). Empirical model for estimating drop size distributions of aerosols generated by inductively coupled plasma nebulizers. *Spectrochimica Acta Part B: Atomic Spectroscopy, 43*(9–11), 1321–1335. https://doi.org/10.1016/0584-8547(88)80172-1.

Clifford, R. H., Ishii, I., Montaser, A., & Meyer, G. A. (1990). Dual-beam, light-scattering interferometry for simultaneous measurements of droplet-size and velocity distributions of aerosols from commonly used nebulizers. *Analytical Chemistry, 62*(4), 390–394. https://doi.org/10.1021/ac00203a015.

Domann, R., & Hardalupas, Y. (2003). Quantitative measurement of planar droplet Sauter mean diameter in sprays using planar droplet sizing. *Particle and Particle Systems Characterization, 20*(3), 209–218. https://doi.org/10.1002/ppsc.200390027.

Fulk, S. M., Beaudry, M. R., & Rochelle, G. T. (2017). *Energy procedia. 114* (pp. 939–951). Elsevier Ltd. https://doi.org/10.1016/j.egypro.2017.03.1237.

Gras, L., Alvarez, M. L., & Canals, A. (2002). Evaluation of new models for drop size distribution prediction of aerosols in atomic spectrometry: Pneumatic nebulizers. *Journal of Analytical Atomic Spectrometry, 17*(5), 524–529. https://doi.org/10.1039/b109114j.

Groombridge, A. S., Inagaki, K., Fujii, S. I., Nagasawa, K., Okahashi, T., Takatsu, A., et al. (2012). Modified high performance concentric nebulizer for inductively coupled plasma optical emission spectrometry. *Journal of Analytical Atomic Spectrometry*, *27*(10), 1787–1793. https://doi.org/10.1039/c2ja30118k.

Gupta, A., & Yan, D. (2016). *Particle size estimation and distributions* (pp. 33–69). Elsevier. https://doi.org/10.1016/b978-0-444-63589-1.00002-2.

Gustavsson, A. (1984). Prediction of nebulizer characteristics for concentric nebulizer systems with a mathematical model. *Analytical Chemistry*, *56*(4), 815–817. https://doi.org/10.1021/ac00268a053.

Hieftje, G. M., & Malmstadt, H. V. (1968). Unique system for studying flame spectrometric processes. *Analytical Chemistry*, *40*(12), 1860–1867. https://doi.org/10.1021/ac60268a024.

Hieftje, G. M., & Malmstadt, H. V. (1969). New approach to flame spectrometric analysis utilizing isolated droplets of sample solution. *Analytical Chemistry*, *41*(13), 1735–1744. https://doi.org/10.1021/ac60282a022.

Inagaki, K., Fujii, S., Miyashita, S., Nagasawa, K., Okahashi, T., Groombridge, A. S., et al. (2014). A novel concentric grid nebulizer for inductively coupled plasma optical emission spectrometry. *Journal of Analytical Atomic Spectrometry*, *29*(11), 2136–2145. https://doi.org/10.1039/C4JA00233D.

Jorabchi, K., Brennan, R. G., Levine, J. A., & Montaser, A. (2006). Interferometric droplet imaging for in situ aerosol characterization in an inductively coupled plasma. *Journal of Analytical Atomic Spectrometry*, *21*(9), 839–846. https://doi.org/10.1039/b604403d.

Jorabchi, K., Kahen, K., Gray, C., & Montaser, A. (2005). In situ visualization and characterization of aerosol droplets in an inductively coupled plasma. *Analytical Chemistry*, *77*(5), 1253–1260. https://doi.org/10.1021/ac048576k.

Kahen, K., Aeon, B. W., & Montaser, A. (2005). Modified Nukiyama-Tanasawa and Rizk-Lefebvre models to predict droplet size for microconcentric nebulizers with aqueous and organic solvents. *Journal of Analytical Atomic Spectrometry*, *20*(7), 631–637. https://doi.org/10.1039/b501186h.

Kahen, K., Jorabchi, K., Gray, C., & Montaser, A. (2004). Spatial mapping of droplet velocity and size for direct and indirect nebulization in plasma spectrometry. *Analytical Chemistry*, *76*(24), 7194–7201. https://doi.org/10.1021/ac0489541.

Koropchak, J. A., Aryamanya-Mugisha, H., & Winn, D. H. (1988). Effects of capillary diameter on thermospray sample introduction to inductively coupled plasma atomic emission spectrometry. *Journal of Analytical Atomic Spectrometry*, *3*, 799–802.

Lefebvre, A. H., & McDonell, V. G. (2017). Atomization and sprays. In *Atomization and sprays* (pp. 1–284). CRC Press. https://doi.org/10.1201/9781315120911.

Liu, H., & Montaser, A. (1994). Phase-Doppler diagnostic studies of primary and tertiary aerosols produced by a high-efficiency nebulizer. *Analytical Chemistry*, *66*(19), 3233–3242. https://doi.org/10.1021/ac00091a037.

Matusiewicz, H., Ślachciński, M., Almagro, B., & Canals, A. (2009). Evaluation of various types of micronebulizers and spray chamber configurations for microsamples analysis by microwave induced plasma optical emission spectrometry. *Chemia Analityczna*, *54*(6), 1219–1244. http://www.chem.uw.edu.pl/chemanal.

Mohamed, N., Fry, R. C., & Wetzel, D. L. (1981). Laser Fraunhofer diffraction studies of aerosol droplet size in atomic spectrochemical analysis. *Analytical Chemistry*, *53*(4), 639–645. https://doi.org/10.1021/ac00227a017.

Monnig, C. A., & Koirtyohann, S. R. (1985). Mie scattering from sample aerosol inside the inductively coupled plasma. *Analytical Chemistry*, *57*(13), 2533–2536. https://doi.org/10.1021/ac00290a022.

Novak, J. W., & Browner, R. F. (1980). Characterization of droplet sprays produced by pneumatic nebulizers. *Analytical Chemistry*, *52*(6), 792–796. https://doi.org/10.1021/ac50056a005.

Olesik, J. W. (1997). Investigating the fate of individual sample droplets in inductively coupled plasmas. *Applied Spectroscopy*, *51*(5), 158A–175A. https://doi.org/10.1366/0003702971940792.

Olesik, J. W., & Bates, L. C. (1995). Characterization of aerosols produced by pneumatic nebulizers for inductively coupled plasma sample introduction: effect of liquid and gas flow rates on volume based drop size distributions. *Spectrochimica Acta Part B: Atomic Spectroscopy, 50*(4–7), 285–303. https://doi.org/10.1016/0584-8547(94)00151-K.

Olesik, J. W., & Fister, J. C. (1991). Incompletely desolvated droplets in argon inductively coupled plasmas: their number, original size and effect on emission intensities. *Spectrochimica Acta Part B: Atomic Spectroscopy, 46*(6–7), 851–868. https://doi.org/10.1016/0584-8547(91)80086-I.

Qieni, L., Wenhua, J., Tong, L., Xiang, W., & Yimo, Z. (2014). *Optics Communications, 312*, 312–318. https://doi.org/10.1016/j.optcom.2013.09.049.

Russo, R. E., & Hieftje, G. M. (1980). An accurate model for sample droplet acceleration in an air—acetylene flame. *Analytica Chimica Acta, 118*(2), 293–299. https://doi.org/10.1016/S0003-2670(01)93605-0.

Schaldach, G., Berndt, H., & Sharp, B. L. (2003). *Journal of Analytical Atomic Spectrometry, 18*(7), 742–750. https://doi.org/10.1039/b302052e.

Sharp, B. L. (1988). Pneumatic nebulisers and spray chambers for inductively coupled plasma spectrometry: A review: Part 1. Nebulisers. *Journal of Analytical Atomic Spectrometry, 3*(5), 613–652. https://doi.org/10.1039/JA9880300613.

Skogerboe, R. K., & Freeland, S. J. (1985a). Experimental characterization of aerosol production, transport, vaporization, and atomization systems. Part I: factors controlling aspiration rates. *Applied Spectroscopy, 39*, 916–920.

Skogerboe, R. K., & Freeland, S. J. (1985b). Experimental characterization of aerosol production, transport, vaporization, and atomization systems. Part II: factors controlling aerosol size distributions produced. *Applied Spectroscopy, 39*, 920–925.

Wang, M., Si, J., Tan, W., Zheng, Y., & Yang, Y. (2019). *Optics Express, 27*(11), 16080–16087. https://doi.org/10.1364/OE.27.016080.

Williams, O. J. H., Nguyen, T., Schreyer, A.-M., & Smits, A. J. (2015). *Physics of Fluids, 27*.

SECTION 2

Nebulizer designs

CHAPTER

5

Nanonebulizers

Miguel Ángel Aguirre[a] *and Mazaher Ahmadi*[b]

[a]Department of Analytical Chemistry and Food Science and University Institute of Materials, Faculty of Science, University of Alicante, Alicante, Spain, [b]Department of Analytical Chemistry, Faculty of Chemistry, Bu-Ali Sina University, Hamedan, Iran

5.1 Introduction

The analytical nebulizers are commonly used as a part of the liquid sample introduction systems for introducing the analyte into atomic spectrometric detectors. However, analytical nebulizers are also employed as an ionization source [e.g., electrospray ionization (Ho et al., 2003), and as interface for liquid chromatography (Becker, Jochmann, Teutenberg, & Schmidt, 2020) or for capillary electrophoresis (Todebush, He, & De Haseth, 2003)]. The main disadvantages of the traditional analytical nebulizers are their low transport efficiency (<5%), high sample consumption, and wide droplet size distribution. In order to increase the efficiency of sample introduction and to reduce the consumption rate, nanonebulizers operating in the nanoliter sample flow rate range have been developed. The development of nanonebulizers working at very low liquid flow rates has opened the possibility of easily analyzing samples when the amount of sample is limited. Such a situation is encountered in many areas such as forensic, biological, clinical, geological, among others.

5.2 Design and fundamentals

5.2.1 Nanonebulizers in atomic spectrometric detectors

As early as 2006, Giusti, Lobinski, Szpunar, and Schaumlöffel (2006) designed the first dedicated nanonebulizer as a sheathless interface of nano-HPLC with ICP-MS by fabricating a pneumatic nanoflow nebulizer.

Analytical Nebulizers
https://doi.org/10.1016/B978-0-323-91181-8.00010-2

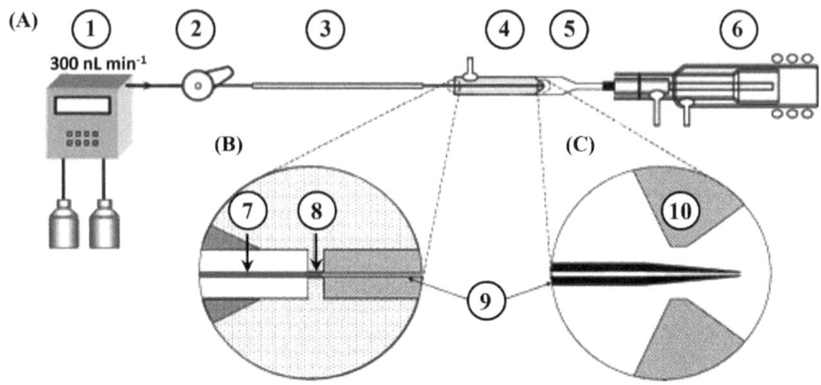

FIG. 5.1 (A) Instrumental setup of the nano-HPLC-ICP-MS coupling: (1) nanoflow HPLC pump, (2) nanoliter-volume injection valve, (3) nano-HPLC column, (4) nanoflow nebulizer nDS-200, (5) spray chamber, and (6) ICP torch. (B) Enlargement of the zero-dead-volume connection in the nebulizer: (7) nanocolumn outlet, (8) zero-dead-volume connection of the nanocolumn outlet to the nebulizer needle, and (9) fused-silica hollow needle. (C) Enlargement of the nebulizer tip: (9) fused-silica hollow needle and (10) sapphire orifice. *Reprinted with permission from: Giusti, P., Lobinski, R., Szpunar, J., & Schaumlöffel, D. (2006). Development of a nebulizer for a sheathless interfacing of nano-HPLC and ICPMS.* Analytical Chemistry, 78, *965–971. Copyright 2006 American Chemical Society.*

The nebulizer (nDS-200) worked at sample uptake rates of less than 500 nL min^{-1} and it was made of a hollow fused-silica needle centered in a sapphire orifice (Fig. 5.1).

The developed interface entirely preserved the chromatographic resolution, allowing the detection of low-abundant selenopeptides at the femtomole level (i.e., the absolute detection limit was 25 fg ^{80}Se). Later, the same research group further developed an improved robust nanonebulizer as an interface for nano-HPLC and ICP-MS to the elution of a lutetium-labeled peptide (Rappel & Schaumlöffel, 2009, 2010). In the first work, the nano-HPLC-ICP-MS was combined with isotope dilution analysis and the nanonebulizer was modified in order to cope with higher flow rates resulting from the added isotopically labeled spike solution (i.e., 1500 nL min^{-1}) (Rappel & Schaumlöffel, 2009). In the second work, the authors improved the nanonebulizer design employing a nebulizer capillary with nontapered tip. In addition, the large cross section minimized the risk of clogging and facilitated flushing, working at liquid flow rates as low as 50 nL min^{-1} and as high as 4000 nL min^{-1}, when the postcolumn addition of an internal standard is required (Rappel & Schaumlöffel, 2010). More recently, Shen et al. (2015) fabricated a demountable nanoflow nebulizer using a tip-tapered narrow-bore capillary by grinding for successful sheathless interfacing nano-HPLC with ICP-MS for arsenic speciation. Three years later, the same research group improved the nanonebulizer using nanoliter connectors, narrow-bore tubes, and injection nanovalves in order to downscale the dead volume of the instrumentation (Cheng, Shen, Liu, Xu, & Wang, 2018).

Another promising design is the combination of a chip device and the nanonebulizer. Chip-based nebulizers provide an excellent chance for rapid on-chip sample preparation or separation for speciation studies and the analysis of extremely low amount of samples. There are few studies combining the benefits of chip technology with atomic spectrometric detectors (e.g., ICP-MS). In 2003, chip-based devices were used for the electrophoretic separation of inorganic species of Cr, Cu, and As, with ICP-MS detection (Song, Greenway, & McCreedy, 2003, 2004). Although the chip-based device was successfully interfaced with ICP-MS via a commercial low flow micronebulizer (i.e., MicroMist), some problems arose by the use of commercial sample introduction systems, including (i) the need for a transfer capillary between the chip-based device and the nebulizer and (ii) the free aspiration rate produced by pneumatic concentric nebulizers induced noncontrolled laminar flow in the chip. Another problem is the liquid flow incompatibility (typically one order of magnitude lower in chip-based device than the micronebulizer). These problems were overcome in the past by a makeup flow. Nevertheless, the introduction of the makeup flow led to a sample dilution, increasing the LOD values. Taking into account all these limitations, Pearson and Greenway (2007) reported a highly efficient interface for ICP-MS, which integrated a chip-based device with a low flow rate cross-flow nebulization. The authors demonstrated that the lowest sample injection volume that could reproducibly give a usable signal was 0.5 μL min^{-1}, corresponding to 42 nL of sample volume (i.e., 5 s injection). Later, in 2012, Chen et al. (Cheng, Liu, Xu, & Yin, 2012; Cheng, Xu, Liu, Wang, & Yin, 2012) fabricated a chip-based nebulizer providing reproducible sample injection even at 0.18–0.4 nL of sample volumes, and high accuracy considering the matrix complexity (i.e., Chinese rice wine and serum sample). The majority of chip-based nebulizers are based on the conversion of a liquid into a polydisperse aerosol, with the high sample consumption and the low transport efficiency being the main drawbacks. One approach to overcome those drawbacks was reported by Verboket, Borovinskaya, Meyer, Günther, and Dittrich (2014). In this work, monodisperse aqueous droplets are produced using the highly volatile carrier phase perfluorohexane (PFH) in a custom-designed liquid-assisted droplet ejection (LADE) chip made of poly (dimethylsiloxane) (PDMS). The chip-based nebulizer can produce droplets in the size range from 40 to 60 μm and the aqueous droplets ejected from the chip in the stream of PFH remain intact and can be transported into the ICP-MS via a custom-built transport system (i.e., a membrane desolvator, which permitted the PFH vapor removal) (Fig. 5.2).

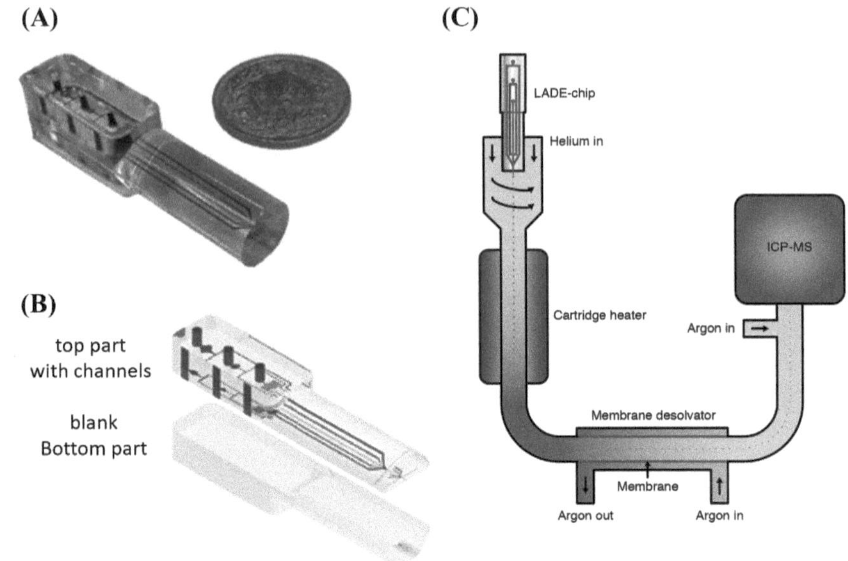

FIG. 5.2 (A) Photograph of the LADE chip next to a one Swiss franc coin for scale. All fluidic channels are filled with blue food dye for visualization. (B) Rendering of the two components of the LADE chip. (C) Schematic drawing of the experimental setup (not to scale). The system consists of the LADE chip, a heater, a membrane desolvator, and an ICP-MS. *Reprinted with permission from: Verboket, P. E., Borovinskaya, O., Meyer, N., Günther, D., & Dittrich, P. S. (2014). A new microfluidics-based droplet dispenser for ICPMS.* Analytical Chemistry, *86, 6012–6018. Copyright 2014 American Chemical Society.*

5.2.2 Nanonebulizers as an ionization source

The electrospray ionization (ESI) is a soft ionization technique extensively used for converting solution-phase analytes into gas-phase ions (Kebarle & Verkerk, 2009) prior to mass spectrometric analysis. Nowadays, nanoelectrospray ionization (nESI) is the evolution of conventional ESI, which has been widely employed in several fields of mass spectrometry analysis (Lee, Huang, Tseng, & Chen, 2021; Tan et al., 2022; Xiao et al., 2020). The introduction of the nESI by Wilm and Mann (1996) has shed new light on the nebulization process. In conventional ESI (i.e., μL/min flow rate), the diameters of the initially formed droplets are in the μm range. In contrast, in nESI the droplet sizes are in the 200 nm range or below, and the ion signal intensities being equal or even higher than traditional ESI (Wilm & Mann, 1996). In addition, nESI has also been found to be highly tolerant of the suppression effects caused by salts and matrixes (Schmidt, Karas, & Dülcks, 2003). With these benefits, nESI is now considered as one of the most promising alternatives for omics studies (Yu et al., 2021).

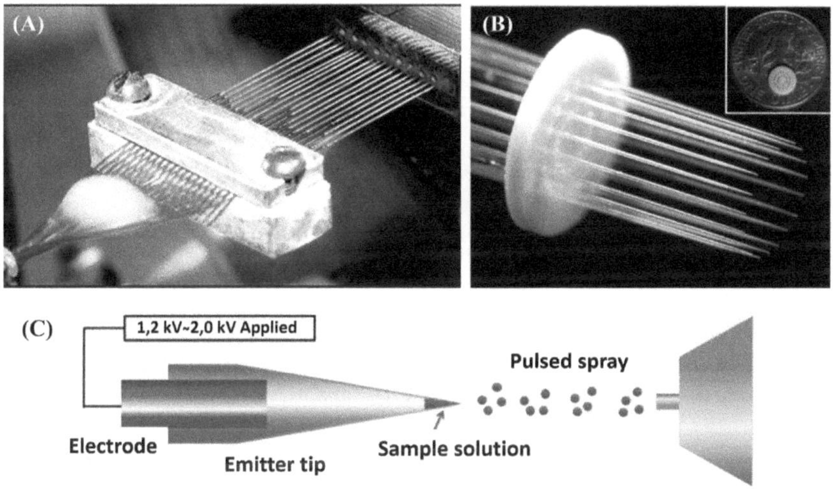

FIG. 5.3 Different designs of nESI. (A) Photograph of an array of 19 emitters positioned in front of a heated multicapillary inlet. (B) Photographs of a circular array of nESI emitters. It is also shown the disk used to arrange the emitters in the inset with a dime for size comparison. (C) Setup of pulsed-dc-ESI. The electrode was set contactless from the sample solution. The dc voltage was applied to the electrode to generate single polarity pulsed electrospray of the sample solution. *(A) Reprinted with permission from: Kelly, R. T., Page, J. S., Zhao, R., Qian, W. J., Mottaz, H. M., Tang, K., & Smith, R. D. (2008) Capillary-based multi-nanoelectrospray emitters: improvements in ion transmission efficiency and implementation with capillary reversed-phase LC-ESI-MS.* Analytical Chemistry, 80, *143–149. Copyright 2008 American Chemical Society. (B) Reprinted with permission from: Kelly, R. T., Page, J. S., Marginean, I., Tang, K., & Smith, R. D. (2008) Nanoelectrospray emitter arrays providing interemitter electric field uniformity.* Analytical Chemistry, 80, *5660–5665. Copyright 2008 American Chemical Society. (C) Reprinted with permission from: Wei, Z., Xiong, X., Guo, C., Si, X., Zhao, Y., He, M., et al. (2015). Pulsed direct current electrospray: enabling systematic analysis of small volume sample by boosting sample economy.* Analytical Chemistry, 87, *11242–11248. Copyright 2015 American Chemical Society.*

Several variants have been published that further enhanced the detection capability of nESI. Kelly et al. (2008) developed, for the first time, an nESI multiemitter (i.e., an array of 19 nanoemitters), which achieved better sensitivity and quantitation performance compared with a single nanoemitter (Fig. 5.3A). The authors reported a sensitivity enhancement averaging 11-fold for peptides in depleted human plasma due to increased ionization efficiency of the nESI multiemitters, along with efficient transfer of ions into the mass spectrometer. A modification of the nESI multiemitters from the linear array into a circular array was also carried out by the same research group (Kelly, Page, Marginean, Tang, & Smith, 2008). The new design enabled more flexibility for nanoemitter positioning and guaranteed that all of the nanoemitters could operate optimally at a given electric potential (Fig. 5.3B). Another improvement of nESI was developed by Wei et al. (2015) for extremely low sample volume (i.e., picoliter). In this work, a pulsed high voltage was applied to the electrode to create a pulsed aerosol. Conventionally, nESI generated a continuous electrospray while the mass analyzer worked intermittently. The authors demonstrated that if the pulsed electrospray was perfectly synchronized with the mass analyzer, the sample economy would boost, avoiding meaningless sample loss between the scan events of the mass analyzer. The pulsed aerosol was created placing the electrode several millimeters away from the solution (Fig. 5.3C). Using this configuration, the sensitivity was improved, as well as sample economy, due to the even lower flow rate (Wei et al., 2015).

Several efforts have also been made for coupling separation techniques [e.g., liquid chromatography (Sanders & Edwards, 2020) and electrophoresis capillary (Zhang, Hong, Xu, & Qu, 2020)] with nESI. Nevertheless, the integration of microfluidics device with nESI represents a promising alternative. Bedair and Oleschuk (2006) developed an nESI utilizing a porous polymer monolith (PPM) fabricated at the end of a capillary for coupling to a mass spectrometer. The aerosol is created directly from the end of the chip, having zero dead volume. Moreover, the PPM could be customized to act as a separation medium. Sun, Kelly, Tang, and Smith (2011) developed an integrated poly (dimethylsiloxane) (PDMS) membrane-based microfluidic emitter for nESI-MS. The sharp emitter was fabricated by cutting a thin PDMS membrane into a triangle shape, the fabrication process is illustrated in Fig. 5.4.

Fig. 5.4A shows the different PDMS slabs for nESI fabrication. A thick PDMS slab (1) was bonded on the top of the patterned PDMS membrane (2), leaving approximately 5-mm-long section of the channel uncovered. The substrates were then gently peeled from the patterned template and bonded with the unpatterned PDMS membrane (3) to enclose the microchannel (Fig. 5.4B). After removing the assembly from the bare wafer, a through-hole was created at the end of the microchannel by punching the substrate with a manually sharpened syringe needle, and a triangle-shaped tip was generated by cutting the two-layer membrane vertically (Fig. 5.4C). Finally, a PDMS slab (4) was bonded together with the assembly, leaving the sharp tip protruding (Fig. 5.4D). Fig. 5.4E and F are photographs of the nESI from the top view and side view, respectively.

In this work, the electric field at the tip was greatly enhanced by the reduced surface area at the emitter; and therefore, the aerosol created by the emitter was stable at lower ESI potentials. While sample preparation in a microfluidic platform coupled to nESI was greatly demonstrated (Hao, Bao, Huang, Hu, & Xiong, 2018; Li, Xiao, Ou, McCullum, & Liu, 2013), the efficient coupling of segmented microfluidic flows to nESI is far from trivial. Segmentation of water-dissolved analytes flowing in a microfluidic channel into an immiscible carrier liquid allows precise control of chemical reactions in these femto to nanoliter volume reaction chambers.

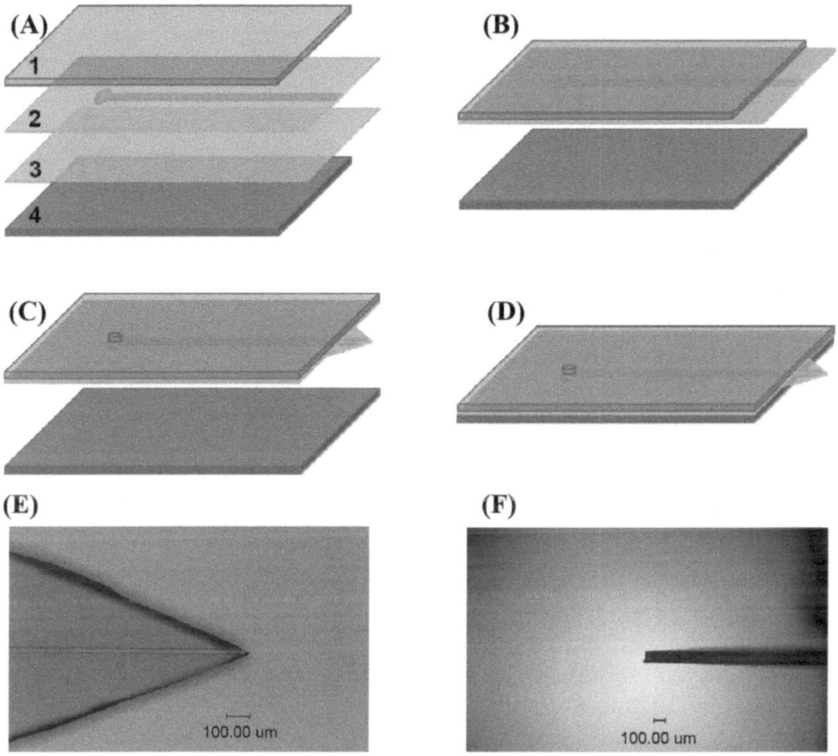

FIG. 5.4 (A–D) Fabrication of PDMS membrane-based microfluidic ESI emitter. Numbers 1 and 4 are PDMS slabs; 2 and 3 are PDMS membranes with and without microstructures, respectively. (E, F) Photographs of a PDMS membrane-based microfluidic ESI emitter from top view and side view, respectively. *Reprinted with permission from: Sun, X., Kelly, R. T., Tang, K., & Smith, R. D. (2011). Membrane-based emitter for coupling microfluidics with ultrasensitive nanoelectrospray ionization-mass spectrometry.* Analytical Chemistry, 83, *5797–5803. Copyright 2011 American Chemical Society.*

In this kind of analysis, the use of an immiscible carrier liquid disrupts the formation of the aerosol, resulting in increased instability of the signal. In order to overcome this limitation, Zhang et al. (2022) designed a novel nESI emitter that can spatially and temporally separate oil and aqueous phases on a droplet-by-droplet basis during electrospray, enabling the detection of gamma aminobutyric acid in pL-scale droplets with high stability.

5.3 Applications

Regarding elemental determination, the speciation of metals and metalloids is of vital importance in many fields of analytical chemistry since their properties (e.g., the toxicity) show a higher correlation with its chemical species than with its total content. For instance, organomercury compounds (e.g., methylmercury) usually present greater toxicity toward a wide range of organisms than the metallic mercury (Natasha et al., 2020). In addition, differences in toxic properties are also observed between As (III), As (V), and organic compounds of As (Ardini, Dan, & Grotti, 2020). As these elements can enter into the food chain through different environmental routes, it is particularly important to determine different chemical species. Among all analytical instruments for ultra-trace elemental speciation, the hyphenation of nano-HPLC and ICP-MS is the most versatile technique with numerous advantages (Bustos, 2020) and the development of new nanonebulizers as an interface addressed the problems of flow incompatibility and low aerosol transport efficiency. Several nanonebulizers including nDS-200 (Giusti et al., 2006; Rappel & Schaumlöffel, 2009), nDS-200e (Holste, Tholey, Hung, & Schaumlöffel, 2013; Rappel & Schaumlöffel, 2010), a demountable direct injection high efficiency nebulizer (d-DIHEN) (Brennan et al., 2007), a demountable nanoflow nebulizer (d-NN) (Cheng, Xu, et al., 2012; Shen et al., 2015), and a modified Teledyne CETAC DS-5 nebulizer (Bishop et al., 2017) have been employed. More recently, Cheng et al. (Cheng, Zhang, Wang, & Liu, 2018) developed a sheathless interface combining a demountable high-pressure nanonebulizer with a heated spray chamber with a makeup gas inlet (Fig. 5.5). This arrangement ensured the independent optimization of the nebulizer gas flow for maximum nebulization efficiency and of the makeup flow for maximum transport efficiency. The authors compared the developed nanonebulizer to commercial nebulizers, and they stated that the nanonebulizer fabrication was simple, rapid, reproducible, and inexpensive.

Regarding organic analysis, successful applications describing the separation and analysis of chiral compounds using nESI-MS are described in the bibliography. Li et al. (2013) utilized an innovative chip design for fast chiral analysis. Enantiomeric separation of 3,4-dihydroxyphenylalanine (DOPA), glutamic acid (Glu), and serine (Ser) were achieved within 130 s. In addition, the chip design was suitable for stereochemical analysis in living cells since it integrated cell culture, sample injection, chiral separation, and detection into a single platform. The limit of detection of the proposed method was 10 times lower than the traditional approach (LODs: 0.04 μM vs 0.50 μM), mainly because of the use of nESI. The results suggested that only L-DOPA was metabolized by human SH-SU5Y cells while D-DOPA was left intact. In further work from the same research group, the authors employed the same chip platform for the simultaneous separation and quantification of extracellular and intracellular levels of D-Ser and L-Ser after incubating PC-12 neuronal cells. The authors demonstrated that PC-12 neuronal cells have a stereochemical preference for D-Ser over L-Ser and they speculated about the mechanism for D-Ser release from PC-12 cells (Li, Mccullum, Zhao, Hu, & Liu, 2015). In another publication, the same authors further

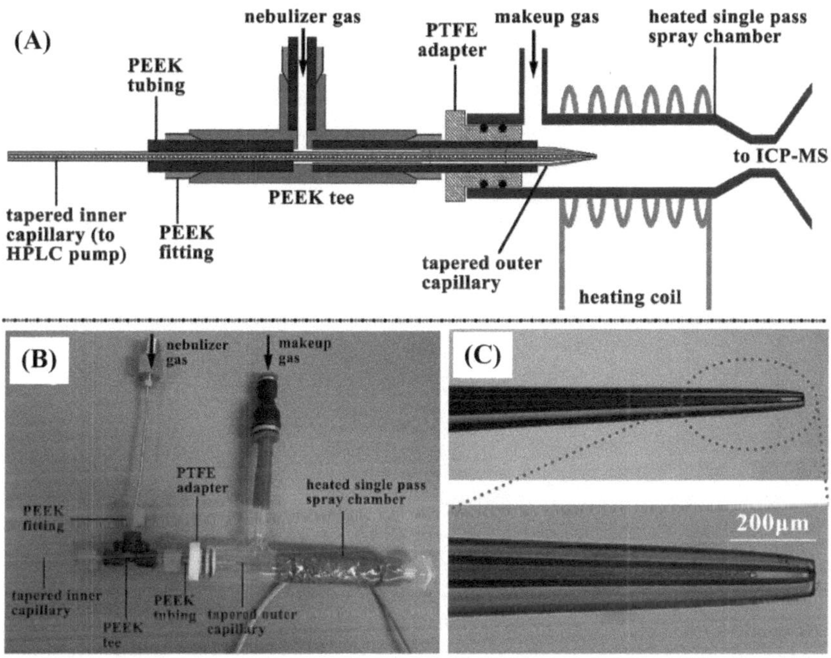

FIG. 5.5 (A) Schematic diagram of the in-column high-pressure nebulizer combined with a heated single pass spray chamber. (B) Photograph of the interface. (C) The enlarged nebulizer tip. *Reprinted from: Cheng, H., Zhang, W., Wang, Y., & Liu, J. (2018). Interfacing nanoliter liquid chromatography and inductively coupled plasma mass spectrometry with an in-column high-pressure nebulizer for mercury speciation.* Journal of Chromatography A, 1575, *59–65. Copyright 2018 Elsevier.*

improved the system, incorporating a chemical stimulation chamber to study the release of chiral serine from single PC-12 cell after stimulation (Li, Zhao, & Liu, 2016). The authors have shown that the proposed method allowed real-time chiral monitoring neurotransmitter release events in single nerve cell, giving novel insights into communication between neurons.

Another important application of nanonebulizer in organic analysis is the development of microfluidic devices compatible with nESI. Chip-based devices could facilitate the improvement of omics analysis since those devices offered significant advantages in high throughput and automatic sample preparation (Feng, Liu, Li, & Liu, 2015). Therefore, microfluidic devices coupled with nESI are considered as an ideal tool to investigate metabolites and peptides (Hao et al., 2018). Moreover, monolithic nESI nozzles have revealed an enhanced reproducibility compared with commercial nESI tips coupled with microfluidic devices, and their excellent compatibility with various microfluidic applications (Wang et al., 2015). In this sense, monolithic nESI would be a useful component

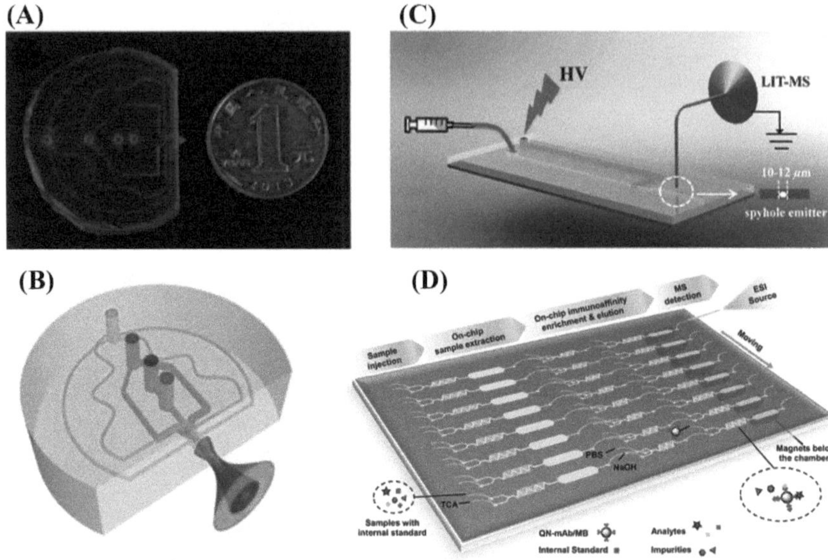

FIG. 5.6 (A, B) Photograph and scheme of a monolithic nESI, respectively. Sample and reference solutions were in red and blue (light gray and gray in print version), respectively. (C) Schematic representation of the spyhole-nanoESI device. (D) Schematic diagram of the immunoaffinity microfluidic chip-mass spectrometry system for online extraction and quantitative analysis of quinolones. *(A, B) Adapted from: Wang, L., Wang, Y., Jiang, S., Ye, M., Su, P., Xiong, B. (2016). Microfluidic nitrogen-assisted nanoelectrospray emitter: A monolithic interface for accurate mass measurements based on a single nozzle.* Journal of Chromatography A, 1470, *1–8. Copyright 2016 Elsevier. (C) Reprinted with permission from: Zhong, X., Qiao, L., Stauffer, G., Liu, B., Girault, H. H. (2018). On-chip spyhole nanoelectrospray ionization mass spectrometry for sensitive biomarker detection in small volumes.* Journal of the American Society for Mass Spectrometry, 29, *1538–1545. Copyright 2018 American Chemical Society. (D) Reprinted with permission from: Zhao, Y., Tang, M., Liu, F., Li, H., Wang, H., Xu, D. (2019). Highly integrated microfluidic chip coupled to mass spectrometry for online analysis of residual quinolones in milk.* Analytical Chemistry, 91, *13418 – 13426. Copyright 2019 American Chemical Society.*

in a versatile microfluidic platform for metabolic and proteomic identifications (Sun, Kelly, Tang, & Smith, 2010; Wang et al., 2016; Fig. 5.6A and B). Recently, several researchers have developed integrated microfluidic chip coupled to mass spectrometry. Chun (2018) reported a microfluidic glass chip for performing analyte-selective electropreconcentration with nESI for peptide analysis. In order to perform the electropreconcentration the authors coated the chip with positively charged polyamine. Therefore, when the electric double layer (EDL) is overlapped in the nanochannel, the nanochannel becomes a charge-selective filter; and then, only counterions to the nanochannel surface charge can pass through it while co-ion (to the nanochannel surface charge) flow is blocked. In this study, the nanochannel is positively charged and the EDL is overlapped, anions can pass through it while cations are blocked. The author demonstrated that the developed electropreconcentration-nESI microchip showed linear sample collecting rate and a signal enhancement by one order of magnitude only for cationic peptides [i.e., angiotensin III, GFR (Gly-Phe-Arg), and FMRF (Phe-Met-Arg-Phe)] (Chun, 2018). Zhong, Qiao, Stauffer, Liu, and Girault (2018) developed a new type of nESI called spyhole-nESI. This novel interface is characterized by drilling a microhole with a diameter of 10–12 μm directly on the top of a microchannel (Fig. 5.6C). To take advantages of the high sensitivity and the low dead volume, the spyhole-nESI was used to analyze a small cell lung cancer biomarker protein, obtaining an LOD of 50 pM for Ang I, which was 600 times lower than standard ESI (Zhong et al., 2018). More recently, Zhao et al. (2019) developed an immunoaffinity microfluidic chip coupled to mass spectrometry for qualitative and quantitative analysis of seven different regulated quinolones in a milk matrix. In this work, sample loading, sample extraction, immunoaffinity reaction, magnetic separation, and online elution are integrated into the same platform and fully automated (Fig. 5.6D). The proposed microfluidic platform offered several advantages such as small reagent consumption, simplified sample pretreatment, high-throughput, and online detection (Zhao et al., 2019).

Over the past decade, the development of miniature mass spectrometers allowed the application of conventional mass spectrometry techniques from laboratory into different on-field analysis areas (Ouyang & Cooks, 2009; Snyder, Pulliam, Ouyang, & Cooks, 2016). The Mini 12 (Li et al., 2014) was developed as a point-of-care prototype as a self-contained system to produce quantitative results for unprocessed samples of small volumes including nonvolatile analytes (Fig. 5.7A–C). The integrated Mini 12 system of 25 kg can perform direct mass spectrometry analysis on whole blood, food, and environmental samples, without sample preparation or chromatographic separation. The entire process required no intervention by the operator, or any spectral interpretation or data analysis. With the improved stability in ion current using extraction nano-ESI, quantitation of amitriptyline in blood samples with an LOQ of 7.5 $ng\,mL^{-1}$ and RSD better than 10% was achieved. Another promising combination between nESI and miniature mass spectrometer was developed by Xu et al. (He et al., 2015; Zhai, Feng, Wei, Wang, & Xu, 2015). In the first research, the capillary electrophoresis (CE) is inserted into a glass nESI capillary (Fig. 5.7D and E). The system was able to separate protonated MRFA (Met-Arg-Phe-Ala, m/z 524.3) and doubly protonated angiotensin II (m/z 524.1). In the second study, the authors developed an instrument weighted approximately 6 kg with nESI. After optimization, good stability (i.e., RSD value $<7\%$), reasonable sensitivity [i.e., LOD 1 $\mu g\,mL^{-1}$ for the peptide GPRP (Gly-Pro-Arg-Pro)], relatively broad mass range (200–2500 m/z), and high scan speed (5 Hz) were achieved for the analysis of organic, peptide, and protein samples.

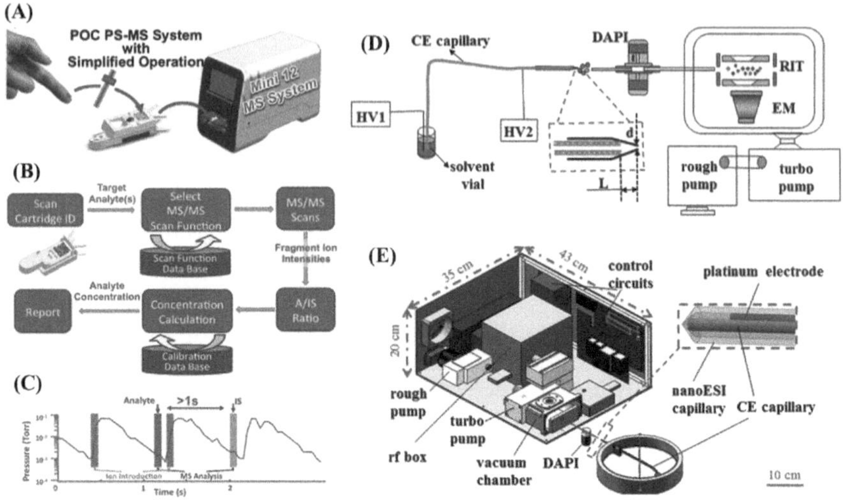

FIG. 5.7 (A) Mini 12 system that provides a simplified operational protocol. (B) Flowchart for automated quantitative analysis of the analyte (A) using an internal standard (IS). (C) Operation of the dual Mini 12 system in the time domain. (D) Schematic structure and (E) 3D structure of the mini CE/MS system. *(A–C) Adapted with permission from: Li, L., Chen, T. C., Ren, Y., Hendricks, P. I., Cooks, R. G., & Ouyang, Z. (2014). Mini 12, miniature mass spectrometer for clinical and other applications—Introduction and characterization.* Analytical Chemistry, 86, *2909–2916. Copyright 2014 American Chemical Society. (D, E) Adapted with permission from: He, M., Xue, Z., Zhang, Y., Huang, Z., Fang, X., Qu, F., et al. (2015). Development and characterizations of a miniature capillary electrophoresis mass spectrometry system.* Analytical Chemistry, 87, *2236–2241. Copyright 2015 American Chemical Society.*

5.4 Conclusions

The recent bibliography surrounding the nESI clearly demonstrates that significant gains in sensitivity can be achieved compared to conventional ESI. Taking the current nESI technology and potential future advances into account, new and exciting possibilities exist to expand impact of omics analysis on human and environmental health and disease. The increased sensitivity of the interface using nESI platforms offer the potential to uncover the roles played by low abundant analytes in health and disease studies. While many aspects of omics analysis are still under progress (e.g., development of more automated analytical methods), the development of analytical methods of higher sensitivity is also a priority to be as comprehensive as possible.

One detector that is considered a major tool in the analytical sciences is the miniature mass spectrometer, especially given the increasing interest in point-of-care measurements. Even though there are some commercial mass spectrometers, it is essential to improve the performance by optimizing design and implementing new advanced technologies. The recent availability of miniature components, such as analyzers and pumping systems, will undoubtedly accelerate developments in this area. Increasing analytical performance of the miniaturized systems endlessly expands their application areas and this trend of development is guaranteed for many years.

Acknowledgments

The authors are grateful to the Spanish Ministry of Science and Innovation (PID2021-126155OB-I00) and the Regional Government of Valencia (Spain) (CIPROM/2021/062) for the financial support. The authors extend their appreciation to Ministry of Science, Innovation and Universities for granting the Spanish Network of Excellence in Sample Preparation (RED2018-102522-T). This chapter is based upon work from the Sample Preparation Study Group and Network, supported by the Division of Analytical Chemistry of the European Chemical Society.

References

Ardini, F., Dan, G., & Grotti, M. (2020). *Journal of Analytical Atomic Spectrometry, 35*(2), 215–237. https://doi.org/10.1039/c9ja00333a.

Becker, C., Jochmann, M. A., Teutenberg, T., & Schmidt, T. C. (2020). *Talanta*, 206. https://doi.org/10.1016/j.talanta.2019.120229.

Bedair, M. F., & Oleschuk, R. D. (2006). *Analytical Chemistry, 78*(4), 1130–1138. https://doi.org/10.1021/ac0514570.

Bishop, D. P., Blanes, L., Wilson, A. B., Wilbanks, T., Killeen, K., Grimm, R., et al. (2017). *Journal of Chromatography A, 1497*, 64–69. https://doi.org/10.1016/j.chroma.2017.03.025.

Brennan, R. G., O'Brien Murdock, S. A. E., Farmand, M., Kahen, K., Samii, S., Gray, J. M., et al. (2007). *Journal of Analytical Atomic Spectrometry, 22*(9), 1199–1205. https://doi.org/10.1039/b703257a.

Bustos, A. R. M. (2020). *Chromatographia, 83*(2), 145–147. https://doi.org/10.1007/s10337-019-03846-2.

Cheng, H., Liu, J., Xu, Z., & Yin, X. (2012). *Spectrochimica Acta-Part B Atomic Spectroscopy, 73*, 55–61. https://doi.org/10.1016/j.sab.2012.07.010.

Cheng, H., Shen, L., Liu, J., Xu, Z., & Wang, Y. (2018). *Journal of Separation Science, 41*(7), 1524–1531. https://doi.org/10.1002/jssc.201701178.

Cheng, H., Xu, Z., Liu, J., Wang, X., & Yin, X. (2012). *Journal of Analytical Atomic Spectrometry, 27*(2), 346–353. https://doi.org/10.1039/C1JA10273G.

Cheng, H., Zhang, W., Wang, Y., & Liu, J. (2018). *Journal of Chromatography A, 1575*, 59–65. https://doi.org/10.1016/j.chroma.2018.09.023.

Chun, H. (2018). *Journal of Chromatography A, 1543*, 67–72. https://doi.org/10.1016/j.chroma.2018.02.037.

Feng, X., Liu, B. F., Li, J., & Liu, X. (2015). *Mass Spectrometry Reviews, 34*(5), 535–557. https://doi.org/10.1002/mas.21417.

Giusti, P., Lobinski, R., Szpunar, J., & Schaumlöffel, D. (2006). *Analytical Chemistry, 78*(3), 965–971. https://doi.org/10.1021/ac051656j.

Hao, Y., Bao, Y., Huang, X., Hu, Y., & Xiong, B. (2018). *RSC Advances, 8*(70), 39811–39817. https://doi.org/10.1039/C8RA08276F.

He, M., Xue, Z., Zhang, Y., Huang, Z., Fang, X., Qu, F., et al. (2015). *Analytical Chemistry, 87*(4), 2236–2241. https://doi.org/10.1021/ac504868w.

Ho, C. S., Lam, C. W. K., Chan, M. H. M., Cheung, R. C. K., Law, L. K., Lit, L. C. W., et al. (2003). Electrospray ionisation mass spectrometry: principles and clinical applications. *The Clinical Biochemist. Reviews, 24*(1), 3–12.

Holste, A., Tholey, A., Hung, C. W., & Schaumlöffel, D. (2013). *Analytical Chemistry, 85*(6), 3064–3070. https://doi.org/10.1021/ac303618v.

Kebarle, P., & Verkerk, U. H. (2009). *Mass Spectrometry Reviews, 28*(6), 898–917. https://doi.org/10.1002/mas.20247.

Kelly, R. T., Page, J. S., Marginean, I., Tang, K., & Smith, R. D. (2008). *Analytical Chemistry, 80*(14), 5660–5665. https://doi.org/10.1021/ac800508q.

Kelly, R. T., Page, J. S., Zhao, R., Qian, W. J., Mottaz, H. M., Tang, K., et al. (2008). *Analytical Chemistry, 80*(1), 143–149. https://doi.org/10.1021/ac701647s.

Lee, C. H., Huang, H. C., Tseng, M. C., & Chen, C. J. (2021). *Journal of Food and Drug Analysis, 29*(3), 468–489. https://doi.org/10.38212/2224-6614.3368.

Li, L., Chen, T. C., Ren, Y., Hendricks, P. I., Cooks, R. G., & Ouyang, Z. (2014). *Analytical Chemistry, 86*(6), 2909–2916. https://doi.org/10.1021/ac403766c.

Li, X., Mccullum, C., Zhao, S., Hu, H., & Liu, Y. M. (2015). *ACS Chemical Neuroscience, 6*(4), 582–587. https://doi.org/10.1021/cn5003122.

Li, X., Xiao, D., Ou, X.-M., McCullum, C., & Liu, Y.-M. (2013). *Journal of Chromatography A, 1318*, 251–256. https://doi.org/10.1016/j.chroma.2013.10.020.

Li, X., Zhao, S., & Liu, Y. M. (2016). *Chemistry Select, 1*(17), 5554–5560. https://doi.org/10.1002/slct.201600748.

Natasha, S. M., Khalid, S., Bibi, I., Bundschuh, J., Khan Niazi, N., & Dumat, C. (2020). *Science of the Total Environment, 711*, 134749. https://doi.org/10.1016/j.scitotenv.2019.134749.

Ouyang, Z., & Cooks, R. G. (2009). *Annual Review of Analytical Chemistry, 2*, 187–214. https://doi.org/10.1146/annurev-anchem-060908-155229.

Pearson, G., & Greenway, G. (2007). *Journal of Analytical Atomic Spectrometry, 22*(6), 657–662. https://doi.org/10.1039/b702624b.

Rappel, C., & Schaumlöffel, D. (2009). *Analytical Chemistry, 81*(1), 385–393. https://doi.org/10.1021/ac801814a.

Rappel, C., & Schaumlöffel, D. (2010). *Journal of Analytical Atomic Spectrometry, 25*(12), 1963–1968. https://doi.org/10.1039/c0ja00050g.

Sanders, K. L., & Edwards, J. L. (2020). *Analytical Methods, 12*(36), 4404–4417. https://doi.org/10.1039/d0ay01194k.

Schmidt, A., Karas, M., & Dülcks, T. (2003). *Journal of the American Society for Mass Spectrometry, 14*(5), 492–500. https://doi.org/10.1016/S1044-0305(03)00128-4.

Shen, L., Sun, J., Cheng, H., Liu, J., Xu, Z., & Mu, J. (2015). *Journal of Analytical Atomic Spectrometry, 30*(9), 1927–1934. https://doi.org/10.1039/c5ja00198f.

Snyder, D. T., Pulliam, C. J., Ouyang, Z., & Cooks, R. G. (2016). *Analytical Chemistry, 88*(1), 2–29. https://doi.org/10.1021/acs.analchem.5b03070.

Song, Q. J., Greenway, G. M., & McCreedy, T. (2003). *Journal of Analytical Atomic Spectrometry, 18*(1), 1–3. https://doi.org/10.1039/b209530k.

Song, Q. J., Greenway, G. M., & McCreedy, T. (2004). *Journal of Analytical Atomic Spectrometry, 19*(7), 883–887. https://doi.org/10.1039/b401657b.

Sun, X., Kelly, R. T., Tang, K., & Smith, R. D. (2010). *Analyst, 135*(9), 2296–2302. https://doi.org/10.1039/c0an00253d.

Sun, X., Kelly, R. T., Tang, K., & Smith, R. D. (2011). *Analytical Chemistry, 83*(14), 5797–5803. https://doi.org/10.1021/ac200960h.

Tan, S., Wang, J., Li, M., Yin, X., Gao, X., Li, C., et al. (2022). *Rapid Communications in Mass Spectrometry, 36*(5). https://doi.org/10.1002/rcm.9237.

Todebush, R. A., He, L. T., & De Haseth, J. A. (2003). *Analytical Chemistry, 75*(6), 1393–1399. https://doi.org/10.1021/ac025858g.

Verboket, P. E., Borovinskaya, O., Meyer, N., Günther, D., & Dittrich, P. S. (2014). *Analytical Chemistry, 86*(12), 6012–6018. https://doi.org/10.1021/ac501149a.

Wang, L., Wang, Y., Jiang, S., Ye, M., Su, P., & Xiong, B. (2016). *Journal of Chromatography A, 1470*, 1–8. https://doi.org/10.1016/j.chroma.2016.09.065.

Wang, X., Yi, L., Mukhitov, N., Schrell, A. M., Dhumpa, R., & Roper, M. G. (2015). *Journal of Chromatography A, 1382*, 98–116. https://doi.org/10.1016/j.chroma.2014.10.039.

Wei, Z., Xiong, X., Guo, C., Si, X., Zhao, Y., He, M., et al. (2015). *Analytical Chemistry, 87*(22), 11242–11248. https://doi.org/10.1021/acs.analchem.5b02115.

Wilm, M., & Mann, M. (1996). Analytical properties of the nanoelectrospray ion source. *Analytical Chemistry, 68*(1), 1–8. https://doi.org/10.1021/ac9509519.

Xiao, X., Chen, C., Deng, J., Wu, J., He, K., Xiang, Z., et al. (2020). *Talanta, 217*, 121064. https://doi.org/10.1016/j.talanta.2020.121064.

Yu, D., Zheng, F., Wang, L., Li, C., Lu, X., Lin, X., et al. (2021). *Analytical Chemistry, 93*(41), 13765–13773. https://doi.org/10.1021/acs.analchem.1c01507.

Zhai, Y., Feng, Y., Wei, Y., Wang, Y., & Xu, W. (2015). *Analyst, 140*(10), 3406–3414. https://doi.org/10.1039/c5an00462d.

Zhang, F., Hong, J., Xu, W., & Qu, F. (2020). *Talanta, 206*, 120183. https://doi.org/10.1016/j.talanta.2019.120183.

Zhang, Y., Kim, S., Shi, W., Zhao, Y., Park, I., Brenden, C., et al. (2022). *Lab on a Chip, 22*(1), 40–46. https://doi.org/10.1039/d1lc00758k.

Zhao, Y., Tang, M., Liu, F., Li, H., Wang, H., & Xu, D. (2019). *Analytical Chemistry, 91*(21), 13418–13426. https://doi.org/10.1021/acs.analchem.9b01844.

Zhong, X., Qiao, L., Stauffer, G., Liu, B., & Girault, H. H. (2018). *Journal of the American Society for Mass Spectrometry, 29*(7), 1538–1545. https://doi.org/10.1007/s13361-018-1937-7.

CHAPTER

6

Multinebulizers

Mariusz Ślachciński

Poznan University of Technology, Faculty of Chemical Technology, Poznan, Poland

6.1 Introduction

The most commonly used approach for introducing liquid samples into plasma sources in analytical optical emission (OES) and mass spectrometry (MS) techniques is based on the injection of sample aerosols generated usually by pneumatic nebulization (PN) of aqueous solutions (Todoli & Mermet, 2005; Todolí & Mermet, 2006; Canals & Aguirre, 2015; Hou, Amais, Jones, & Donati, 2016; Donati, Amais, & Williams, 2017; Browner & Boorn, 1984).

The aerosol produced by a nebulizer/spray chamber assembly should be fine to be efficiently transported and finally atomized/excited/ionized by a plasma discharge. On the other hand, introducing large amounts of sample aerosol into the plasma can lead to a reduction in plasma temperature, incomplete atomization, and increased matrix-induced spectroscopic and nonspectroscopic interference effects (Canals & Aguirre, 2015).

Pneumatic nebulizers show some drawbacks such as low transport efficiency (<5%), high sample consumption, and wide size distribution of aerosol droplets. In order to eliminate these disadvantages, micronebulizers operating in a microliter sample flow rate range have been developed (Canals & Aguirre, 2015; Hou et al., 2016). The reduction in dimensions (e.g., solution capillary diameter and wall thickness, gas annulus area) results in enhanced gas-liquid interactions and a finer primary aerosol compared to conventional devices. Coupling the micronebulizers with miniaturized spray chambers of a low inner volume leads the improvement in analytical figures of merit, better analyte transport efficiency, low liquid solvent loading (the avoidance of plasma cooling), and reduction in memory and acid effects (Todolí & Mermet, 2002a, 2002b, 2003,

Analytical Nebulizers
https://doi.org/10.1016/B978-0-323-91181-8.00009-6

2006; Mora, Maestre, Hernandis, & Todolí, 2003; Cairns et al., 2004; Ślachciński, 2016).

An alternative way of increasing the sample introduction efficiency in plasma spectrometric techniques is to use ultrasonic nebulizers (USN) (Priego-Capote & de Castro, 2004). USNs have been shown to be more efficient at aerosol formation and can produce aerosols with smaller and more uniform droplet size. Generating a finer primary aerosol with a very small mean drop size (ideally $\leq 5\,\mu m$) is an alternative to direct injection to the plasma, as such fine aerosol drops should pass to the plasma with high efficiency. This results in more efficient atomization, higher sensitivity, and reduced matrix effects.

The modification of conventional single-channel nebulizers has been made to allow the introduction of two or several separate streams of solutions, e.g., for applying the online calibration methodology and/or chemical vapor generation. These arrangements mainly consist in adding additional channels to provide solutions of the necessary regents.

Many alternative sample introduction systems were developed to improve the analytical performance of plasma spectrometric techniques, including chemical vapor generation (CVG) by reaction with sodium or potassium tetrahydroborate (III) ($NaBH_4$ or KBH_4, THB) in acidic medium of a strong mineral acid, which involves efficient generation of volatile hydrides and cold element vapors, separation of the gaseous phase from the liquid phase, and transportation of the gases into atomization and excitation sources and to the detector (Robbins & Caruso, 1979;Nakahara, 1983; R. E. Sturgeon & Mester, 2002; Pohl, 2004; Brindle, 2007; Pohl & Jamroz, 2011).

By introducing the analyte as a volatile species, some disadvantages of PN can be avoided, and the efficiency of sample transport grows significantly, which translates into higher sensitivity and lower detection limits. Furthermore, separation of the volatile hydrides from the matrix minimizes potential spectroscopic and nonspectroscopic interferences. However, the drawbacks of CVG are also obvious. These traditional CVG approaches often use relatively expensive and/or unstable reagents such as THB, as well as other derivatization reagents, and are a potential source of contamination. In addition, the traditional THB-based CVG methods show strong dependence of the oxidation state of the element in the solution and are prone to interference by transition metals due to cogeneration of volatile species with the same reagent. The large amount of hydrogen generated by the THB-acid system may reduce the stability of a plasma source or even extinguish it, as, for example, when a low-power microwave-induced plasma is used for detection. Efforts to develop new vapor-generation systems that may replace or reduce the use of chemical reagents remain an important research area in CVG-atomic spectrometry (Ślachciński, 2014, 2019).

Therefore, a new CVG method was developed which can amend or substitute for the conventional ones. As one of the most promising new CVG techniques, photochemical vapor generation (PVG) retains the principal advantages of conventional CVG as well as provides several unique advantages. A less expensive, efficient, and more environmentally friendly arrangement for vapor generation can be realized with use of ultraviolet (UV) irradiation (He, Hou, Zheng, & Sturgeon, 2007; Wu, He, Zheng, Hou, & Sturgeon, 2010; Bendicho, Pena, Costas, Gil, & Lavilla, 2010; Yin, Liu, & Jiang, 2011; Sturgeon & Grinberg, 2012).

The modern multichannel sample introduction systems, described in this chapter, allow to take advantage of all the above-mentioned sample/analyte conversion techniques in one measuring cycle in a single device.

6.2 Design and fundamentals

6.2.1 Liquid solution pneumatic nebulization in multichannel systems

For over a dozen years, two or multichannel different nebulization systems were investigated based on concentric, cross-flow, parallel-path, flow blurring/flow-focusing (micro)nebulizers as well as (micro)ultrasonic nebulizers. These modern nebulizer configurations, often leading to increase in the sensitivity of the analytical signals, were applied to extend the analytical performance of the instruments used and to facilitate, often time-consuming and being a source of important errors, procedures (Carrión, Murillo, Montiel, & Díaz, 2003; Feng, Lam, & Sturgeon, 2001; Moor, Lam, & Sturgeon, 2000; Ślachciński, 2019; Tao & Sturgeon, 1999).

It was proposed to modify the nebulizer itself by introducing a common liquid inlet splitting the sample stream into three streams multimicrospray nebulizer (MMSN). The three orifices form the points of an equilateral triangle and are 2 mm apart. Each of the three capillaries is 80 mm long. The end of each capillary is 0.1 mm from the orifice. The MMSN was designed to be directly inserted into the endcap of the spray chamber and the sample solution can be precisely fed, at an uptake rate as low as 5 $\mu L\,min^{-1}$, by using a syringe pump (Huang, Hirabayashi, Shirasaki, & Koizumi, 2000).

An example of an application of the pneumatic cross-flow, dual-channel nebulizer is the trihedral cross-flow (Th-CFN) and T-shaped cross-flow (T-CFN) nebulizers, where two solutions were nebulized simultaneously by a single gas flow. In Th-CFN, the three capillaries, i.e., two for liquids and one for gas, are positioned orthogonally to each other, while in T-CFN the capillary for gas introduction is placed orthogonally to both capillaries for liquid, which face each other. This device was a modification of a conventional cross-flow nebulizer achieved by adding an additional capillary of 300 μm for liquid introduction.

The indicated weaknesses of this arrangement was the necessity to calibrate at least once a day the response of both channels of the dual sample-loading nebulizers (Bauer & Broekaert, 2007).

In 2009, a new multiple nebulizer system (flow-focusing multinebulizer –FFPN), based on flow-focusing (FFPN) and blurring (FBPN) principles originally described by Gañán-Calvo (1998), was developed and characterized by the Canals group for liquid sample introduction in ICP-based spectrometers (Kovachev et al., 2009; Fig. 6.1). The critical dimensions for FF/FB are the inner capillary of the nebulizer to exit orifice placed on a platinum sheet located inside the PTFE body, the orifice diameter, and the jet diameter. But, the capillary wall and plate thickness as well as their shape are also of importance. The liquid vein is focused on the center of the nebulizer exit hole, so that the sample never touches the hole edge. This fact eliminates or at least decreases the possibility of generating salt deposit growth nuclei and clogging (Almagro, Gañán-Calvo, Hidalgo, & Canals, 2006b).

The FF nebulizer enables the introduction of microliter volumes of solutions and, what is more, lower liquid uptakes lead to finer and, especially, more monodisperse primary aerosols (Almagro, Gañán-Calvo, & Canals, 2004; Almagro, Gañán-Calvo, Hidalgo, & Canals, 2006a; Almagro et al., 2006b; (Matusiewicz, Ślachciński, Hidalgo, & Canals, 2007).

These advantages were used in the construction of a multinebulizer consisting of various independent nozzles in a single body for the simultaneous introduction to the spray chamber of aerosols from different solutions (sample, solvent, and standard solutions), opening a new way of performing "in spray chamber" analytical procedures (i.e., dilution, derivatization, addition of internal standard, standard addition, etc.) that

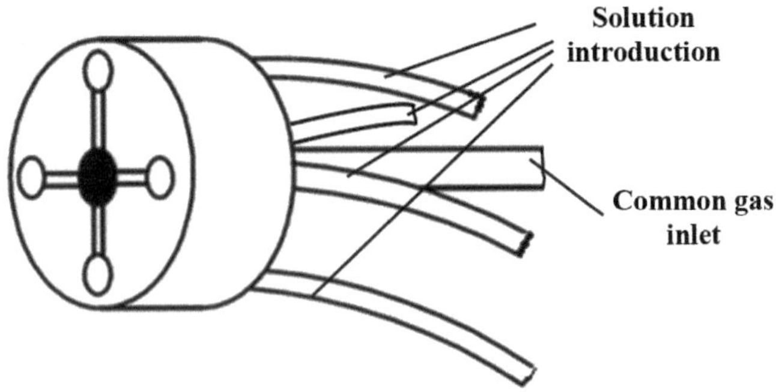

FIG. 6.1 Four-channel flow focusing multinebulizer (FFPN) based on Kovachev et al. (2009). *No permission required.*

normally are the most time-consuming part of the sample preparation process in many methods of analysis by plasma-based techniques (Bauer & Broekaert, 2008; Gómez, Márquez, & Chirinos, 2006; Hamier & Salin, 1998; Huxter, Hamier, & Salin, 2003; Krachler, Rausch, Feuerbacher, & Klemens, 2005; Maldonado, Chirinos, Benzo, Gómez, & Marcano, 2006; Olofsson, Rodushkin, & Axelsson, 2000). The main body of the nebulizer is a cylindrical piece with a 38 mm diameter and 10 mm length made of Teflon with 25% graphite, providing chemical resistance to commonly used inorganic acids and organic solvents. Inside the main body, four FF nebulization nozzles, a common gas inlet, and four liquid uptake inlets are incorporated. The common nebulization gas inlet is cross-divided to four channels inside the nebulizer body that lead it to the individual gas inlets of the four nebulization nozzles. Each nozzle, situated at 10 mm from the nebulizer's axis, has a separate liquid uptake tube. A shorter distance was not possible to achieve due to technical difficulties in the construction of the prototype. The inlet tube is connected to a standard, fused silica capillary of 50 μm id, which is precisely positioned with respect to the exit hole at a distance of 100 μm. The exit hole has a diameter of 100 mm and is drilled on a platinum plate (9) that forms the front part on the nebulization nozzle (Kovachev et al., 2009).

It was found that the four nozzles produce different aerosols regarding both, Sauter mean diameter and mean velocity and each of the four nozzles does not equally contribute to the total aerosol composition. The highest share of approximately 60% was found for one individual nozzle, which coincides with the smallest Sauter mean diameter and fastest mean velocity compared to the remaining nozzles (Kovachev et al., 2009).

Subsequently, novel flow blurring dual nebulization (FBDN) systems for online calibration methodologies were tested by the same research group. The effect of different geometries (parallel to 30° convergent) of this system was examined. Three different designs of the dual flow-blurring system as well as a standard nebulizer as a reference system were applied. It was found that a 30° convergent setup of both nebulizers showed the best achievable accuracy while precision was somewhat degraded for all geometries compared to the reference arrangement (Aguirre, Kovachev, Almagro, Hidalgo, & Canals, 2010).

The significant modification of the above-described nebulization system was the flow blurring multinebulizer (FBMN), which has a common nebulization gas inlet, and independent liquid inlets for each nozzle. The FBMN directly was connected with a commercial spray chamber (Aguirre, Kovachev, Hidalgo, & Canals, 2012).

The drawbacks of two nebulizers with one or two spray chambers and multiple-nozzle nebulizer systems are coarse aerosol generation and difficulty in synchronizing the performances of individual sprays (Aguirre et al., 2010, 2012; Hamier & Salin, 1998; Huxter et al., 2003; Ido et al., 2020; Kovachev et al., 2009; Maldonado et al., 2006). Aerosols generated with these systems would be coarser compared with a standard nebulizer/spray chamber, because the total gas flow for sample introductions has to be split between two nebulizers or nozzles for controlling the residence time of aerosols inside the plasma. In addition, each nebulizer or nozzle works independently, resulting in variances of the loading amounts of two liquids into the plasma. High-performance nebulizers such as high-efficiency nebulizer (HEN) (Maldonado et al., 2006), FBN, and multispray-type FBN, described above, which can generate finer aerosols than conventional nebulizers, were applied to improve the performance of these systems. Even though the performance of each spray with a low gas flow rate was comparable to conventional concentric nebulizers with high gas flow rate, the sprays in these systems would not synchronize, which was a potentially large source of variance of analytical results (Aguirre et al., 2010, 2012; Hamier & Salin, 1998; Huxter et al., 2003; Ido et al., 2020; Kovachev et al., 2009; Maldonado et al., 2006).

An advanced version of described systems above is the MultiNeb, working in the dual mode. This nebulizer incorporates two separate liquid inlets placed in a single nebulization body with a common nebulization gas inlet and a unique outlet orifice. The nebulizer was coupled to a standard 50 mL internal-volume cyclonic-type spray chamber and the liquid streams were mixed at the tip of the nebulizer (in the aerosol phase inside the nebulizer at high pressure) and the aerosol of the liquid mixture exits by the unique hole (García, Aguirre, & Canals, 2020).

Much finer aerosol (D_{50}: 2.9 and 3.1 μm for two channels) in comparison with Meinhard nebulizer type C (D_{50}: 19.5 μm) was achieved when a two-channel, concentric-grid nebulizer (m-CGrid) was applied for ICP-OES analysis. This device consists of two independent liquid-flow capillaries (i.d./o.d.: 150/220 μm) inside a single nozzle and a PEEK grid screen (356 mesh) placed on the hole of the nozzle (i.d.: 200 μm). Both capillary tips were set at a position recessed by *ca.* 20–100 μm from the grid screen. Due to this unique structure, excellent spray performance was achieved (with sample uptake rate of 0.5 mL min^{-1} for each channel) in comparison with a conventional pneumatic nebulizer, resulting in two- to threefold higher signal intensities in ICP OES with good stability. In addition, the spray performances for the two liquid channels were almost the same. However, a correction of the difference of liquid flow rates of the channels that originated from an instrumental bias of the LC pumps was required and it could be performed by applying the internal standard correction (Ido et al., 2020).

6.2.2 Liquid solution ultrasonic nebulization in multichannel systems

An alternative way of increasing the sample introduction efficiency in plasma spectrometric techniques is to use ultrasonic nebulizers (USNs) (Canals & Aguirre, 2015; Hou et al., 2016; Matusiewicz & Ślachciński, 2017a, 2017b). USNs have been shown to be more efficient at aerosol formation and can produce aerosols with smaller and more uniform droplet size. Generating a primary aerosol with a very small mean drop size (ideally ≤5 μm) is an alternative to direct injection to the plasma, as such fine aerosol drops should pass to the plasma with high efficiency. This results in more efficient atomization, higher sensitivity, and reduced matrix effects. It has been reported that ultrasonic nebulizers are capable of operating at flow rates down to 20 μL min^{-1}, without the need for any makeup solvent, and to provide very high transport efficiency (Štrupar, 1989; Tarr, Zhu, & Browner, 1993).

Two, three, or even four-channel USNs are used for sample, standard solution, and reagents for chemical vapor generation introduction to the vibrating quartz plate where aerosols/volatile spices are generated. The main advantage of using multichannel ultrasonic nebulizers in comparison with pneumatic ones is that gas flow is not required for the generation of aerosol but is required for the transport of the aerosol to the plasma source as the aerosol is generated immediately after it comes into contact with the vibrating quartz surface of the transducer. Then, the performance of the individual nebulizer channels and aerosol quality is not affected by the carrier gas flow (Matusiewicz & Ślachciński, 2010a, 2010b, 2010c, 2010d, 2017a, 2017b).

The dual-type device contains two, identical, narrow PFA tubes with 0.4 mm orifices for solution injection positioned at an angle of *ca.* 45° and placed in almost direct contact with the vibrating quartz surface of the ultrasonic transducer. The precise position is not important; it makes no difference whether the end of the capillaries touches the quartz plate or not. In this manner, two independent solutions might be pumped simultaneously to the quartz piezoelectric transducer (1.65 MHz resonant frequency and a forward power of 45 W) at a flow rate in the range of 5–50 μL min^{-1} (Matusiewicz & Ślachciński, 2010a, 2010c).

The most advanced prototype of continuous-microflow ultrasonic nebulizer quadruple-mode microcapillary system (μ-USN-QCS) was successfully coupled on a MIP OES spectrometer *via* a dedicated spray chamber with optimized geometry providing a rapid and versatile methodology for the detection of inorganic elements (Hou et al., 2016; Matusiewicz & Ślachciński, 2017a, 2017b). The microflow principle of nebulization has been successfully transferred to a four, independent nebulization (injection) capillary platform (Fig. 6.2). Those capillaries were used for delivery of the samples, standard element solutions, reagents for CVG, and internal standard. Nebulization was achieved by placing four identical solution injection capillaries (narrow PFA tubes with 0.5 mm orifices) in contact with the nebulizer quartz vibrating plate surface. The capillaries were cut straight and placed close to each other in almost direct contact with the quartz surface of the ultrasonic transducer. In this manner, four independent solutions might be pumped simultaneously to the quartz piezoelectric transducer (1.65 MHz resonant frequency and forward power of 80 W) at a flow rate in the range 10–80 μL min^{-1} using a peristaltic pump,

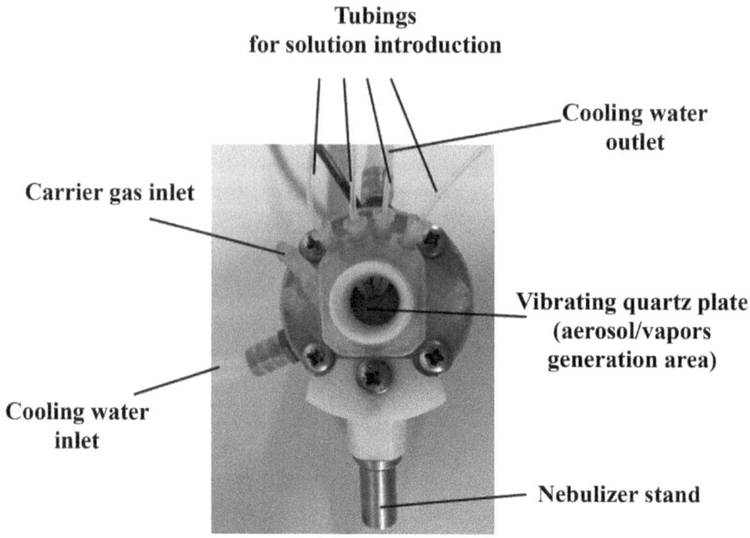

FIG. 6.2 Four-channel, ultrasonic, micronebulizer/vapor generator. *No permission required.*

where they were effectively mixed and dispersed into droplets of *ca.* 2.0–2.5 μm in mean droplet diameter (0.1–4.5 μm size distribution) (Jin, Zhu, Brushwyler, & Hieftje, 1990).

The ultrasonic nebulizer was directly connected to one of the glass spray chambers tested: two cyclonic (with internal volumes of 15 and 50 mL) and two "on axis" (single pass and double pass with internal volumes of 8 mL). The cyclonic spray chamber made it possible to introduce argon in a pseudoconcentric manner, thereby minimizing turbulence. The system was used without desolvation, unlike most higher flow ultrasonic nebulization systems. The transducer quartz plate was cooled with a closed-circuit water cooling system (Matusiewicz & Ślachciński, 2017a, 2017b).

Modifications to the multichannel systems not only concern the nebulizers themselves, but also the spray chambers. In such systems, it is also possible to use a multinebulization mode by the possibility of adding two nebulizers to the same spray chamber or applying the arrangement consisting of two nebulizer/spray chamber sets. The cross-flow nebulizer/Scott spray chamber (sample aspiration at flow rate of 1.62 mL min^{-1}) and a classical Meinhard nebulizer/Scott spray chamber (standard solution aspiration at a flow rate of 4.25 mL min^{-1}) were applied for dual nebulization. The aerosol from the two nebulizers was combined before the plasma at a Y-joint. The dual nebulizer configuration was chosen over a merging flow system, where the Y joint would have been placed before a single nebulizer and spray chamber, because it allowed adjustment of one of the nebulizer parameters with very little effect on the other. The

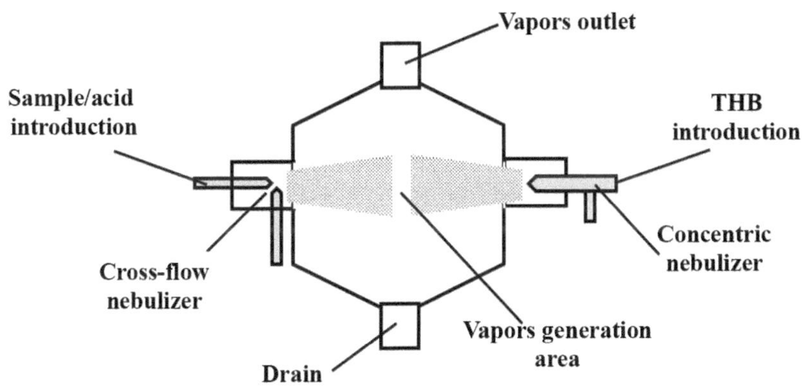

FIG. 6.3 Dual nebulization system based on Gómez et al. (2006). *No permission required.*

relative efficiency of the two nebulizers was determined by alternately introducing a standard into the sample and standard nebulizers while the remaining nebulizer was fed with the blank. Finally, the same standard was run simultaneously in both nebulizers to monitor short-term drift, since the signal obtained should be the sum of the two signals after blank subtraction (Hamier & Salin, 1998; Huxter et al., 2003).

The system consists of both cross-flow and concentric conventional Meinhard pneumatic nebulizers, mounted oppositely in a conventional spray chamber of the instrument via a simple adaptor for hydride generation and other elements' determination (Fig. 6.3). In contrast, analytical figures of merit for the nebulization of elements that do not form volatile hydrides were lower than found for conventional nebulization systems, mainly due to the fact that both pneumatic nebulizers were operated far from their optimal gas flow conditions (Gómez et al., 2006).

In another arrangement, two high-efficiency nebulizers were inserted into a modified cyclonic spray chamber of 50mL inner volume. The dual system allowed efficient hydride generation for the elements at lower sample and reagent uptake rates with a precision lower than 2% (slightly poorer than with the conventional system). This device is useful for handling samples with limited volume because it uses a sample uptake rate of 240 μL min^{-1}.The long-term stability obtained indicated that the system was stable for at least two and a half hours of continuous operation (Maldonado et al., 2006).

6.3 Applications

Matrix effects are major technical issues that require attention in optical emission spectrometry techniques, and a standard addition method is a robust way to compensate for the effects (Montaser & Golightly, 1992; Nölte, 2003).

However, a classical standard addition method is time and labor consuming, because individual calibration curves for each sample should be made, requiring preparation and measurement of numerous solutions (Ido et al., 2020).

The way to overcome this drawback is to use multichannel systems with separate nozzles (pneumatic nebulizers) or capillaries (ultrasonic nebulizers) for "in spray chamber" analytical procedures, such as dilution, derivatization, addition of internal standard, and standard addition, which normally are the most time-consuming part of the sample preparation process in many methods of analysis by plasma-based techniques (Bauer & Broekaert, 2008; Gómez et al., 2006; Hamier & Salin, 1998; Huxter et al., 2003; Krachler et al., 2005; Maldonado et al., 2006; Matusiewicz & Ślachciński, 2017b; Olofsson et al., 2000).

In this case, the addition of a standard is not performed in the sample itself, but it is introduced simultaneously during the nebulization, prior to introduction to the plasma source. Such a procedure can be carried out, e.-g., using a dual nebulizer sample introduction system, consisting of a cross-flow nebulizer/Scott spray chamber (sample aspiration) and a classical Meinhard nebulizer/Scott spray chamber (standard solution introduction) to an ICP-MS instrument. This calibration methodology gave better results than those obtained by the method of external calibration and were statistically equivalent to those obtained by standard additions, which is the favored technique for analytical calibration with difficult samples because it provides the highest accuracy independent of the matrix (Hamier & Salin, 1998; Huxter et al., 2003).

An increase on the sensitivities of the analytical signals of As, Cd, Cr, V, Co, and Mn in comparison with a single-orifice concentric nebulizer was also achieved, when a new nebulizer called the multimicrospray nebulizer (MMSN) was applied to the MIP-MS apparatus. The nebulizer has several orifices, each containing a capillary for delivering the sample solution. The nebulizer gas was ejected from the area between the orifice and the capillary. A fine aerosol was produced due to collision between the high-speed gas and the sample solution. An increase in the sensitivities of the analytical signals in comparison with a single orifice nebulizer using the same mechanism for nebulization was reported (Huang et al., 2000).

Pneumatic nebulization systems [trihedral cross-flow nebulizer (Th-CFN) and T-shaped cross-flow nebulizer (T-CFN)], where two solutions were nebulized simultaneously by a single gas flow (modifications of a conventional cross-flow nebulizer) was proposed.

It was tested to what extent these systems can be used to correct for matrix effects, induced by four elements having low first (Na, K, Rb, and Ba) and second (Ba) ionization potentials, by applying internal standardization and calibration by standard addition in an online way. The use of standard addition was shown to be often more accurate than internal standardization.

The detection limits and the precision obtained for dual mode were lower than those achieved with a commercially available cross-flow nebulizer. The indicated weakness of this arrangement was the necessity to calibrate at least once a day the response of both channels of the dual sample-loading nebulizers (Bauer & Broekaert, 2007). The T-CFN has been described for Ba, Ca, Mg, P, and Zn determinations in organic solutions by ICP OES using aqueous solutions for calibration. With this nebulizer, the simultaneous generation of an aerosol from kerosene and from an aqueous solution was possible; a mixed aqueous/organic solution aerosol can be directly formed in the spray chamber and carried to the plasma without the use of a surfactant. The additive properties of the analyte signals were demonstrated, as well as the constancy of the intensity ratios for analyte and reference element lines (Bauer & Broekaert, 2008).

The new multiple nebulizer, the FFMN for liquid sample introduction to an ICP OES spectrometer, was developed and characterized. The prototype described has four, separate, liquid sample ducts for each nebulization nozzle. Since the FFMN has dimensions that do not permit its direct connection to the conventional spray chambers used in the sample introduction systems, a dedicated spray chamber of different designs were evaluated. This new system was successfully applied to the ICP OES instrument for element determination in certified reference material (LGC 6016 estuarine water). However, some drawbacks like the different individual nozzle performance, relatively low transport efficiency, and incompatibility with standard spray chambers were indicated (Gañán-Calvo, 1998; Kovachev et al., 2009).

Subsequently, novel flow blurring dual nebulization (FBDN) systems for online calibration methodologies was tested by the same research group. Synthetic samples prepared in four different matrices of Ba, K, Na, and Rb were analyzed by using external calibration, online internal standard calibration, and online standard addition methodologies. With all the calibration methodologies, the accuracy and uncertainty of the obtained results were very similar for both standard (conical nebulizer/cyclonic spray chamber) and FBDN (in different nebulizer arrangement) systems. The main difference was a significant reduction in resource consumption (i.e., samples, reagents, and time) when FBDN were used. For most of the emission lines (of Cd, Co, Cr, Cu, Fe, Mn, Ni, and Pb) evaluated, the dual nebulization mode gave higher precision values and lower limits of detection than the standard system. Online internal standard calibration and online standard addition calibration, carried out with FBDN, led to a substantial reduction in matrix effects (Aguirre et al., 2010).

The significant modification of the above-described nebulization system was the flow blurring multinebulizer (FBPN), which has a common nebulization gas inlet, and independent liquid inlets for each nozzle. The multinebulizer directly connected with commercial spray chamber was successfully applied for biodiesel and oil sample analysis by online calibration to ICP OES. The aqueous calibration standards were used, which allowed avoiding the use of organic calibration standards and reducing carbon deposition formation (spectral interference elimination and avoidance of injector clogging) (Aguirre et al., 2012).

The FBPN successfully corrects matrix effects in As and Se determination by ICP OES. The multinebulizer makes feasible the implementation of an online internal standard strategy by mixing the internal standard solution with analytical reference solutions or a sample solution directly in the spray chamber. It was demonstrated that the combination of FBMN and In as an internal standard leads to better accuracy for determining As in a lake sediment (Pereira, Aguirre, Nóbrega, Hidalgo, & Canals, 2012).

The FBMN was also applied for the compensation of inorganic acid matrix effects in ICP OES and ICP-MS using an online standard addition calibration. The applicability of the proposed method has been evaluated analyzing two whole milk powders, a certified reference material, and a commercial product, showing excellent recovery values. Compared with other calibration strategies and experimental setups used, the online standard addition calibration using multichannel system was faster, easier to handle, and significantly reduced reagent and sample consumption (Aguirre, Fialho, Nóbrega, Hidalgo, & Canals, 2014).

Online internal standard calibration was also applied for determining Na, K, Ca, Mg, P, and 20 heavy metals in biodiesel samples using the FBMN in the dual nebulization mode and ICP OES. The standards and samples were introduced through one of the multinebulizer nozzles, while the aqueous solution containing yttrium as an internal standard was introduced through the second nozzle. Thus, the spectral interferences were compensated and the formation of carbon deposits on the ICP torch was prevented (Avila Orozco et al., 2014).

An advanced version of described system above, the MultiNeb, incorporates two independent liquid inlets into a single nebulization body with a common nebulization gas inlet and a unique outlet orifice. The nebulizer was coupled to a 50-mL internal volume cyclonic-type spray chamber. The liquid streams were mixed at the tip of the nebulizer (in the aerosol phase inside the nebulizer at high pressure) and the aerosol of the mixture of the liquids exits by the unique hole. This system for the online standard dilution analysis to ICP OES was successfully applied for the determination of Cd, Cr, Cu, Fe, Mn, Ni, and Pb in used lubricating oils. The methodology proposed allows, *inter alia*, for simultaneous correcting for matrix effects, reduces the time of the analysis as a minimal sample pretreatment is needed, requires only one standard solution per sample and there is no need of preparing mixtures, provides lower organic solvent consumption, as only aqueous standards are used, and is suitable for samples of different matrices. Successful elimination of the matrix effects caused by high concentration of easily ionized elements using multinebulization systems based was reported (García, Aguirre, & Canals, 2020).

The MultiNeb coupled to a quartz double-pass Peltier-cooled (2°C) spray chamber was also successfully employed for trace element determination (Cd, Cu, Hg, Ni, Pb, and Zn) in marine biota-certified reference material (biota IAEA-476) using online isotope dilution (OID-ICP-MS). The online sample spiking was achieved by using both liquid inlets (with flow rate of 0.3 mL min^{-1} for each channel) of the MultiNeb for the continuous multielement spike solution addition and the other one for the sequential introduction of the natural isotopic composition reference standard, the procedural blanks, and the liquid sample solutions. The solutions were efficient mixed at the inner cavity of the multinebulizer tip prior to the introduction to the plasma (García, Aguirre, Vassileva, & Canals, 2020).

A two-channel, concentric-grid nebulizer (m-CGrid) was developed for realizing efficient online standard addition in ICP OES without premixing of liquids before nebulization. This device consists of two independent liquid-flow capillaries inside a single nozzle and a grid screen placed on the hole of the nozzle. The dual systems nebulizer provided almost the same or better spray performance compared with a conventional nebulizer (Meinhard nebulizer), which led to achieve two- to threefold higher signal intensities for 23 elements. The performance of m-CGrid in an online standard addition was demonstrated in the analysis of NIST SRM1577b bovine liver and NIES No. 3 *Chlorella* reference-certified materials. The analytical results were in good agreement with their certified values (Ido et al., 2020). Operating parameters of selected chemical/photochemical vapor generation multichannel systems to atomic emission spectrometry detection are listed in Table 6.1.

6.3.1 Vapor generation in multichannel systems

The modification of conventional single-channel nebulizers have been made to allow a chemical vapor generation to be carried out. These arrangements mainly consist in adding additional channels to provide solutions of the necessary regents. An example of such an assembly is a standard, concentric, pneumatic nebulizer with inserted capillary into the existing channel (Tao & Sturgeon, 1999). The THB solution was pumped through the standard solution introduction channel of the nebulizer, while the acidified sample was injected *via* an additional capillary. Other pneumatic nebulizers were also used for this purpose, e.g., cross-flow, V-groove, and parallel path. Direct hydride generation nebulizer

TABLE 6.1 Operating parameters of selected chemical/photochemical vapor generation multichannel systems to atomic emission spectrometry detection.

Multichannel system	CVG/PVG conditions	Sample flow rate (mL min^{-1})	Element/ species	Matrix sample type	DL (ng mL^{-1})	Detection	Ref.
Meinhard nebulizer modified to dual mode	Sample acidity: 4.8 M HCl, 0.5% ($m v^{-1}$) $NaBH_4$	1.2	Se	Nickel oxide	2	ICP OES	Tao and Sturgeon (1999)
V-groove Babington-type nebulizer modified to dual mode	Sample acidity: 2.3 M HCl, 0.3% ($m v^{-1}$) $NaBH_4$	1.0	Se	Not-fat milk powder, oyster tissue, human urine, sediment	0.2	ICP OES	Carrión et al. (2003)
V-groove nebulizer modified to dual mode	Sample acidity: 30% v/v HCl (As, Bi, Sb, Se, Te); 5% v/v HCl (Sn), 2.0% (m/v) $NaBH_4$	4.0	As, Bi, Sb, Se, Sn, Te	Standard solution	0.5 (As), 0.4 (Bi), 2 (Sb), 1 (Se), 4 (Sn), 5 (Te)	ICP OES	Hutton and Preston (1983)
Direct hydride generation nebulizer (DHGN)	Sample acidity: 2.3 M HCl, 0.3% ($m v^{-1}$) $NaBH_4$	2.2	As, Sb, Se, Ba, Sb, Cd, Cu, Co, Fe, Ni, Cr, Mn, Mg, Zn	Oyster tissue	0.4 (As), Sb (2), 0.1 (Se), 18 (Ba), 3 (Cd), 11 (Cu), 6 (Ni), 6 (Cr), 2 (Mn), 3 (Mg), 3 (Zn)	ICP OES	Rojas, Murillo, Carrión, and Chirinos (2003)

nebulization system	H_2SO_4, 10% (m/v) $NaBH_4$		Se, Te, Ag, Al, Be, Ca, Cd, Co, Cr, Cu, Fe, K, Mg, Mn, Mo, Ni P, Sr, Ti, Zn	flour, corn, potatoes, soybeans	20 (Ge), 0.7 (Sb), 0.5 (Se), 0.7 (Te), 5 (Ag), 10 (Al), <1 (Be), 6 (Ca), 9 (Cd), 3 (Co), 3 (Cr), 5 (Cu), 6 (Fe), 60 (K), 1 (Mg), <1 (Mn), 5 (Mo), 6 (Ni) 100 (P), 4 (Sr), 1 (Ti), 7 (Zn)		Fricke, Hahn, and Caruso (1981)
Integrated nebulizer-hydride generation system with concentric nebulizer	Sample acidity 1% (v/v) HNO_3, 2% (m/v) KBH_4, 1:9 (v/v) H_2O_2	1.5	As, Sb, Bi, Sn, Pb, Hg, Co, Mn, Fe, Cu	Standard solution	0.4 (As), 0.55 (Sb), 0.9 (Bi), 0.65 (Sn), 7.7 (Pb), 0.3 (Hg), 8 (Co), 1 (Mn), 4 (Fe), 10 (Cu)	ICP OES	Fengzhou, Guoying, and Chun (1991)
Two high efficiency nebulizers (HEN) coupled to a modified cyclonic spray chamber	10% (v/v) HCl, 0.5% (m/v) $NaBH_4$	0.06	As, Sb, Se, Ba, Cd, Mg, Mn, Sr, Zn	Bovine liver, oyster tissue	5.3 (As), 2.9 (Sb), 4.9 (Se), 1.1 (Ba), 1.1 (Cd), 0.6 (Mg), 0.4 (Mn), 0.4 (Sr), 2.8 (Zn)	ICP OES	Maldonado et al. (2006)
Dual sample introduction system: two Meinhard nebulizers mounted at angel of 45° in a modified cyclonic chamber	Sample acidity: 15% (v/v) HCl, 0.5% (m/v) $NaBH_4$	0.9 for each nebulizer	As, Be, Cd, Co, Cr, Cu, Fe, Mg, Mn, Mo, Ni, Pb, Se, Sb, Sr, V, Zn	Urban particulate matter	2.2 (As), 0.4 (Be), 1.3 (Cd) 4.4 (Co), 5.9 (Cr), 3.3 (Cu), 3.6 (Fe), 1.3 (Mg), 0.7 (Mn), 12.8 (Mo), 8.4 (Ni), 23.1 (Pb), 9.3 (Se), 3.8 (Sb), 0.3 (Sr) 3.8 (V), 1.6 (Zn)	ICP OES	Benzo et al. (2008)

Continued

TABLE 6.1 Operating parameters of selected chemical/photochemical vapor generation multichannel systems to atomic emission spectrometry detection—cont'd

Multichannel system	CVG/PVG conditions	Sample flow rate (mL min^{-1})	Element/ species	Matrix sample type	DL (ng mL^{-1})	Detection	Ref.
Dual sample introduction system: two Meinhard nebulizers mounted at angel of 45° in a modified cyclonic chamber	Sample acidity: 15% (v/v) HCl, 0.4% (m/v) $NaBH_4$	0.9 for each nebulizer	As, Be, Cd, Co, Cr, Cu, Fe, Mg, Mn, Mo, Ni, Pb, Se, Sb, Sr, V, Zn	Tomato leaves	2.2 (As), 0.4 (Be), 1.3 (Cd) 4.4 (Co), 5.9 (Cr), 3.3 (Cu), 3.6 (Fe), 1.3 (Mg), 0.7 (Mn), 12.8 (Mo), 8.4 (Ni), 23.1 (Pb), 9.3 (Se), 3.8 (Sb), 0.3 (Sr) 3.8 (V), 1.6 (Zn)	ICP OES	Benzo et al. (2009)
Flow blurring multinebulization (FBMN) systems	1.39% $NaBH_4$, 2.97 M HCl	0.936	As, Se, Hg	Fuel	25 (As), 140 (Se), 13 (Hg) μg kg^{-1}	ICP OES	García, Aguirre, and Canals (2017)
Dual-flow ultrasonic nebulizer	2.0 M HCl, 1.0% (m/v) $NaBH_4$	0.2	As, Sb, Se	Mineral water, hard drinking water	1.5 (As) conventional continuous hydride generation, 1.3 (Sb), 0.7 (Se)	ICP OES	Tyburska et al. (2010)

Dual-flow ultrasonic nebulizer	Sample acidity: 1.2 M HNO_3, 0.6% (m/v) $NaBH_4$	0.011	As, Bi, Ge, Sb, Se Sn, Ba, Ca, Fe, Li, Mg, Sr	Dogfish livers, stream sediment, soil	1.3 (As), 5.9 (Bi), 6.6 (Ge), 1.8 (Sb), 3.6 (Se), 2.6 (Sn), 41 (Ba), 8.1(Ca), 11 (Fe), 7.5 (Li), 9.2 (Mg), 12 (Sr)	MIP OES	Matusiewicz and Ślachciński (2010a)
Dual-flow ultrasonic nebulizer	Sample acidity: 0.6 M HNO_3,1.0% (m/v) $NaBH_4$	0.015	Au, Ag, Cd, Cu, Mn, Ni, Pb, Pd, Pt, Rh, Zn	Lobster hepatopancreas, human hair, soil, soya bean flour	2.49 (Au), 1.75 (Ag), 3.39 (Cd), 2.13 (Cu), 4.80 (Mn), 6.21 (Ni), 4.26 (Pb), 2.01 (Pd), 7.65 (Pt), 3.92 (Rh), 4.65 (Zn	MIP OES	Matusiewicz and Ślachciński (2010c)
Triple-mode microcapillary ultrasonic nebulizer	3.5 M H_2SO_4, 3.0 M H_2O_2, 10 mM $NaNO_2$ (in sample), 1.0 mg L^{-1} Fe^{3+} (in sample)	0.015	I	Oyster tissue, nonfat milk powder, milk powder, iodized salt, iodid tablet, mineral water	1.6 (I)	ICP OES	Matusiewicz and Ślachciński (2010b)
Triple-mode microcapillary ultrasonic nebulizer	1.0 M HCl, 2.0% (m/v) $NaBH_4$, 1 M thiourea (in sample), 100 ng mL^{-1} of Ge (internal standard)	0.05	As, Bi, Sb, Se, Sn	Dogfish liver, soil, water	1(As), 5 (Bi), 2 (Sb), 3 (Se), 2 (Sn)	MIP OES	Matusiewicz and Ślachciński (2017b)

Continued

TABLE 6.1 Operating parameters of selected chemical/photochemical vapor generation multichannel systems to atomic emission spectrometry detection—cont'd

Multichannel system	CVG/PVG conditions	Sample flow rate (mL min^{-1})	Element/ species	Matrix sample type	DL (ng mL^{-1})	Detection	Ref.
Dual-mode ultrasonic nebulizer with 50-mL cyclonic spray chamber and 6 W Hg pen lamp	40% ($v\ v^{-1}$) acetic acid	0.03	As, Bi, Sb, Se, Sn, Hg, Cd	Dogfish liver, water	11 (As), 22 (Bi), 16 (Sb), 19 (Se), 20 (Sn), 11 (Cd), 15 (Hg)	MIP OES	Matusiewicz and Ślachciński (2013)
Dual-mode ultrasonic nebulizer with 50-mL cyclonic spray chamber and 6 W Hg pen lamp	20% v/v solution of acetic acid	0.04	Hg	Dogfish liver, marine sediment	4.1	RFP[a] OES	Matusiewicz, Ślachciński, Pawłowski, and Portalski (2015)

[a] *RFP, rotating field plasma dual-mode ultrasonic nebulizer with 50-mL cyclonic spray chamber and 6 W Hg pen lamp.*

(DHGN), which was a modification of the V-groove nebulizer, was attached to a double-pass Scott-type spray chamber.

Two additional solution channels with 70 or 120 μm orifices were successfully used for the simultaneous determination of As, Sb, and Se and other elements in ICP OES (Carrión et al., 2003; Rojas et al., 2003).

Two new introduction channels, mounted perpendicular to the sample introduction channel, were incorporated in the cross-flow nebulizer coupled with a conventional, Scott, double-pass, spray chamber. The HCl and THB solutions were pumped independently by two peristaltic pumps. This system was used for the determination of Sb and Sn (hydride-forming) together with two conventional elements (V and Zn) (Zoltan et al., 2005).

Another arrangement of the cross-flow nebulizer, into which a narrow tube was inserted for the introduction of both sample and $NaBH_4$ solutions, coupled with a Scott spray chamber, was also used by Moor et al. (2000). A chemical vapor-generation technique was also suitable for noble and transition metal determination using a modified Burgener parallel-path nebulizer (Feng et al., 2001).

6.3.1.1 Chemical vapor generation in multichannel systems with pneumatic nebulization

In recent years, several multichannel systems have been developed for the simultaneous determination of conventional hydride- and nonhydride-forming elements to inductively coupled plasma optical emission spectrometry (ICP OES), microwave-induced plasma optical emission spectrometry (MIP OES), or inductively coupled plasma-mass spectrometry (ICP MS).

The first system of this type was a tandem nebulization generator, consisting of both concentric and cross-flow nebulizers connected with two separate spray chambers. The THB solution was pumped through the concentric nebulizer to the Teflon spray chamber, while acidified sample nebulization was achieved in the second nebulizer/spray chamber configuration, where the CVG reaction proceeded (Wolnik et al., 1981). Such a system made it possible to carry out simultaneous nebulization and hydride generation for the determination of hydride/vapor forming and nonhydride forming elements. In addition, the contact time between reagents was reduced, which increases the tolerance to the different interferences present in the sample solution. This arrangement also gave the analyst more option to perform measurements by enabling switching from conventional nebulization to a hydride generation mode within a single experiment (Pohl & Sturgeon, 2010; Bings, Orlandini Von Niessen, & Schaper, 2014). Similarly, Hutton and Preston described a modification to the V-groove nebulizer configuration that allows either sensitive hydride or conventional solution analysis to be performed with no downtime. The additional tubing used was positioned directly in front of the nebulizer gas jet, in the position previously occupied by the aerosol impactor (Hutton & Preston, 1983). Huang et al. developed a nebulizer hydride generator systems in which large droplets of the acid sample aerosol formed during pneumatic nebulization were trapped by the smoking-pipe-shaped hydride generator for further reaction with the reductant. This system allowed for the simultaneous determination of the hydride-forming and other elements with detection limits improved by more than 20 times (for HG) compared to those obtained with the conventional nebulizing system, while those of the other elements remained practically unaffected (Huang, Zeng, Zhang, & Liu, 1988; Huang, Zhang, & Zeng, 1987).

Modifications to the arrangement of the spray chambers allowing multiple nebulization were also applied. This direction of development of multichannel sample introduction assembles has become very important in optical emission and mass spectrometry, leading to the manufacture of commercially available systems that are standard equipment mounted in the spectrometers. An example of such a system is an integrated nebulizer-hydride generator in which the spray chamber was modified for HG. The upright spray chamber served as a phase separator for hydride generation, and, on the other hand, permitted the small droplets of the aerosol to pass through. Thus, both volatile hydrides and conventional pneumatically nebulized solutions were performed individually or simultaneously without the need to change or disassemble the sample introduction system. Detection limits for the elements forming volatile hydrides were improved by one to two orders of magnitude, while the detection limits for other elements remain essentially unchanged or were slightly degraded (Fengzhou et al., 1991).

A double-pass Scott spray chamber, with additional gas inlet added and equipped with cross-flow nebulizer for the introduction of both sample and reductant solutions (THB), was also successfully employed for Se determination in biological-certified reference materials and for detection of other hydride-forming elements and mercury vapors by ICP-MS (Moor et al., 2000).

Modification of the above systems are concomitant metals analyzer (CMA) (Hutton & Preston, 1983; da Luz et al., 2009; Pohl & Zyrnicki, 2000) from Jobin Yvon and a generator that has become the basic equipment of modern spectrometers: multimode sample introduction system

(MSIS) (Asfaw & Wibetoe, 2005, 2006; Kiryakova, Varbanova, Simitchiev, & Kmetov, 2019; Kmetov, Kiryakova, Varbanova, Simitchiev, & Canals, 2021; Matusiewicz & Ślachciński, 2007; McLaughlin & Brindle, 2002, 2005; Proch & Niedzielski, 2021; Wiltsche, Brenner, Prattes, & Knapp, 2008). These systems are not based on the multinebulization principle, but sample, acid, and reducing reagent solutions are pumped in separate tubings (CMA), while in the MSIS hydride formation takes between two glass capillaries placed centrally in the chamber, into which the acidified sample and THB solutions are introduced. Additionally, the sample is pumped through the nebulizer for nonhydride/vapor-forming element detection.

Gómez et al. mounted opposite cross-flow and concentric (Meinhard) nebulizers in a conventional spray chamber for simultaneous determination of volatile element hydrides and other elements. This configuration not only permitted hydride generation but also allowed other elements to be determined by pneumatic nebulization (Gómez et al., 2006).

Dual-sample introduction designs used for the determination of volatile and nonvolatile elements have been also compared. Four cyclonic spray chamber arrangements with entrance angles of 0°, 45°, 90°, and 180° were tested with connection to conventional concentric pneumatic nebulizers. Sample and $NaBH_4$ aerosols were generated and a reaction occurred in the cyclonic spray chamber to produce the volatile hydrides. Two peristaltic pumps were used: one delivered the acidified sample and the sodium tetrahydroborate solutions and the second removed the waste. The dual configuration allows the aerosols used for hydride generation to be optimized by changing the solution uptake rate and gas flow of both nebulizers, independently (Benzo et al., 2009).

Maldonado and colleagues proposed a system based on a modified cyclonic spray chamber in which two high-efficiency (HEN) nebulizers were inserted (Maldonado et al., 2006).

Very effective invention in multichannel introduction systems, not only for nebulization but also for CVG, is an arrangement based on three independent flow-blurring nebulizers (flow-blurring multiple nebulizer-FBMN). This system was applied for the hydride generation of As and Se and online internal standardization (germanium solution) followed by determination by ICP OES. Higher recoveries and lower standard deviations were obtained when using a Ge(I) 303.906 nm emission line in comparison with external calibration. Addition–recovery experiments indicated that a solution containing $100\,\mu g\,L^{-1}$ of Ge presented the best performance for correcting for matrix effects and improving accuracy. When online internal standardization was applied, the mean recoveries for As emission lines were 96% for Ca, 94% for Mg, 9% for K, and 99% for their mixture. For the most-sensitive As emission line (i.e., As(I) 228.812 nm), recoveries were 97% for Ca and 92% for Mg. In addition, this system does not change appreciably from a conventional sample introduction arrangement used in ICP OES and allows achieving lower detection limits for As in comparison with two separate nebulizer setup (Pereira, Aguirre, Nóbrega, Hidalgo, & Canals, 2014).

The analysis of As, Se, and Hg in commercial fuel samples by ICP OES without oxygen supply was also performed using the same aerosol/hydrides generation system. Aqueous solutions of THB and HCl and organic calibration standards/samples were introduced separately as aerosols into the spray chamber through the FBMN. The primary aerosols generated by the nozzles then interacted inside the spray chamber generating the conditions for CVG. This simple design and low-cost (i.e., the use of valves, mix coils, and gas-liquid phase separator was avoided) generator provided low sample and reagent consumptions, decreasing the total analytical cost and capability to measure simultaneously forming and nonforming volatile species (García et al., 2017).

6.3.1.2 Chemical vapor generation in multichannel systems with ultrasonic nebulization

In most cases, multimode vapor generators equipped with pneumatic nebulizers have been used. However, pneumatic nebulizers suffer from a relatively low rate of conversion of a liquid sample into aerosol. Conventional liquid sample introduction systems, using a pneumatic nebulizer and a spray chamber, are generally only about 1%–5% efficient. In order to improve the transport efficiency, a great variety of nebulizers have been reported (Canals & Aguirre, 2015).

Another possibility is the introduction of a very fine aerosol generated by ultrasonic (micro)nebulizers for chemical and photochemical vapor generation taking place at a transducer vibrating plate/spray chamber area. Aerosol with a very small mean drop size (ideally $\leq 5\,\mu m$) can be effectively involved in the CVG reaction or UV-irradiated during a PVG procedure. In addition, such fine aerosol drops should pass the spray chamber and reach the plasma discharge with high efficiency (Karnicky, Zitelli, & Van der Wal, 1987; Tarr et al., 1993).

A two-channel ultrasonic nebulizer was applied for simultaneous aerosol/vapor generation. The feasibility of the use of dual-flow ultrasonic nebulizer for the determination of Se, Sb, and As in waters by hydride-generation ICP OES has been investigated (Tyburska et al., 2010).

The range of applications for such a nebulizer/vapor generator was extended to the simultaneous determination of hydride-forming (As, Bi, Ge, Sb, Se, and Sn) and nonhydride-forming elements (Ba, Ca, Fe, Li, Mg, and Sr) (Matusiewicz & Ślachciński, 2010a). An acidified sample and THB solutions were pumped through two separate capillaries to the vibrating quartz plate, where volatile hydrides/aerosols were generated. The same system for the simultaneous determination of Au, Ag, Cd, Cu, Mn, Ni, Pb, Zn, Pd, Pt, and Rh volatile metal species by MIP OES have been applied (Matusiewicz & Ślachciński, 2010c).

A triple-mode ultrasonic nebulization/vapor generator system was used for the determination of iodine with ICP OES detection. Sulfuric acid, sample (containing sodium nitrite and Fe^{3+} ions), and hydrogen peroxide were pumped to the ultrasonic nebulizer through three separate capillaries. The use of this system made it possible to generate iodine vapors directly on the nebulizer vibrating plate, without the need for prior acidification of the sample, which would oxidize iodide ions to iodine (Matusiewicz & Ślachciński, 2010b). The same apparatus for simultaneous multielement determination of Au, Ag, Pd, Pt, and Rh metal volatile species by microwave-induced plasma spectrometry has also been adopted (Matusiewicz & Ślachciński, 2010d). Operation of the triple-mode system consisted of pumping the THB solution with a peristaltic pump through one capillary, the sample solution delivered with the same pump through the second capillary, and the acidic solution delivered by the third capillary. The reaction occurred *in situ* on the very localized aerosol generation area and in the quartz piezoelectric transducer. The sample can be introduced without having to be acidified and it is easier to perform speciation analysis.

A proposed quadruple-mode microcapillary system (μ-USN/QCS-HG) proposed (Matusiewicz & Ślachciński, 2017a, 2017b), allowed not only for simultaneous efficient generation of As, Bi, Ge, Sb, Se, and Sn hydrides but also online internal standardization using a germanium solution with detection by microwave-induced plasma optical emission spectrometry (MIP OES).

6.3.1.3 Photochemical vapor generation in multichannel systems

The conventional CVG technique has some limitations related to the decrease of sensitivity and reproducibility due to cogeneration of volatile species with the same reducing reagent during the generation step:

coproduction of active metals that catalytically decompose $NaBH_4$ or adsorb and/or decompose the organometallic hydrides. It is well established that efficiency of CVG strongly depends on oxidation states of hydride-forming elements taking part in the reaction with THB in acid media. The large amount of hydrogen generated by the THB-acid system may reduce the stability of a plasma sources or even extinguish it, as for example, when a low-power microwave-induced plasma (MIP) is used for detection (He et al., 2007).

A technique that allow eliminating, to some extent, the disadvantageous aspects of CVG is photochemical vapor generation (PVG) associated with spectrometric determination consisting of a very sensitive approach, which aims at the conversion of inorganic analytes into volatile chemical species upon sample exposure to UV radiation (Guo, Sturgeon, Mester, & Gardner, 2004).

It has been shown that volatile species of several hydride-forming elements (As, Bi, Sb, Se, Sn, Pb, and Te), cold vapor (Cd and Hg), transition metals (Co, Cu, Fe, and Ni), noble metals (Au, Pd, Pt, Rh, and Ag), and nonmetals (I and S) have also been detected following UV irradiation of their aqueous solutions containing low molecular-weight acids (LMW). Radical-induced reactions in irradiated solutions of low molecular-weight organic (LMW) acids provide small ligands capable of reacting with many elements to yield volatile products (Bendicho et al., 2010; Guo et al., 2004; He et al., 2007; Lei, Chen, Yang, & Wen, 2017; Nóbrega et al., 2011; Suzuki et al., 2012; Yin et al., 2011; Zhang, Shen, Li, Zheng, & Hou, 2016). Organic and inorganic compounds, when exposed to UV radiation, tend to decompose and form radicals such as $OH^{\bullet}$, $O^{\bullet}$, $CO^{\bullet}$, and some smaller organic radicals, which may reduce the analyte to its elemental state or lead to the formation of volatile species containing the analyte. The efficiency of radical production is typically higher when low molecular-weight organic acids, such as formic acid, acetic acid, or propionic acid, are added in an aqueous medium (Golimowski & Golimowska, 1996; Sturgeon, Willie, & Mester, 2006; Zheng, Ma, Wu, Hou, & Sturgeon, 2010). The mechanisms of photochemical vapor generation were proposed by many authors (He et al., 2007; Ralph E. Sturgeon & Grinberg, 2012; Yin et al., 2011).

The use of multichannel nebulizers enables delivering the sample and organic acid through separate capillaries to the ultrasonic transducer, low-pressure channels for PVG. An example of such arrangement is the combination of a cyclonic spray chamber equipped with a lamp emitting UV radiation connected with an ultrasonic nebulizer.

The use of a very fine aerosol, produced by USN, exposed to UV radiation serves to mitigate the effects of extraneous absorption of

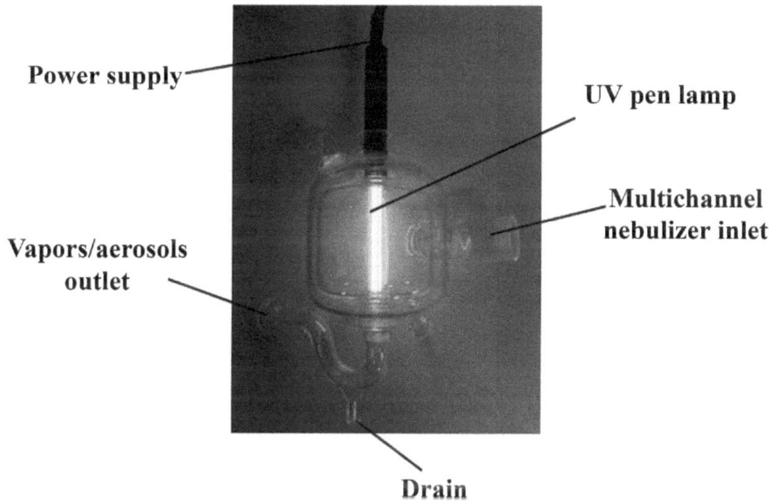

FIG. 6.4 Photogenerator. *No permission required.*

source intensity as it penetrates the aerosol (liquid) medium, thereby delivering the UV intensity with high efficiency and minimizing the need for enhanced irradiation time for the processing of absorbing samples. The generator consisting of an ultrasonic nebulizer equipped with cyclonic spray chamber with inserted 6 W Hg UV pen lamp (Fig. 6.4) was successfully applied for the determination of the conventional hydrides (As, Bi, Sb, Se, and Sn) and vapors (Hg and Cd) using MIP OES in two reference materials: dogfish liver (NRCC DOLT-2) and water (NIST 1643e). Acetic acid was added as a precursor of reducing species upon UV irradiation and as solubilizer of biological samples (Matusiewicz & Ślachciński, 2013). A similar generator was used for total mercury determination in reference materials using a five-phase digitally controlled rotating field plasma source (RFP) in optical emission spectrometry (OES). The synergic effect of ultrasonic nebulization in a dual-mode arrangement and ultraviolet-visible light (UV-Vis) irradiation when used in combination has been exploited for efficient Hg vapor generation. The most advantageous design was an ultrasonic nebulizer fitted with a 6-W mercury lamp supplying a microliter sample to a quartz oscillator, converting liquid into aerosol at the entrance of the UV spray chamber (Matusiewicz, Ślachciński, Pawłowski, & Portalski, 2015). Selected operating parameters and detection limits for chemical/photochemical vapor generation in multichannel systems with atomic emission and mass spectrometry detection are listed in Table 6.2.

TABLE 6.2 Operating parameters of selected multichannel pneumatic and ultrasonic nebulization systems to atomic emission and mass spectrometry detection.

Multichannel system	Sample flow rate (mL min^{-1})	Element	Matrix sample type	DL (ng mL^{-1})	Detection	Ref.
Multimicrospray nebulizer (MMSN) with double-pass spray chamber	0.005 (0.2, max. sensitivity)	Cr, Mn, Co, Cu	Standard solution	–	MIP-MS	Huang et al. (2000)
Cross flow dual-channel nebulizer are the trihedral cross-flow (Th-CFN) and T-shaped cross-flow (T-CFN) with online standard addition	1.1–1.5	Cd, Co, Cr, Cu, Fe, Mn, Ni, Pb	Alumina sample	0.2 (Cd), 1 (Co), 0.8 (Cr), 1 (Cu), 1 (Fe), 0.3 (Mn), 2 (Ni) and 5 (Pb)	ICP OES	Bauer and Broekaert (2007)
Flow focusing multinebulization (FFMN) with dedicated spray chamber	0.5	K, Cu, Li, Zn, Mg, Cd, Sr, Ba, Mn, Cr, Fe, Ni, Pb	Estuarine water	35 (K), 5 (Cu), 17 (Li), 22 (Zn), 6 (Mg), 43 (Cd), 1 (Sr), 8 (Ba), 1 (Mn), 11 (Cr), 7 (Fe), 65 (Ni), 191 (Pb)	ICP OES	Kovachev et al. (2009)
Flow blurring dual nebulization (FBDN) with online calibration	0.2	Cr, Ni, Cu, Mn, Cd, Fe, Co, Pb	Estuarine water	0.04 (Cr), 0.1(Ni), 0.3 (Cu), 0.01 (Mn), 0.13 (Cd), 0.08 (Fe), 0.08 (Co), 2 (Pb)	ICP OES	Aguirre et al. (2010)
Flow blurring multinebulizer (FBMN) with online calibration	0.2	Ag, Al, Ba, Ca, Cd, Cr, Cu, Fe, K, Mg, Mn, Mo, Na, Ni, Pb, P, Si, Sn, Ti, V, Zn	Biodiesel and oil	LOQ: 0.7–2 (Ag), 18–26 (Al), 0.4–0.6 (Ba), 3–16 (Ca), 1–2 (Cd), 6–7 (Cr), 4–9 (Cu), 2 (Fe), 4–7 K, 0.5–1.2 (Mg), 0.13–0.3 9 (Mn), 14–28 (Mo), 5–7 (Na), 3–4 (Ni), 22–28 (Pb), 14–20 (P), 7 (Si), 5–32 (Sn), 0.4–0.8 (Ti), 2–5 (V), 1.5–4 (Zn)	ICP OES	Aguirre et al. (2012)

MultiNeb, dual mode, with line standard dilution analysis	2.5 (total liquid uptake rate)	Cd, Cr, Cu, Fe, Mn, Ni, Pb	Used lubricating oils	20 (Cd), 10 (Cr), 6 (Cu), 20 (Fe), 4 (Mn), 30 (Ni), 200 (Pb) ng g^{-1}	ICP OES	García et al. (2020)
MultiNeb, with online isotope dilution	0.6 (total liquid uptake rate)	Cd, Cu, Hg, Ni, Pb, Zn	Marine biota	0.006 (Cd), 0.4 (Cu), 0.09 (Hg), 30 (Ni), 0.2 (Pb), 4 (Zn) ng g^{-1}	ICP-MS	Zoltan et al. (2005)
Concentric grid nebulizer (m-CGrid), two-channel, with online standard addition	0.5	P, K, Na, Mg, Fe, Cu, Zn, Ca, Mn, Sr (23 elements)	Bovine liver, chlorella	–	ICP OES	Ido et al. (2020)
Quadruple-mode microflow ultrasonic nebulizer	0.04	Ba, Ca, Cd, Cu, Fe, Mg, Mn, Pb, Sr, Zn	Dogfish liver, stream sediment, soil, water	30 (Ba), 6.0 (Ca), 6.5 (Cd), 2.8 (Cu), 12 (Fe), 7.1 (Mg), 5.3 (Mn), 8.5 (Pb), 10 (Sr), 8.8 (Zn)	MIP OES	Matusiewicz and Ślachciński (2017a)
Dual nebulizer system: nebulizer/Scott spray chamber (sample) and Meinhard nebulizer/Scott spray chamber (standards solution), tandem calibration method	1.62 (sample) 4.25 (standard solution)	Cu, Y, Pt, Pb	Standard solution	–	ICP-MS	Huxter et al. (2003)

6.4 Conclusions

Recent multichannel apparatus are used for nebulization as well as chemical/photo-chemical vapor generation in the analysis of biological materials, clinical, environmental, food, toxicology, and petrochemicals. The variety and number of samples indicate the wide application of these systems in today's analytical applications for detection in optical emission and mass spectrometry.

The great advantages of using multichannel sample introduction systems include: low sample/reagent consumption (microliter volumes), possibility of introducing reagents through separate channels (reaction proceeding in the aerosol phase), the use of online calibration and standardization, lower reagent consumption, easy to change the operating mode from standard nebulization to CVG/PVG or combined nebulization-CVG/PVG, and application of time- and cost-effective methodologies. In addition, a photo-CVG technique that allows replacing or reducing the use of chemical reagents seems to have great potential for the generation of a broad spectrum of elements as volatile species for detection in optical emission spectrometry and mass spectrometry.

In recent decades, there have been many studies on the use of volatile species generations (vapors, hydrides, etc.) in analytical atomic spectrometry and mass spectrometry. The CVG technique is a highly selective method of separation of the element of the matrix, often complex and interfering, allowing lowering detection limits. The chemical generation technique enables introducing analytes with very high efficiency to the atomization/excitation/ionization sources. The progress in CVG and PVG is evident over the past several years and major improvement has been made to simultaneous vapor generation and nebulization. Multimode systems open a widespread area for real multielement analyses of hydride/vapor- and nonhydride/vapor-forming elements in a variety of samples, also prepared as a slurry. However, further research is needed to increase the efficiency of PVG and to enable the hyphenation of PVG and CVG in one measurement cycle with the use of multichannel systems (Shen, Zheng, Jiang, Wu, & Hou, 2015) as especially, the multimode systems are becoming a standard equipment in the apparatus used in spectrochemical laboratories and this trend will not only continue, but will also grow intensively in the near future.

Acknowledgment

This work was supported by the Polish Ministry of Science and Higher Education.

References

Aguirre, M. A., Fialho, L. L., Nóbrega, J. A., Hidalgo, M., & Canals, A. (2014). Compensation of inorganic acid interferences in ICP-OES and ICP-MS using a flow blurring® multinebulizer. *Journal of Analytical Atomic Spectrometry, vol. 29*(7), 1218–1227. Royal Society of Chemistry https://doi.org/10.1039/c4ja00079j.

Aguirre, M. A., Kovachev, N., Almagro, B., Hidalgo, M., & Canals, A. (2010). Compensation for matrix effects on ICP-OES by on-line calibration methods using a new multi-nebulizer based on Flow Blurring® technology. *Journal of Analytical Atomic Spectrometry, 25*, 1724–1732. https://doi.org/10.1039/C004854B.

Aguirre, M. A., Kovachev, N., Hidalgo, M., & Canals, A. (2012). Analysis of biodiesel and oil samples by on-line calibration using a Flow Blurring® multinebulizer in ICP OES without oxygen addition. *Journal of Analytical Atomic Spectrometry, 27*(12), 2102–2110. https://doi.org/10.1039/c2ja30202k.

Almagro, B., Gañán-Calvo, A. M., & Canals, A. (2004). Preliminary characterization and fundamental properties of aerosols generated by a flow focusing pneumatic nebulizer. *Journal of Analytical Atomic Spectrometry, 19*(10), 1340–1346. https://doi.org/10.1039/B400502N.

Almagro, B., Gañán-Calvo, A. M., Hidalgo, M., & Canals, A. (2006a). Behaviour of a flow focusing pneumatic nebulizer with high total dissolved solids solution on radially- and axially-viewed inductively coupled plasma atomic emission spectrometry. *Journal of Analytical Atomic Spectrometry, 21*(10), 1072–1075. https://doi.org/10.1039/B606621F.

Almagro, B., Gañán-Calvo, A. M., Hidalgo, M., & Canals, A. (2006b). Flow focusing pneumatic nebulizer in comparison with several micronebulizers in inductively coupled plasma atomic emission spectrometry. *Journal of Analytical Atomic Spectrometry, 21*(8), 770–777. https://doi.org/10.1039/b518282d.

Asfaw, A., & Wibetoe, G. (2005). Simultaneous determination of hydride (Se) and non-hydride-forming (Ca, Mg, K, P, S and Zn) elements in various beverages (beer, coffee, and milk), with minimum sample preparation, by ICP-AES and use of a dual-mode sample-introduction system. *Analytical and Bioanalytical Chemistry, 382*(1), 173–179. https://doi.org/10.1007/s00216-005-3188-2.

Asfaw, A., & Wibetoe, G. (2006). Dual mode sample introduction for multi-element determination by ICP-MS: The optimization and use of a method based on simultaneous introduction of vapor formed by NaBH4 reaction and aerosol from the nebulizer. *Journal of Analytical Atomic Spectrometry, 21*(10), 1027–1035. https://doi.org/10.1039/b604116g.

Avila Orozco, F. D., Kovachev, N., Aguirre Pastor, M. A., Domini, C. E., Fernández Band, B. S., & Hernández, A. C. (2014). Analysis of metals and phosphorus in biodiesel B100 from different feedstock using a Flow Blurring® multinebulizer in inductively coupled plasma-optical emission spectrometry. *Analytica Chimica Acta, 827*, 15–21. https://doi.org/10.1016/j.aca.2014.04.016.

Bauer, M., & Broekaert, J. A. C. (2007). Investigations on the use of pneumatic cross-flow nebulizers with dual solution loading including the correction of matrix effects in elemental determinations by inductively coupled plasma optical emission spectrometry. *Spectrochimica Acta - Part B Atomic Spectroscopy, 62*(2), 145–154. https://doi.org/10.1016/j.sab.2007.02.006.

Bauer, M., & Broekaert, J. A. C. (2008). The use of dual-channel simultaneous pneumatic nebulization of two non-miscible solutions for the analysis of organic media using aqueous calibration solutions applied to elemental determinations in oils by inductively coupled plasma-optical emission spectrometry. *Journal of Analytical Atomic Spectrometry, 23*(4), 479–486. https://doi.org/10.1039/b718270h.

Bendicho, C., Pena, F., Costas, M., Gil, S., & Lavilla, I. (2010). Photochemistry-based sample treatments as greener approaches for trace-element analysis and speciation. *TrAC, Trends in Analytical Chemistry, 29*(7), 681–691. https://doi.org/10.1016/j.trac.2010.05.003.

Benzo, Z., Maldonado, D., Chirinos, J., Marcano, E., Gómez, C., Quintal, M., et al. (2009). Evaluation of dual sample introduction systems by comparison of cyclonic spray chambers with different entrance angles for ICP-OES. *Microchemical Journal, 93*(2), 127–132. https://doi.org/10.1016/j.microc.2009.05.009.

Benzo, Z., Maldonado, D., Chirinos, J., Marcano, E., Gómez, C., Quintal, M., et al. (2008). Performance of a dual sample introduction system with conventional concentric nebulizers for simultaneous determination of hydride and non hydride forming elements by ICP-OES. *Instrumentation Science & Technology, 36,* 598–610. https://doi.org/10.1080/10739140802448275.

Bings, N. H., Orlandini Von Niessen, J. O., & Schaper, J. N. (2014). Liquid sample introduction in inductively coupled plasma atomic emission and mass spectrometry - Critical review. *Spectrochimica Acta - Part B Atomic Spectroscopy, 100,* 14–37. https://doi.org/10.1016/j.sab.2014.08.011.

Brindle, I. D. (2007). Vapour-generation analytical chemistry: From Marsh to multimode sample-introduction system. *Analytical and Bioanalytical Chemistry, 388*(4), 735–741. https://doi.org/10.1007/s00216-007-1161-y.

Browner, R. F., & Boorn, A. W. (1984). Sample introduction techniques for atomic spectroscopy. *Analytical Chemistry, 56*(7), 875–888. https://doi.org/10.1021/ac00271a003.

Cairns, W. R. L., Barbante, C., Capodaglio, G., Cescon, P., Gambaro, A., & Eastgate, A. (2004). Performance characteristics of a low volume spray chamber with a micro-flow nebulizer for ICP-MS. *Journal of Analytical Atomic Spectrometry, 19*(2), 286–291. https://doi.org/10.1039/b314255h.

Canals, A., & Aguirre, M. A. (2015). Roles of nebulizers in analytical chemistry. In R. A. Meyers (Ed.), *Encyclopedia of analytical chemistry* John Wiley & Sons, Ltd.

Carrión, N., Murillo, M., Montiel, E., & Díaz, D. (2003). Development of a direct hydride generation nebulizer for the determination of selenium by inductively coupled plasma optical emission spectrometry. *Spectrochimica Acta - Part B Atomic Spectroscopy, 58*(8), 1375–1389. https://doi.org/10.1016/S0584-8547(03)00078-8.

da Luz, W., Lopes, R. E., Santelli, E. P., Oliveira, M., de Carvalhoma, B., & Bezerra. (2009). *Talanta, 79,* 1276–1282.

Donati, G. L., Amais, R. S., & Williams, C. B. (2017). Recent advances in inductively coupled plasma optical emission spectrometry. *Journal of Analytical Atomic Spectrometry, 32*(7), 1283–1296. https://doi.org/10.1039/c7ja00103g.

Feng, Y. L., Lam, J. W., & Sturgeon, R. E. (2001). Expanding the scope of chemical vapor generation for noble and transition metals. *Analyst, 126*(11), 1833–1837. https://doi.org/10.1039/b107494f.

Fengzhou, Q., Guoying, G., & Chun, X. (1991). A compact, versatile, integrated nebulizer-hydride generator system for simultaneous determination of volatile elemental hydrides and other elements by ICP-AES. *Applied Spectroscopy, 45*(2), 287–292. https://doi.org/10.1366/0003702914337597.

Gañán-Calvo, A. M. (1998). Generation of steady liquid microthreads and micron-sized monodisperse sprays in gas streams. *Physical Review Letters, 80*(2), 285–288. https://doi.org/10.1103/PhysRevLett.80.285.

García, M., Aguirre, M. A., & Canals, A. (2017). Determination of As, Se, and Hg in fuel samples by in-chamber chemical vapor generation ICP OES using a Flow Blurring® multinebulizer. *Analytical and Bioanalytical Chemistry, 409*(23), 5481–5490. https://doi.org/10.1007/s00216-017-0484-6.

García, M., Aguirre, M. A., & Canals, A. (2020). A new multinebulizer for spectrochemical analysis: Wear metal determination in used lubricating oils by on-line standard dilution analysis (SDA) using inductively coupled plasma optical emission spectrometry (ICP

OES). *Journal of Analytical Atomic Spectrometry, 35*, 265–272. https://doi.org/10.1039/C9JA00255C.

García, M., Aguirre, M. A., Vassileva, E., & Canals, A. (2020). A multinebulization technique for the determination of trace metals in a marine biota sample by on-line isotope dilution inductively coupled plasma mass spectrometry (OID-ICP-MS). *Journal of Analytical Atomic Spectrometry, 35*(11), 2509–2516. https://doi.org/10.1039/d0ja00262c.

Golimowski, J., & Golimowska, K. (1996). UV-photooxidation as pretreatment step in inorganic analysis of environmental samples. *Analytica Chimica Acta, 325*(3), 111–133. https://doi.org/10.1016/0003-2670(96)00034-7.

Gómez, L. R., Márquez, G. D., & Chirinos, J. R. (2006). Dual nebulizer sample introduction system for simultaneous determination of volatile elemental hydrides and other elements. *Analytical and Bioanalytical Chemistry, 386*(1), 188–195. https://doi.org/10.1007/s00216-006-0578-z.

Guo, X., Sturgeon, R. E., Mester, Z., & Gardner, G. J. (2004). Vapor generation by UV irradiation for sample introduction with atomic spectrometry. *Analytical Chemistry, 76*(8), 2401–2405. https://doi.org/10.1021/ac0353536.

Hamier, J., & Salin, E. D. (1998). Tandem calibration methodology: dual nebulizer sample introduction for inductively coupled plasma atomic emission spectrometry. *Journal of Analytical Atomic Spectrometry, 13*(6), 497–505. https://doi.org/10.1039/a801689e.

He, Y., Hou, X., Zheng, C., & Sturgeon, R. E. (2007). Critical evaluation of the application of photochemical vapor generation in analytical atomic spectrometry. *Analytical and Bioanalytical Chemistry, 388*(4), 769–774. https://doi.org/10.1007/s00216-006-1044-7.

Hou, X., Amais, R. S., Jones, B. T., & Donati, G. L. (2016). Inductively coupled plasma optical emission spectrometry. In R. A. Meyers (Ed.), *Encyclopedia of analytical chemistry* John Wiley & Sons.

Huang, M., Hirabayashi, A., Shirasaki, T., & Koizumi, H. (2000). A multimicrospray nebulizer for microwave-induced plasma mass spectrometry. *Analytical Chemistry, 72*(11), 2463–2467. https://doi.org/10.1021/ac991379e.

Huang, B., Zeng, X., Zhang, Z., & Liu, J. (1988). Recent studies on inductively coupled plasma sample introduction-1. Hydride generation. *Spectrochimica Acta Part B: Atomic Spectroscopy, 43*(4–5), 381–389. https://doi.org/10.1016/0584-8547(88)80066-1.

Huang, B., Zhang, Z., & Zeng, X. (1987). A new nebulizer-hydride generator system for simultaneous multielement inductively coupled plasma-atomic emission spectrometry. *Spectrochimica Acta Part B: Atomic Spectroscopy, 129–137*. https://doi.org/10.1016/0584-8547(87)80056-3.

Hutton, R. C., & Preston, B. (1983). A simple versatile hydride-generation configuration for inductively coupled plasmas. *The Analyst, 108*(1292), 1409–1411. https://doi.org/10.1039/an9830801409.

Huxter, V., Hamier, J., & Salin, E. D. (2003). Tandem calibration methodology: dual nebulizer sample introduction for ICP-MS. *Journal of Analytical Atomic Spectrometry, 18*(1), 71–75. https://doi.org/10.1039/b110427f.

Ido, K., Matsushita, R., Fujii, S. I., Miyashita, S. I., Umemura, T., Hokura, A., et al. (2020). Multiple-channel concentric grid nebulizer for online standard addition in inductively coupled plasma optical emission spectrometry. *Analytical Sciences, 36*(6), 717–722. https://doi.org/10.2116/ANALSCI.19P385.

Jin, Q., Zhu, C., Brushwyler, K., & Hieftje, G. M. (1990). Efficient and inexpensive ultrasonic nebulizer for atomic spectrometry. *Applied Spectroscopy, 44*(2), 183–186. https://doi.org/10.1366/0003702904085570.

Karnicky, J. F., Zitelli, L. T., & Van der Wal, S. (1987). Ultrasonic micronebulizer inferface for high-performance liquid chromatography with flame photometric detector. *Analytical Chemistry, 59*(2), 327–333. https://doi.org/10.1021/ac00129a023.

Kiryakova, M. T., Varbanova, E. K., Simitchiev, K. K., & Kmetov, V. J. (2019). Nanoparticles-assisted MSIS-MP-AES hydride generation determination of As and Sb. *Bulgarian*

Chemical Communications, 51, 58–63. https://bcc.iche.bas.bg/BCC_Volumes/Volume_51_Special_D_2019/BCC-51-D-2019-58-63-Kiryakova-14.pdf.

Kmetov, V., Kiryakova, M., Varbanova, E., Simitchiev, K., & Canals, A. (2021). MSIS S determination of As and Sb in complex matrices by magnetic nanoparticles-assisted hydride generation. *Analytical Methods, 13*(9), 1172–1180. https://doi.org/10.1039/d0ay02260h.

Kovachev, N., Almagro, B., Aguirre, M. A., Hidalgo, M., Gañán-Calvo, A. M., & Canals, A. (2009). *Journal of Analytical Atomic Spectrometry.*

Krachler, M., Rausch, N., Feuerbacher, H., & Klemens, P. (2005). A new HF-resistant tandem spray chamber for improved determination of trace elements and Pb isotopes using inductively coupled plasma-mass spectrometry. *Spectrochimica Acta - Part B Atomic Spectroscopy, 60*(6), 865–869. https://doi.org/10.1016/j.sab.2005.04.002.

Lei, Z., Chen, L., Yang, S., & Wen, X. (2017). Effects of four organic reagents on non-aqueous phase cold vapor generation of mercury. *Microchemical Journal, 132,* 89–92. https://doi.org/10.1016/j.microc.2017.01.011.

Maldonado, D., Chirinos, J., Benzo, Z., Gómez, C., & Marcano, E. (2006). Analytical evaluation of a dual micronebulizer sample introduction system for inductively coupled plasma spectrometry. *Journal of Analytical Atomic Spectrometry, 21*(8), 743–749. https://doi.org/10.1039/b604044f.

Matusiewicz, H., & Ślachciński, M. (2007). Simultaneous determination of hydride forming (As, Bi, Ge, Sb, Se, Sn) and Hg and non-hydride forming (Ca, Fe, Mg, Mn, Zn) elements in sonicate slurries of analytical samples by microwave induced plasma optical emission spectrometry with dual-mode sample introduction system. *Microchemical Journal, 86*(1), 102–111. https://doi.org/10.1016/j.microc.2006.12.002.

Matusiewicz, H., & Ślachciński, M. (2010a). Analytical evaluation of an integrated ultrasonic nebulizer-hydride generator system for simultaneous determination of hydride and non-hydride forming elements by microwave induced plasma spectrometry. *Spectroscopy Letters, 43*(6), 474–485. https://doi.org/10.1080/00387010903360040.

Matusiewicz, H., & Ślachciński, M. (2010b). In situ vapor generation inductively coupled plasma spectrometry for determination of iodine using a triple-mode microflow ultrasonic nebulizer after alkaline solubilization. *Analytical Methods, 2*(10), 1592–1598. https://doi.org/10.1039/c0ay00420k.

Matusiewicz, H., & Ślachciński, M. (2010c). Method development for simultaneous determination of transition (Au, Ag, Cd, Cu, Mn, Ni, Pb, Zn) and noble (Pd, Pt, Rh) metal volatile species by microwave- induced plasma spectrometry using ultrasonic micronebulizer dual capillary sample introduction system. *Spectroscopy Letters, 43*(3), 172–182. https://doi.org/10.1080/00387010903284323.

Matusiewicz, H., & Ślachciński, M. (2010d). Method development for simultaneous multi-element determination of transition (Au, Ag) and noble (Pd, Pt, Rh) metal volatile species by microwave induced plasma spectrometry using a triple-mode microflow ultrasonic nebulizer and in situ chemical vapor generation. *Journal of Analytical Atomic Spectrometry, 25*(8), 1324–1333. Royal Society of Chemistry https://doi.org/10.1039/c002886j.

Matusiewicz, H., & Ślachciński, M. (2013). Ultrasonic nebulization/UV photolysis vapor generation sample introduction system for the determination of conventional hydride (As, Bi, Sb, Se, Sn) and cold vapor (Hg, Cd) generation elements in reference materials in the presence of acetic acid by microwave-induced plasma spectrometry. *Spectroscopy Letters, 46*(5), 315–326. https://doi.org/10.1080/00387010.2012.725702.

Matusiewicz, H., & Ślachciński, M. (2017a). Analytical evaluation of a quadruple-mode micro-flow ultrasonic nebulizer for sample introduction in microwave induced plasma spectrometry. *Microchemical Journal, 130,* 345–352. https://doi.org/10.1016/j.microc.2016.10.010.

Matusiewicz, H., & Ślachciński, M. (2017b). Simultaneous determination of As, Bi, Sb, Se and Sn by microwave induced plasma spectrometry using a quadruple-mode microflow ultrasonic nebulizer for in situ hydride generation with internal standardization. *Microchemical Journal, 131*, 70–78. https://doi.org/10.1016/j.microc.2016.11.017.

Matusiewicz, H., Ślachciński, M., Hidalgo, M., & Canals, A. (2007). Evaluation of various nebulizers for use in microwave induced plasma optical emission spectrometry. *Journal of Analytical Atomic Spectrometry, 22*(9), 1174–1178. https://doi.org/10.1039/b704612j.

Matusiewicz, H., Ślachciński, M., Pawłowski, P., & Portalski, M. (2015). Evaluation of five phase digitally controlled rotating field plasma source for photochemical mercury vapor generation optical emission spectrometry. *Analytical Sciences, 31*(10), 987–995. https://doi.org/10.2116/analsci.31.987.

McLaughlin, R. L. J., & Brindle, I. D. (2002). A new sample introduction system for atomic spectrometry combining vapour generation and nebulization capacities. *Journal of Analytical Atomic Spectrometry, 17*(11), 1540–1548. https://doi.org/10.1039/b208011g.

McLaughlin, R. L. J., & Brindle, I. D. (2005). *S Patent number 6.*

Montaser, A., & Golightly, D. W. (1992). *Inductively Coupled Plasmas in Analytical Atomic Spectrometry.*

Moor, C., Lam, J. W. H., & Sturgeon, R. E. (2000). A novel introduction system for hydride generation inductively coupled plasma mass spectrometry: Determination of selenium in biological materials. *Journal of Analytical Atomic Spectrometry, 15*(2), 143–149. https://doi.org/10.1039/a909296j.

Mora, J., Maestre, S., Hernandis, V., & Todolí, J. L. (2003). Liquid-sample introduction in plasma spectrometry. *TrAC, Trends in Analytical Chemistry, 22*(3), 123–132. https://doi.org/10.1016/S0165-9936(03)00301-7.

Nakahara, T. (1983). Applications of hydride generation techniques in atomic absorption, atomic fluorescence and plasma atomic emission spectroscopy. *Progress in Analytical Atomic Spectroscopy, 6*, 163–223.

Nóbrega, J. A., Sturgeon, R. E., Grinberg, P., Gardner, G. J., Brophy, C. S., & Garcia, E. E. (2011). UV photochemical generation of volatile cadmium species. *Journal of Analytical Atomic Spectrometry, 26*(12), 2519–2523. https://doi.org/10.1039/c1ja10252d.

Nölte, J. (2003). *ICP emission spectrometry: A practical guide.* Weinheim: Wiley-VCH Verlag GmbH & Co.

Olofsson, R. S., Rodushkin, I., & Axelsson, M. D. (2000). Performance characteristics of a tandem spray chamber arrangement in double focusing sector field ICP-MS. *Journal of Analytical Atomic Spectrometry, 15*(6), 727–729. https://doi.org/10.1039/b002609n.

Pereira, C. D., Aguirre, M. A., Nóbrega, J. A., Hidalgo, M., & Canals, A. (2012). Correction of matrix effects for As and Se in ICP OES using a Flow Blurring® multiple nebulizer. *Journal of Analytical Atomic Spectrometry, 27*(12), 2132–2137. https://doi.org/10.1039/c2ja30231d.

Pereira, C. D., Aguirre, M. A., Nóbrega, J. A., Hidalgo, M., & Canals, A. (2014). Aerosol generation of As and Se hydrides using a new Flow Blurring® multiple nebulizer for sample introduction in inductively coupled plasma optical emission spectrometry. *Microchemical Journal, 112*, 82–86. https://doi.org/10.1016/j.microc.2013.09.006.

Pohl, P. (2004). *TrAC, Trends in Analytical Chemistry, 23*, 87–101.

Pohl, P., & Jamroz, P. (2011). Recent achievements in chemical hydride generation inductively coupled and microwave induced plasmas with optical emission spectrometry detection. *Journal of Analytical Atomic Spectrometry, 26*(7), 1317–1337. https://doi.org/10.1039/c1ja10038f.

Pohl, P., & Sturgeon, R. E. (2010). Simultaneous determination of hydride- and non-hydride-forming elements by inductively coupled plasma optical emission spectrometry. *TrAC, Trends in Analytical Chemistry, 29*(11), 1376–1389. https://doi.org/10.1016/j.trac.2010.07.015.

Pohl, P., & Zyrnicki, W. (2000). Simultaneous determination of As, Bi, Se and Sn and non-hydride forming elements by HG-ICP-AES without liquid phase separation. *Chemia Analityczna, 45*(5), 699–708.

Priego-Capote, F., & de Castro, L. (2004). *Trends in Analytical Chemistry, 23*, 829–841.

Proch, J., & Niedzielski, P. (2021). Multi-mode Sample Introduction System (MSIS) as an interface in the hyphenated system 2 HPLC–MSIS–ICP–OES in simultaneous determination of metals and metalloids species. *Analytica Chimica Acta, 1147*, 1–14. https://doi.org/10.1016/j.aca.2020.12.047.

Robbins, W. B., & Caruso, J. A. (1979). *Analytical Chemistry, 51*.

Rojas, I., Murillo, M., Carrión, N., & Chirinos, J. (2003). Investigation of the direct hydride generation nebulizer for the determination of arsenic, antimony and selenium in inductively coupled plasma optical emission spectrometry. *Analytical and Bioanalytical Chemistry, 376*(1), 110–117. https://doi.org/10.1007/s00216-003-1856-7.

Shen, Y., Zheng, C., Jiang, X., Wu, X., & Hou, X. (2015). Integration of hydride generation and photochemical vapor generation for multi-element analysis of traditional Chinese medicine by ICP-OES. *Microchemical Journal, 123*, 164–169. https://doi.org/10.1016/j.microc.2015.06.002.

Ślachciński, M. (2014). Recent achievements in sample introduction systems for use in chemical vapor generation plasma optical emission and mass spectrometry: From macro- to microanalytics. *Applied Spectroscopy Reviews, 49*(4), 271–321. https://doi.org/10.1080/05704928.2013.823547.

Ślachciński, M. (2016). Slurry micro-sampling technique for use in argon-helium microwave induced plasma optical emission spectrometry. *Talanta, 161*, 812–818. https://doi.org/10.1016/j.talanta.2016.09.043.

Ślachciński, M. (2019). Modern chemical and photochemical vapor generators for use in optical emission and mass spectrometry. *Journal of Analytical Atomic Spectrometry, 34*(2), 257–273. https://doi.org/10.1039/c8ja00383a.

Štrupar, J. (1989). *Vestnik Slovenskega Kemijskega Drustva, 36*, 75–91.

Sturgeon, R. E., & Grinberg, P. (2012). Some speculations on the mechanisms of photochemical vapor generation. *Journal of Analytical Atomic Spectrometry, 27*(2), 222–231. https://doi.org/10.1039/c2ja10249h.

Sturgeon, R. E., & Mester, Z. (2002). *Applied Spectroscopy, 56*.

Sturgeon, R. E., Willie, S. N., & Mester, Z. (2006). UV/spray chamber for generation of volatile photo-induced products having enhanced sample introduction efficiency. *Journal of Analytical Atomic Spectrometry, 21*(3), 263–265. https://doi.org/10.1039/b517884c.

Suzuki, T., Sturgeon, R. E., Zheng, C., Hioki, A., Nakazato, T., & Tao, H. (2012). Influence of speciation on the response from selenium to UV-photochemical vapor generation. *Analytical Sciences, 28*(8), 807–811. https://doi.org/10.2116/analsci.28.807.

Tao, G. H., & Sturgeon, R. E. (1999). Sample nebulization for minimization of transition metal interferences with selenium hydride generation ICP-AES. *Spectrochimica Acta, Part B: Atomic Spectroscopy, 54*(3), 481–489. https://doi.org/10.1016/S0584-8547(99)00003-8.

Tarr, M. A., Zhu, G., & Browner, R. F. (1993). Microflow ultrasonic nebulizer for inductively coupled plasma atomic emission spectrometry. *Analytical Chemistry, 65*(13), 1689–1695. https://doi.org/10.1021/ac00061a010.

Todolí, J. L., & Mermet, J. M. (2002a). Towards a new compact, low consumption liquid sample introduction system for ICP-AES. *Canadian Journal of Analytical Sciences and Spectroscopy, 47*(6), 164–170.

Todolí, J.-L., & Mermet, J.-M. (2002b). New torch design with an in-built chamber for liquid sample analysis by ICP-AES. *Journal of Analytical Atomic Spectrometry, 17*(4), 345–351. https://doi.org/10.1039/B111416F.

Todolí, J. L., & Mermet, J. M. (2003). Optimization of the evaporation cavity in a torch integrated sample introduction system based ICP-AES system. Applications to matrix and transient effects, analysis of microsamples and analysis of certified solid samples. *Journal of Analytical Atomic Spectrometry*, *18*(10), 1185–1191. https://doi.org/10.1039/B212698B.

Todoli, J. L., & Mermet, J. M. (2005). *TrAC, Trends in Analytical Chemistry*, *24*, 107–114.

Todolí, J. L., & Mermet, J. M. (2006). Sample introduction systems for the analysis of liquid microsamples by ICP-AES and ICP-MS. *Spectrochimica Acta - Part B Atomic Spectroscopy*, *61*(3), 239–283. https://doi.org/10.1016/j.sab.2005.12.010.

Tyburska, A., Jankowski, K., Ramsza, A., Reszke, E., Strzelec, M., & Andrzejczuk, A. (2010). Feasibility study of the determination of selenium, antimony and arsenic in drinking and mineral water by ICP-OES using a dual-flow ultrasonic nebulizer and direct hydride generation. *Journal of Analytical Atomic Spectrometry*, *25*(2), 210–214. https://doi.org/10.1039/B916729C.

Wiltsche, H., Brenner, I. B., Prattes, K., & Knapp, G. (2008). Characterization of a multimode sample introduction system (MSIS) for multielement analysis of trace elements in high alloy steels and nickel alloys using axially viewed hydride generation ICP-AES. *Journal of Analytical Atomic Spectrometry*, *23*(9), 1253–1262. https://doi.org/10.1039/b803943g.

Wolnik, K. A., Fricke, F. L., Hahn, M. H., & Caruso, J. A. (1981). Sample introduction system for simultaneous determination of volatile elemental hydrides and other elements in foods by inductively coupled argon plasma emission spectrometry. *Analytical Chemistry*, *53*(7), 1030–1035. https://doi.org/10.1021/ac00230a025.

Wu, P., He, L., Zheng, C., Hou, X., & Sturgeon, R. E. (2010). Applications of chemical vapor generation in non-tetrahydroborate media to analytical atomic spectrometry. *Journal of Analytical Atomic Spectrometry*, *25*(8), 1217–1246. https://doi.org/10.1039/c003483e.

Yin, Y., Liu, J., & Jiang, G. (2011). Photo-induced chemical-vapor generation for sample introduction in atomic spectrometry. *TrAC, Trends in Analytical Chemistry*, *30*(10), 1672–1684. https://doi.org/10.1016/j.trac.2011.04.021.

Zhang, R., Shen, B., Li, C., Zheng, C., & Hou, X. (2016). Integrating photochemical vapor generation with photo-oxidation trapping for effective mercury removal from polluted water and its on-line monitoring. *Microchemical Journal*, *17*(1), 98–103. https://doi.org/10.1016/j.microc.2016.06.014.

Zheng, C., Ma, Q., Wu, L., Hou, X., & Sturgeon, R. E. (2010). UV photochemical vapor generation-atomic fluorescence spectrometric determination of conventional hydride generation elements. *Microchemical Journal*, *95*(1), 32–37. https://doi.org/10.1016/j.microc.2009.09.010.

Zoltan, T., Benzo, Z., Murillo, M., Marcano, E., Gómez, C., Salas, J., et al. (2005). Performance of a new nebulizer system for simultaneous determination of Sb, Sn (hydride generation), V, and Zn by ICP-OES. *Analytical and Bioanalytical Chemistry*, *382*(6), 1419–1430. https://doi.org/10.1007/s00216-005-3278-1.

CHAPTER

7

Discrete droplet generation

Mazaher Ahmadi[a] *and Miguel Ángel Aguirre*[b]

[a]Department of Analytical Chemistry, Faculty of Chemistry, Bu-Ali Sina University, Hamedan, Iran, [b]Department of Analytical Chemistry and Food Science and University Institute of Materials, Faculty of Science, University of Alicante, Alicante, Spain

7.1 Introduction

Using a nebulizer, samples for inductively coupled plasma (ICP) analysis are normally delivered as an aerosol into the plasma (Montaño, Olesik, Barber, Challis, & Ranville, 2016). The aerosol is commonly generated using pneumatic nebulizers operating at sonic gas velocities, with droplet around 20 μm in diameter (Olesik & Bates, 1995). If the entire sample were to reach the plasma, the solution load could be too large to keep the temperature of plasma constant in the center channel. Furthermore, in the ICP, droplets of 20 μm sizes are not entirely vaporized under routinely employed plasma settings (Olesik & Hobbs, 1994). As a result, a spray chamber is employed to limit the number of droplets that reach the plasma and to reduce their size.

Droplet evaporation, droplet-droplet coagulation, and droplet impact on the walls are three major mechanisms in the spray chamber which decide whether a droplet reaches the plasma. The larger the evaporation of a droplet, the smaller it gets, and the more probable it is to be maintained in the percentage of gas entering the plasma (Canals & Aguirre, 2015; Montaño et al., 2016). Larger droplets are more likely to collide with the spray chamber's walls and so be eliminated. While the spray chamber removes the biggest droplets, the aerosol that exits the spray chamber remains polydisperse. As the sample uptake rate is raised, the analyte transport efficiency falls (Canals & Aguirre, 2015; Montaño et al., 2016). Transport efficiency in open, double pass spray chambers, or conical spray chambers can be as high as 100% at sample uptake rates of less than 10 μL min^{-1} (Montaño et al., 2016). Some spray chamber designs, such

Analytical Nebulizers
https://doi.org/10.1016/B978-0-323-91181-8.00007-2

as unheated spray chambers (Holland et al., 2001), and heated spray chambers employed with modest sample uptake rates (Miyashita et al., 2014), can achieve transport efficiencies of around 100% at flow rates up to $0.5\,mL\,min^{-1}$. Droplet evaporation in the spray chamber is more extreme at sample uptake rates less than $10\,\mu L\,min^{-1}$, and the chance of droplet-droplet coagulation/collision reduces at a great rate as the number of droplets formed per second declines. Because the spray is aimed at the spray chamber wall, Baffled cyclonic spray chambers usually exhibit poorer transport efficiency at sample uptake rates of $10\,\mu L\,min^{-1}$ than double pass or open conical spray chambers. Furthermore, if the sample matrix alters the surface tension of the solution, influencing the rate of droplet evaporation in the spray chamber or droplet size distribution, the transport efficiency may differ from that of pure, deionized water.

Microdroplet generators have been utilized to produce approximately 100% efficient aerosols from nanoparticle suspensions. This allows for the measurement of inductively coupled plasma-mass spectrometry (ICP-MS) sensitivity without the use of known-size nanoparticle standards (Borovinskaya, Hattendorf, Tanner, Gschwind, & Günther, 2013; Gschwind et al., 2011; Gschwind, Hagendorfer, Frick, & Günther, 2013). Monodisperse droplet producers often create droplets that are too big to be entirely vaporized in the ICP under conventional plasma conditions. As a result, a desolvation mechanism is frequently utilized before the droplets are introduced into the ICP. He gas is used in a long tube in one case to allow droplets to evaporate at a considerably faster pace than Ar gas without needing to be heated (Gschwind et al., 2011; Koch et al., 2013). A capillary surrounded by a piezoelectric tube that ejects a droplet when a voltage pulse is applied, thermal inkjet printer-like designs (Niessen, Schaper, Petersen, & Bings, 2011), and a microfluidics-based droplet dispenser (Verboket, Borovinskaya, Meyer, Günther, & Dittrich, 2014) are all examples of monodisperse droplet generator designs.

7.2 Design and fundamentals

The monodisperse dried-microparticulate injector (MDMI) (Fig. 7.1) was the first device to allow the droplet formation frequency to be varied within a particular range (French, Etkin, & Jong, 1994; Montaser, 1998), whereas earlier devices had to be set to a specific frequency. Because the MDMI is controlled by a piezo-actuator, the frequency of operation must be kept to a minimum to ensure consistent droplet formation. Olesik et al. concentrated on the influence and fate of discrete droplets in the ICP (Olesik & Fister, 1991; Olesik, Smith, & Williamsen, 1989) and looked at the observed signal induced by isolated droplets on a μs time frame (Olesik & Dziewatkoski, 1996; Olesik & Hobbs, 1994). The MDMI was shown to be a useful diagnostic tool for clarifying fundamental ICP mechanisms,

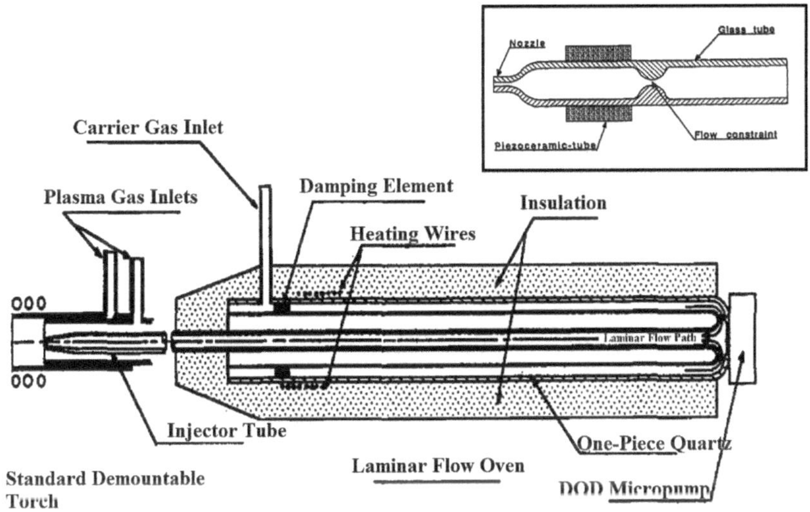

FIG. 7.1 Cross section of the assembled MDMI apparatus (inset: droplet generator design). *Adapted from J. B. French, B. Etkin, & R. Jong (1994). Monodisperse dried microparticulate injector for analytical instrumentation. Analytical Chemistry, 66, 685–691. Copyright 1994 American Chemical Society.*

however, it was not used to introduce and analyze actual materials (Hill, 2008).

In 2006, the alternative thermal-inkjet concept was used for the first time in atomic spectrometry (Fittschen et al., 2006). For calibration purposes in laser ablation ICP-MS and total reflection X-ray fluorescence spectrometry (TXRF), original thermal inkjet printers were utilized to spray droplets of pL volume over solid objects. Fittschen et al. looked at an innovative picoliter pipette system developed by Hewlett-Packard in later experiments (Fittschen & Havrilla, 2010). This automated dispenser allows varying quantities of droplets to be transferred, despite the fact that it suffers from a lot of solvent evaporation via the dosing nozzles. This was found to cause analyte enrichment in the channel leading to the dosing nozzle, which can only be accounted for by washing the relevant nozzle by rejecting the first of a series of droplets expelled during each dosing sequence. As a result, a lesser number of droplets sprayed by this device lead to increased average mass errors, particularly when single droplets are dosed (Verkouteren & Verkouteren, 2009). Petersen et al. demonstrated the use of a humidified shielding gas combined with a mechanical shutter to reject a predetermined number of unwanted droplets, thus considerably improving attainable accuracy, particularly when a single droplet is transferred to a laser ablation substrate (Bings, Niessen, & Schaper, 2014).

Bings and coworkers have described the use of a customized thermal inkjet printer cartridge typically used in commercial inkjet printers as an aerosol generator for sample introduction in mass and atomic spectrometry (Niessen, Krone, & Bings, 2014; Niessen, Petersen, Schaper, & Bings, 2012; Niessen et al., 2011; Schaper, Pfeuffer, Shelley, Bings, & Hieftje, 2012). Unlike previous inkjet systems used as dispensers, the developed "drop-on-demand, DOD" aerosol generator is controlled by a homemade microcontroller, which allows complete access to all electronic parameters vital for influencing and optimizing the thermal inkjet process. Supplementary printer hardware and software support are no longer required. A modified HP45 cartridge with 300 nozzles with a diameter of approximately 28 μm may be adjusted to an ICP-MS through a homemade aerosol transfer chamber (Fig. 7.2). Sensitivities for a significant number of examined elements can be enhanced by a factor of 8–18 compared to a MicroMist-based reference system using the developed DOD sample introduction in ICP-MS (Niessen, Schaper, Petersen, & Bings, 2011). Signal noise was found to be marginally increased, as it is for all of the microflow systems outlined above. It was also found to vary from that produced by other pneumatic low-flow nebulizers, such as the

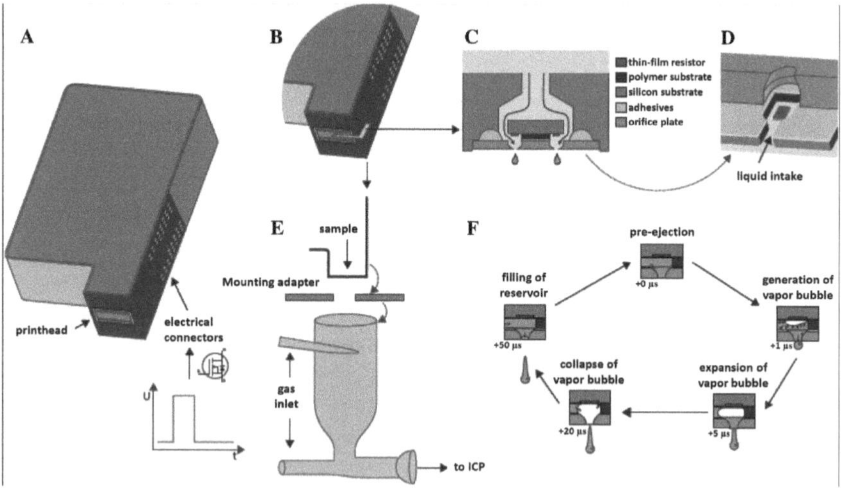

FIG. 7.2 Schematic representation of the DOD aerosol generator. (A) Original ink cartridge, type HP45, (B) modified ink cartridge for analytical applications, (C) cross section of the dosing device showing both liquid channels integrated into the silicon chip, (D) close-up bottom view of a single dosing nozzle, (E) the combination of the modified dosing cartridge with the aerosol transport chamber via a mounting plate, and (F) the principle of thermal inkjet droplet generation process. *Reproduced with permission from N. Bings, J. O. O. V. Niessen, & J. Schaper (2014). Liquid sample introduction in inductively coupled plasma atomic emission and mass spectrometry—Critical review. Spectrochimica Acta Part B: Atomic Spectroscopy, 100, 14–37. Copyright 2014 Elsevier.*

EnyaMist and MicroMist (Bings et al., 2014; Massmann, Petersen, Schaper, Broekaert, & Bings, 2009). Both emit frequency-dependent noise, with the DOD producing more white noise at higher frequencies. The magnitude of this noise was discovered to be related to the span of the drop-to-drop time interval distribution. The major cause of signal noise was identified as the divergence of the corresponding droplet transit time from the dosing nozzle to the ICP (Chan, Zhu, & Hieftje, 2012a, 2012b). The DOD enables separate optimization of the liquid flow and carrier gas rates, plus an excellent estimated aerosol transport efficiency of up to 94 vol.% (Bings et al., 2014; Massmann et al., 2009). For a single nozzle, the latter may be changed from 0.04 to 6.3 $\mu L\,min^{-1}$ without affecting the aerosol properties (Niessen, Schaper, Petersen, & Bings, 2011). Modifications to the DOD's controller circuitry increased the range to 0.005–12.5 $\mu L\,min^{-1}$ (Niessen et al., 2014). Full control of aerosol properties can therefore be obtained simply by adjusting the dosing device's electrical operation conditions, rather than by attaching, for example, a heating chamber for aerosol manipulation.

Orlandini et al. demonstrated a dual-DOD device and used it to analyze urine reference material (Niessen et al., 2012). New calibration procedures based on dosage frequency were developed and appropriately verified using this system, comparable to the methodology previously described by Bastiaans and Hieftje as the "null-point technique" (Bastiaans & Hieftje, 1973). The DOD device was shown to be less vulnerable to the impacts of the urine matrix than the MicroMist reference system, as seen by smaller spatial ion dispersion inside the ICP. Considering the large matrix load, five certified trace elements were determined accurately using dosing frequency-based external calibration without matrix matching, whereas only two of the five elements in the MicroMist system using traditional nonmatrix matching external calibration agreed with the reference data (Niessen et al., 2012). For both the standard addition approach and the external calibration, the dosing frequency-based calibration was shown to dramatically cut on sample usage by at top 12 μL of reference material. All of the tested methodologies (i.e., both dual-DOD calibration techniques and the use of the reference sample introduction system) yield equal accuracies, however, the observed precision is lower than the data obtained with the reference system.

The DOD is suited for usage with low-temperature plasma sources due to its extremely low sample uptake rate. The field of ambient desorption/ionization (ADI) mass spectrometry has received a lot of interest in recent years. The flowing atmospheric-pressure afterglow (FAPA), the low-temperature plasma probe (LTP), and the direct analysis in real-time system (DART) sources are just a few of the innovative ion sources that have been presented (Harris, Galhena, & Fernández, 2011). Each of these sources produces a stream of reactive ions that is typically focused onto the sample surface to desorb and ionize the analyte of interest, which is subsequently sampled into the mass spectrometer. Dissolved materials are evaluated by spotting residues from dried quantities of the relevant liquid onto targets or by directing the plasma source directly onto the liquid surface (Shelley, Wiley, & Hieftje, 2011). The direct introduction of liquids into these ADI sources was previously impossible owing to the low energy of the plasma until the hyphenation of the DOD to the FAPA for direct MS identification and quantification of drugs of abuse and their metabolites in urine samples was reported by Schaper and collaborators (Schaper et al., 2012). When compared to results obtained with alternate liquid sampling methods for FAPA-MS and other ion sources for molecular MS, the attainable detection limits were considerably improved.

7.3 Applications

The development of methods and apparatus for the detection and quantification of discrete particles and droplets, particularly nanoparticles, is a key approach of modern instrumental analytical chemistry. To attain this aim, the utilized detection approach must have a high sensitivity, as well as high-speed data capture.

Bochert and Dannecker (1989) and Kawaguchi, Fukasawa, and Mizuike (1986) were the first to use inductively coupled plasma optical emission spectrometry (ICP-OES) for the size characterization of airborne aerosols. Degueldre et al. modified the approach to ICP-MS and then used it for the analysis of ZrO_2 particles (Degueldre, Favarger, & Bitea, 2004). Since then, this method has been dubbed "single-particle (SP) ICP-MS." In general, there are two techniques for detecting individual events. To begin with, use a fast detector with a high time resolution for data gathering and a repetition rate that is a multiple of the particle introduction frequency to detect each unique transient event, such as each droplet or particle presented. On the contrary, by combining a long detector dwell time with low droplet introduction rates, each recorded signal might reflect an average of only one droplet or particle (Olesik & Gray, 2012). Other than synchronizing droplet/particle formation and data gathering to avoid erroneous particle size measurements owing to wrongly set dwell durations, the latter is simple to execute and does not require significant modifications to the existing ICP-MS apparatus. However, as compared to the first method, background/noise reduction is more difficult. The association of the quality of the acquired SP-ICP-MS data with the detector dead time, the integration time, and the dynamic range in pulse counting or analog mode was also studied by Olesik and Gray (2012). Even with a data repetition rate of $500\,s^{-1}$ and a particle introduction rate of just 10 particles per second, there is still a 1% chance that the detected signal is produced by more than one particle.

Both techniques use the same data processing method: To discriminate between signals originating from the background or particles and signals created by a single and numerous particles, appropriate thresholds must first be evaluated. The detected events must then be classified in order to produce a Gaussian distribution of the number of occurrences versus the measured intensity. The intensity distribution gives the size distribution of the corresponding particles to be defined after suitable calibration using reference particles of known size. Franze et al. presented this data processing technique and demonstrated its ability to discriminate between silver nanoparticles and dissolved silver (Franze, Strenge, & Engelhard, 2012).

Single-particle ICP-MS has already been used to characterize and quantify nanoparticles (Franze et al., 2012; Garcia, Murtazin, Groh, Becker, & Niemax, 2010; Gschwind et al., 2011; Murtazin, Groh, & Niemax, 2012), as well as to create new immunoassays (Hu et al., 2009). The MicroDrop dispenser was employed to generate individual droplets in this context, and Niemax and coworkers further defined and used it (Garcia, Murtazin, Groh, Becker et al., 2010; Garcia, Murtazin, Groh, Horvatic, & Niemax, 2010; Gschwind et al., 2011; Murtazin et al., 2012). It is based on a setup similar to MDMI, and consists of a piezo-actuator that pressurizes a capillary to generate monodisperse droplets of certain sizes. Contrary to the MDMI, no heating/evaporation device is linked to the dispenser, instead a long vertical tube is used to accomplish complete evaporation of the droplets as well as enhanced sampling efficiency. The droplet formation frequency cannot be changed across a large range in a piezo-based device, which is a drawback. In the instance of the method used by Gschwind et al. (2011), only 5 or 100 s^{-1} were relevant.

Franze et al. examined three sample introduction methods, the APEX Q, the PC^3, and a MicroDrop dispenser, and used these methods in ICP-MS investigations to characterize Ag nanoparticles of various sizes (Franze et al., 2012). The latter system was found to be the best for this application since it is the sole discontinuous sample introduction system. The MicroDrop dispenser's ICP-MS sensitivity was better, allowing the detection of nanoparticles as small as 20 nm in diameter.

Utilizing known-volume monodisperse droplets and known-concentration liquids, Niemax and collaborators suggested an atomic spectrometric approach for the size classification of nanoparticles (Garcia, Murtazin, Groh, Becker et al., 2010; Murtazin et al., 2012). Because the work was based on ICP-OES detection, the reported limit of detections (LODs) of nanoparticle size of ca. 470 and 200 nm for SiO_2 and Au, respectively, are relatively high. In subsequent work, size-related LODs of 21 and 33 nm for Au and Ag nanoparticles were reported employing ICP-MS detection using the high-speed data-acquisition technique

(Gschwind et al., 2011). The proposed approach, which relied on manual data capture, was, however, plagued by the substantial analog background noise of around 180 mV. When employing the pulse-counting method, such as that used by Franze et al., Ag nanoparticles as small as 20 nm were easily identified, however, no size-related LOD was found (Franze et al., 2012).

The proficiency of SP-ICP-MS could be increased even further by coordinating the droplet introduction into the plasma with the collection of ICP-MS data, such that the first event serves as a trigger for the measurement procedure to begin (Bings et al., 2014). Significant differences in travel time emerge during the transit of the droplets toward the ICP, as previously indicated, which makes the application of such a trigger pulse difficult. In a different approach, Chan et al. effectively devised a technique for leveraging either a distinct change in plasma impedance or the plasma's H-emission line as a trigger signal, both of which were related to the entrance of a droplet (Chan et al., 2012a, 2012b).

7.4 Conclusions

The idea of discontinuous sample introduction, it may be argued, has several advantages: LODs relating to size can be considerably improved, owing to enhanced transport efficiency and lower background noise in ICP-MS. Also, because less severe dilution of the liquid nanoparticle samples is necessary to guarantee a maximum quantity of one particle per droplet, sample preparation may be simplified and related mistakes can be eliminated. Droplet dispensers can be used to make monodisperse pL droplets from much diluted solutions containing the desired, dissolved ingredient, which can then be utilized to make nanoparticles. Following the evaporation of the solvent, the production of the desired nm-sized residues occurs. The advantages of discontinuous sample introduction techniques are equally beneficial to the relatively young field of nanoparticle analysis. Owing to improved transport efficiencies of the produced aerosol and suppressed background noise in ICP-MS, the smallest observable particle sizes may be greatly lowered.

Acknowledgments

The authors are grateful to the Spanish Ministry of Science and Innovation (PID2021-126155OB-I00) and the Regional Government of Valencia (Spain) (CIPROM/2021/062) for the financial support. The authors extend their appreciation to Ministry of Science, Innovation and Universities for granting the Spanish Network of Excellence in Sample Preparation (RED2018-102522-T). This article is based upon work from the Sample Preparation Study Group and Network, supported by the Division of Analytical Chemistry of the European Chemical Society.

References

Bastiaans, G. J., & Hieftje, G. M. (1973). Evaluation and application of a new high precision null point analytical technique in flame spectrometry. *Analytical Chemistry, 45*, 1994–2001.
Bings, N., Niessen, J. O. O. V., & Schaper, J. (2014). Liquid sample introduction in inductively coupled plasma atomic emission and mass spectrometry—Critical review. *Spectrochimica Acta Part B: Atomic Spectroscopy, 100*, 14–37.
Bochert, U. K., & Dannecker, W. (1989). On-line aerosol analysis by atomic emission spectroscopy. *Journal of Aerosol Science, 20*, 1525–1528.
Borovinskaya, O., Hattendorf, B., Tanner, M., Gschwind, S., & Günther, D. (2013). A prototype of a new inductively coupled plasma time-of-flight mass spectrometer providing temporally resolved, multi-element detection of short signals generated by single particles and droplets. *Journal of Analytical Atomic Spectrometry, 28*, 226–233.
Canals, A., & Aguirre, M.Á. (2015). Roles of nebulizers in analytical chemistry. *Encyclopedia of analytical chemistry* (pp. 1–45). John Wiley & Sons.
Chan, G. C. Y., Zhu, Z., & Hieftje, G. M. (2012a). Effect of single aerosol droplets on plasma impedance in the inductively coupled plasma. *Spectrochimica Acta Part B: Atomic Spectroscopy, 76*, 87–95.
Chan, G. C. Y., Zhu, Z., & Hieftje, G. M. (2012b). Operating parameters and observation modes for individual droplet analysis by inductively coupled plasma-atomic emission spectrometry. *Spectrochimica Acta Part B: Atomic Spectroscopy, 76*, 77–86.
Degueldre, C., Favarger, P. Y., & Bitea, C. (2004). Zirconia colloid analysis by single particle inductively coupled plasma–mass spectrometry. *Analytica Chimica Acta, 518*, 137–142.
Fittschen, U. E. A., & Havrilla, G. J. (2010). Picoliter droplet deposition using a prototype Picoliter pipette: Control parameters and application in micro X-ray fluorescence. *Analytical Chemistry, 82*, 297–306.
Fittschen, U. E. A., Hauschild, S., Amberger, M. A., Lammel, G., Streli, C., Förster, S., et al. (2006). A new technique for the deposition of standard solutions in total reflection X-ray fluorescence spectrometry (TXRF) using pico-droplets generated by inkjet printers and its applicability for aerosol analysis with SR-TXRF. *Spectrochimica Acta Part B: Atomic Spectroscopy, 61*, 1098–1104.
Franze, B., Strenge, I., & Engelhard, C. (2012). Single particle inductively coupled plasma mass spectrometry: Evaluation of three different pneumatic and piezo-based sample introduction systems for the characterization of silver nanoparticles. *Journal of Analytical Atomic Spectrometry, 27*, 1074–1083.
French, J. B., Etkin, B., & Jong, R. (1994). Monodisperse dried microparticulate injector for analytical instrumentation. *Analytical Chemistry, 66*, 685–691.
Garcia, C. C., Murtazin, A., Groh, S., Becker, M., & Niemax, K. (2010). Characterization of particles made by desolvation of monodisperse microdroplets of analyte solutions and particle suspensions for nanoparticle calibration in inductively coupled plasma spectrometry. *Spectrochimica Acta Part B: Atomic Spectroscopy, 65*, 80–85.
Garcia, C. C., Murtazin, A., Groh, S., Horvatic, V., & Niemax, K. (2010). Characterization of single Au and SiO_2 nano- and microparticles by ICP-OES using monodisperse droplets of standard solutions for calibration. *Journal of Analytical Atomic Spectrometry, 25*, 645–653.
Gschwind, S., Hagendorfer, H., Frick, D. A., & Günther, D. (2013). Mass quantification of nanoparticles by single droplet calibration using inductively coupled plasma mass spectrometry. *Analytical Chemistry, 85*, 5875–5883.
Gschwind, S., Flamigni, L., Koch, J., Borovinskaya, O., Groh, S., Niemax, K., & Günther, D. (2011). Capabilities of inductively coupled plasma mass spectrometry for the detection of nanoparticles carried by monodisperse microdroplets. *Journal of Analytical Atomic Spectrometry, 26*, 1166–1174.
Harris, G. A., Galhena, A. S., & Fernández, F. M. (2011). Ambient sampling/ionization mass spectrometry: Applications and current trends. *Analytical Chemistry, 83*, 4508–4538.
Hill, S. J. (2008). *Inductively coupled plasma spectrometry and its applications*. John Wiley & Sons.

Holland, G., Tanner, S. D., Olesik, J. W., Hensman, C., Rabb, S., & Rago, D. (2001). Sample introduction, plasma-sample interactions, ion transport and ion-molecule reactions: Fundamental understanding and practical improvements in ICP-MS. In S. D. Tanner, & G. Holland (Eds.), *Plasma source mass spectrometry: The new millennium* (pp. 3–16). Royal Society of Chemistry.

Hu, S., Liu, R., Zhang, S., Huang, Z., Xing, Z., & Zhang, X. (2009). A new strategy for highly sensitive immunoassay based on single-particle mode detection by inductively coupled plasma mass spectrometry. *Journal of the American Society for Mass Spectrometry, 20*, 1096–1103.

Kawaguchi, H., Fukasawa, N., & Mizuike, A. (1986). Investigation of airborne particles by inductively coupled plasma emission spectrometry calibrated with monodisperse aerosols. *Spectrochimica Acta Part B: Atomic Spectroscopy, 41*, 1277–1286.

Koch, J., Flamigni, L., Gschwind, S., Allner, S., Longerich, H., & Günther, D. (2013). Accelerated evaporation of microdroplets at ambient conditions for the on-line analysis of nanoparticles by inductively-coupled plasma mass spectrometry. *Journal of Analytical Atomic Spectrometry, 28*, 1707–1717.

Massmann, J., Petersen, J., Schaper, J., Broekaert, J., & Bings, N. (2009). Development and characterization of a drop-on-demand aerosol generator for the introduction of small sample volumes in plasma spectrometry. *TraceSpec*.

Miyashita, S.-I., Groombridge, A. S., Fujii, S.-I., Minoda, A., Takatsu, A., Hioki, A., et al. (2014). Highly efficient single-cell analysis of microbial cells by time-resolved inductively coupled plasma mass spectrometry. *Journal of Analytical Atomic Spectrometry, 29*, 1598–1606.

Montaño, M. D., Olesik, J. W., Barber, A. G., Challis, K., & Ranville, J. F. (2016). Single particle ICP-MS: Advances toward routine analysis of nanomaterials. *Analytical and Bioanalytical Chemistry, 408*, 5053–5074.

Montaser, A. (1998). *Inductively coupled plasma mass spectrometry*. John Wiley & Sons.

Murtazin, A., Groh, S., & Niemax, K. (2012). Investigation of sample introduction- and plasma-related matrix effects in inductively coupled plasma spectrometry applying single analyte droplet and particle injection. *Spectrochimica Acta Part B: Atomic Spectroscopy, 67*, 3–16.

Niessen, J. O. O. V., Krone, K. M., & Bings, N. H. (2014). Effect of operation conditions of the drop-on-demand aerosol generator on aerosol characteristics: Pseudo-cinematographic and plasma mass spectrometric studies. *Spectrochimica Acta, 92*, 51–59.

Niessen, J. O. O. V., Petersen, J. H., Schaper, J. N., & Bings, N. H. (2012). Comparison of novel and conventional calibration techniques for the analysis of urine samples using plasma source mass spectrometry combined with a new dual-drop-on-demand aerosol generator. *Journal of Analytical Atomic Spectrometry, 27*, 1234–1244.

Niessen, J. O. O. V., Schaper, J. N., Petersen, J. H., & Bings, N. H. (2011). Development and characterization of a thermal inkjet-based aerosol generator for micro-volume sample introduction in analytical atomic spectrometry. *Journal of Analytical Atomic Spectrometry, 26*, 1781–1789.

Olesik, J. W., & Bates, L. C. (1995). Characterization of aerosols produced by pneumatic nebulizers for inductively coupled plasma sample introduction: Effect of liquid and gas flow rates on volume based drop size distributions. *Spectrochimica Acta Part B: Atomic Spectroscopy, 50*, 285–303.

Olesik, J. W., & Dziewatkoski, M. P. (1996). Time-resolved measurements of individual ion cloud signals to investigate space-charge effects in plasma mass spectrometry. *Journal of the American Society for Mass Spectrometry, 7*, 362–367.

Olesik, J. W., & Fister, J. C. (1991). Incompletely desolvated droplets in argon inductively coupled plasmas: Their number, original size and effect on emission intensities. *Spectrochimica Acta Part B: Atomic Spectroscopy, 46*, 851–868.

Olesik, J. W., & Gray, P. J. (2012). Considerations for measurement of individual nanoparticles or microparticles by ICP-MS: Determination of the number of particles and the analyte mass in each particle. *Journal of Analytical Atomic Spectrometry, 27*, 1143–1155.

Olesik, J. W., & Hobbs, S. E. (1994). Monodisperse dried microparticulate injector: A new tool for studying fundamental processes in inductively coupled plasmas. *Analytical Chemistry, 66*, 3371–3378.

Olesik, J. W., Smith, L. J., & Williamsen, E. J. (1989). Signal fluctuations due to individual droplets in inductively coupled plasma atomic emission spectrometry. *Analytical Chemistry, 61*, 2002–2008.

Schaper, J. N., Pfeuffer, K. P., Shelley, J. T., Bings, N. H., & Hieftje, G. M. (2012). Drop-on-demand sample introduction system coupled with the flowing atmospheric-pressure afterglow for direct molecular analysis of complex liquid microvolume samples. *Analytical Chemistry, 84*, 9246–9252.

Shelley, J. T., Wiley, J. S., & Hieftje, G. M. (2011). Ultrasensitive ambient mass spectrometric analysis with a pin-to-capillary flowing atmospheric-pressure afterglow source. *Analytical Chemistry, 83*, 5741–5748.

Verboket, P. E., Borovinskaya, O., Meyer, N., Günther, D., & Dittrich, P. S. (2014). A new microfluidics-based droplet dispenser for ICPMS. *Analytical Chemistry, 86*, 6012–6018.

Verkouteren, R. M., & Verkouteren, J. R. (2009). Inkjet metrology: High-accuracy mass measurements of microdroplets produced by a drop-on-demand dispenser. *Analytical Chemistry, 81*, 8577–8584.

CHAPTER

8

Electrospray

Alfonso M. Gañán-Calvo[a,b], *José M. López-Herrera*[a], *and Luis B. Modesto-López*[a]

[a]Departamento de Ingeniería Aeroespacial y Mecánica de Fluidos, ETSI, Universidad de Sevilla, Sevilla, Spain, [b]Laboratory of Engineering for Energy and Environmental Sustainability, Universidad de Sevilla, Sevilla, Spain

8.1 Introduction

In this chapter, we deal with an extraordinary class of nebulization processes such that the driver is the application of an electric potential difference: a liquid nebulization mechanism termed as *electrospray* (Cloupeau & Prunet-Foch, 1989; Fernández de la Mora, 2007; Gañán-Calvo, López-Herrera, Herrada, Ramos, & Montanero, 2018; Jaworek & Krupa, 1999; Jaworek & Sobczyk, 2008). This procedure converts electric potential energy into kinetic, heat, and surface energies, producing very fine and highly charged aerosols. Although it has been known since antiquity, its first detailed reports date back to the 18th century (see Fig. 8.1). Its major application was found in mass spectrometry of biomolecules since it allowed their gentle yet extreme electrical charging to the point of detection (maximum mass-to-charge ratio m/z) by mass spectrometers (Fenn, Mann, Meng, Wong, & Whitehouse, 1989; Whitehouse, Dreyer, Yamashita, & Fenn, 1985).

To understand the physics of electrospray, a brief introduction to the general class of liquid nebulization principles to which electrospray belongs is necessary.

8.1.1 Liquid nebulization by steady jetting: Basic physics and main parameters

Liquid nebulization always entails the ejection of a stream of liquid in a predetermined direction. When the stream is continuous, in the form of a

Analytical Nebulizers
https://doi.org/10.1016/B978-0-323-91181-8.00006-0

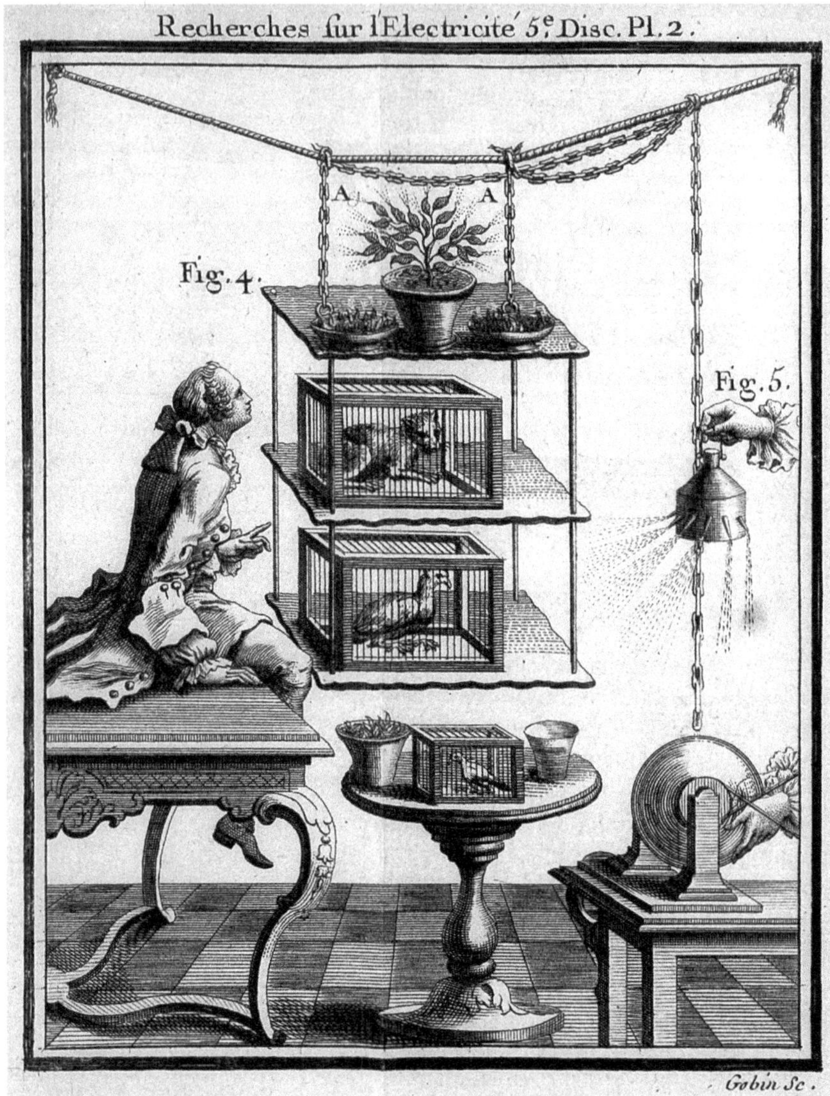

FIG. 8.1 Earliest documented electrospray experiments with a high detail of illustrations showing different ways in which the phenomenon can be observed. *From "Essay sur l'electricité des corps" by M. L'abbe Nollet, Jean-Antoine (1746).*

capillary jet, it can be promoted by the action of surface forces (pressure, tangential stresses, surface electrical stresses, etc.), bulk forces (gravity, acceleration, rotation, electromagnetic forces, etc.), or both. However, when the objective is the generation of small droplets, bulk forces (e.g., gravity acceleration) become negligible compared to surface forces, whose resultant (both normal and tangential) shapes the jet and accelerates it in the axial direction. The jet generally acquires a velocity such that its kinetic energy per unit volume ultimately becomes large compared to the surface tension stress. Indeed, the kinetic energy should be at least of the order of or larger than the surface tension to have a running jet; otherwise, the jet becomes globally unstable. Eventually, following the liquid, the jet breaks up into droplets by the growing action of surface tension forces following the classical Rayleigh-Plateu instability, or in a more chaotic fashion depending on the ratio of kinetic energy to surface tension stress. The relative measure of kinetic energy compared to surface tension is a fundamental ratio in liquid nebulization: the Weber number (We) that can be expressed in general as $We = \rho v^2 a/\sigma$, where ρ and σ are the liquid density and surface tension, respectively, and v and a are the jet local average velocity and the jet radius, respectively.

Thus, one can define a reference velocity, usually called the capillary velocity $v_c = (\sigma/(\rho a))^{1/2}$, as that limiting one for which the Weber number is 1. It measures the typical propagation velocity of capillary waves along the jet. This means that, for a jet to be locally (absolutely) stable the convective velocity v should be generally larger than the capillary velocity v_c to avoid perturbations climbing and growing upstream, eventually affecting the source (Montanero & Gañán-Calvo, 2020). Besides, the viscous forces should be limited to avoid an excessive delay in the capillary breakup of the liquid jet that might prevent the eventual formation of droplets. The ratio of viscous to surface tension stresses can be expressed in relation to the reference velocity v_c, which yields the Ohnesorge number $Oh = \frac{\mu}{(\rho\sigma a)^{1/2}}$ as a measure of the relative weight of those effects, where μ is the liquid viscosity.

Liquid nebulization by capillary jetting operates by the ejection of a continuous supply of liquid, that is, a liquid flow rate Q that is steadily transformed in a spray with a characteristic droplet diameter d_g, related to the jet radius a. The balance of driving and resisting forces in the liquid motion ultimately determines the relationship between Q and d_g, which involves the intermediate dynamical jet breakup process. In the case of We numbers in the range from about 1 to about 20, and very small Oh, the jet breakup is nearly axisymmetric and d_g is proportional to a approximately through a universal constant, given by the most unstable breakup wavelength $\lambda = 2\pi/k$, where $k = 0.697/a$ is the most unstable wave number, which yields $d_g \simeq 3.783a$ by conservation of mass.

The Oh number determines the delay level that the jet breakup may experience by viscous damping. For Oh above unity, that delay may cause the nearly indefinite stretching and winding of the jet without breaking up into droplets, and the process is usually termed *drawing* or *spinning*. In this chapter, we deal with a class of nebulization processes by capillary jet breakup such that We is generally below 100, Oh is smaller than unity, the driver is the application of an electric potential difference, and the outcome are charged droplets: *electrospray*. Its physics, main features, prediction, and control of the outcome are discussed here.

8.1.2 Steady jetting by acceleration from a nearly quiescent source: The Taylor cone

In contrast to those nebulization procedures that operate by the ejection of a relatively rapid capillary stream, for example, from a round orifice, electrospray (Gañán-Calvo & Montanero, 2009; Gañán-Calvo et al., 2018) and flow focusing (Gañán-Calvo, 1998) belong to a class that is often classified as "tip streaming" procedures (Montanero & Gañán-Calvo, 2020). These are characterized by a relatively large and nearly hydrostatic region (liquid meniscus, see Fig. 8.2 and subsequent text for notation) attached to the source of liquid (generally, a capillary tube), which tapers into a capillary jet of diameter d much smaller than the dimension of the source, normally a capillary tube of radius $R_1 \gg a$. The liquid acceleration proceeds by the action of surface forces (normal and tangential) in the

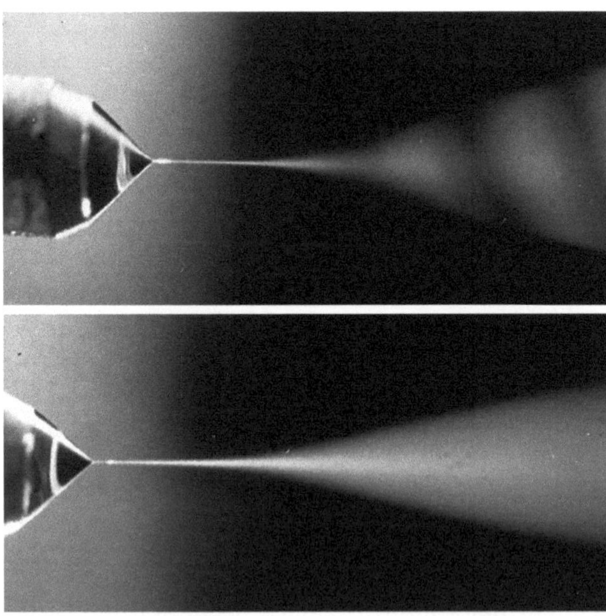

FIG. 8.2 Electrospray of methanol, $K = 4.5 \times 10^{-3}$ S/m, $\rho = 792$ kg/m^3, $\mu = 5.45 \times 10^{-4}$Pa s, $\sigma = 0.0223$ N/m. Parameters: $\beta = 32.7$ (liquid-to-vacuum permittivity ratio $\beta = \varepsilon_i/\varepsilon_o$), $\eta = 1.2275 \times 10^{-4}$ (nondimensional viscosity of the liquid expressed as: $\eta = \left([K\mu^3]/[\rho\sigma^2\varepsilon_o]\right)^{1/3}$). Feeding capillary tube: OD: 1 mm, ID: 0.75 mm. (*Up*) $Q = 1$ μL/min. Observe the optical Tyndal waves indicating homogeneous droplet size. (*Bottom*) $Q = 5$ μL/min.

axial direction. In the case of flow focusing, the axial resultant of normal forces (pressure) and its gradient produces a steep raise of the liquid velocity in the axial direction, which is supplemented by the axial resultant of tangential forces produced by the coflowing stream: usually a gas forced through a small orifice of radius R_o in front of the source tube. One has two degrees of freedom to determine the diameter of the jet, within certain parametrical limits: the liquid flow rate and the applied pressure gradient through the discharge orifice. In electrospray, instead of a coflowing gas, one has an externally applied voltage gradient (electric field) in the direction of the issuing jet. The fortunate existence of an electrohydrostatic conical solution, discovered by Taylor (1964), serves three physical outcomes:

(1) It establishes a perfect geometrical junction between the source capillary tube and the small jet.
(2) It provides a "launching" platform for the liquid that feeds the jet, since the tangential electric field applied on its surface toward the cone apex drags a stream of liquid that facilitates the tapering of the conical meniscus and the emission of the jet.
(3) It produces a component of the electric field in the axial direction along the jet (see Fig. 8.3), further accelerating the jet and contributing to its remarkable stability and robustness.

The fee for these three remarkable effects entails some limitations, though. The existence of the Taylor cone implies that the normal Maxwell stress on the surface of the conical meniscus should balance the surface tension stress. This requires (i) that the surface should become highly charged, that is, it should accumulate a sufficient charge concentration from the liquid bulk to shield it from the outer electric field, and (ii) that the conical liquid surface should acquire nearly the same voltage level as that applied on the source, with at most a very small tangential decay (tangential electric field) along the cone. The first requirement excludes dielectric liquids and liquids with too low electrical conductivity. The second requirement implies that if this tangential electric field was not small, it would produce a significant tangential stress and excessive liquid movement that would destroy the quasihydrostatic nature of the conical geometry. Hence, in first order, neglecting pressure, and tangential fields and stresses, the static balance between surface tension and the normal resultant of the Maxwell stresses on the conical liquid surface with a negligible inner electric field and liquid pressure reads:

$$\frac{\sigma}{R\tan(\alpha_T)} = \frac{\varepsilon_o}{2}E_{n,T}^2, \qquad (8.1)$$

where R is the spherical coordinate radius, ε_o is the permittivity of vacuum, α_T is the cone angle, and $E_{n,T}$ is the normal electric field on the cone. Taylor discovered that the solution of the Laplace equation for the electric

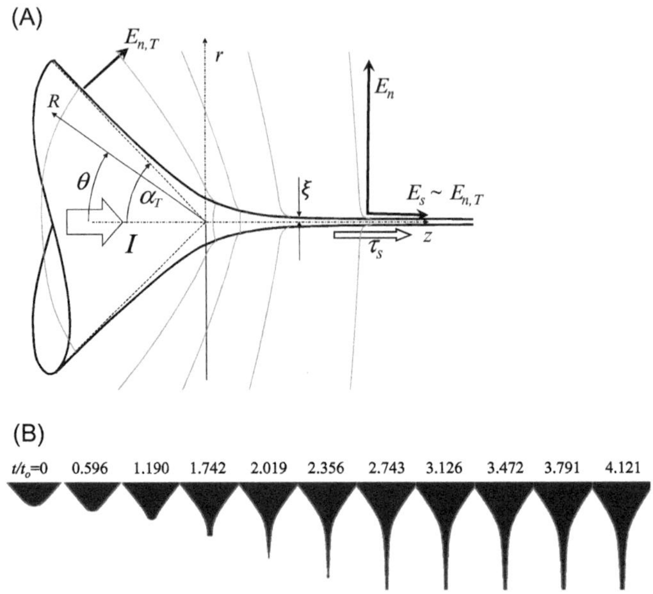

FIG. 8.3 (A) Schematic representation of the Taylor cone-jet around the cone-jet transition with variables involved. Both spherical and cylindrical coordinates to describe the problem are indicated, as well as the underlying Taylor cone angle α_T. *Thin gray curves* represent equipotential lines. The axial electric field on the jet is comparable to the normal electric field on the cone for $z \sim R$. (B) Sequence of the initial ejection that establishes the cone-jet configuration, by numerical simulation (Gañán-Calvo, López-Herrera, Rebollo-Muñoz, & Montanero, 2016). Each frame time is given in units of $t_o = \left(\frac{\rho R_1^3}{\sigma}\right)^{1/2}$. In this sequence, $\beta = 40$ and $\eta = 0.625$.

potential $\nabla^2 \Phi(R, \theta) = 0$ (where θ is the usual angular spherical coordinate) given by the conical Legendre function $Q_{1/2}(\theta)$, has the form:

$$\Phi_T(R,\theta) = A_T R^{1/2} Q_{1/2}(\theta), \tag{8.2}$$

where A_T is a constant with dimensions of electric potential times length$^{-1/2}$. If the cone surface is asymptotically equipotential, there is a specific value of constant A_T and a cone angle α_T for which the surface is conical (i.e., $\theta = \alpha_T$) and equipotential, with a normal electric field as

$$E_{n,T} = -\frac{1}{R}\frac{\partial \Phi_T}{\partial \theta}\bigg|_{\theta=\alpha_T} = -A_T Q'_{1/2}(\alpha_T) R^{-1/2}, \tag{8.3}$$

and thus, according to Eqs. (8.1), (8.3), A_T results

$$A_T = \left(\frac{2\sigma}{\varepsilon_o \tan(\alpha_T) Q'^2_{1/2}(\alpha_T)}\right)^{1/2}, \tag{8.4}$$

where the prime means derivative with respect to θ, and $Q'_{1/2}(\alpha_T)$ means the value of the derivative of $Q_{1/2}(\theta)$ at α_T. This angle has only one possible value: the Taylor cone angle $\alpha_T = 0.860274$ rads.

Consequently, the applied voltage potential on the liquid source cannot be arbitrary. Strictly speaking, Taylor's solution could be thought as an intermediate asymptotic solution between the real cone-jet transition region at the tip of the tapering meniscus, and the region close to the liquid source. Implicitly, this solution should match the specific geometry of that liquid source. A theoretical rigorous analysis of the shape of an electrohydrostatic meniscus with a conical apex, attached to a round capillary tube and in the absence of jet was given for the first time by Pantano, Gañán-Calvo, and Barrero (1994). That work demonstrated that the value of the applied voltage U to have a conical tip-ended equilibrium meniscus with given volume V is unique: that is, there is a continuous function relating U and V, as experimental practice teaches (Cloupeau & Prunet-Foch, 1989), with a relatively limited range of possible values of U for given boundary conditions and electrode geometries.

8.1.3 The effect of ejections

In summary, the existence of the nearly electrohydrostatic meniscus requires that the inner electric field should be negligible. It means that the voltage should be applied for enough time to allow the free charges in the liquid to migrate toward the surface, where they form a charge layer, that becomes nearly equipotential for very long times compared to the characteristic relaxation time $t_e = \beta\varepsilon_o/K$ of charges. β is the relative permittivity of the liquid (i.e., $\beta = \varepsilon_i/\varepsilon_o$, with ε_i and ε_o the electrical permittivities of liquid and vacuum, respectively) and K its electrical conductivity. However, Taylor's solution is singular when $R \to 0$: as one approaches the meniscus tip at $R = 0$, the strength of the electric field increases as $R^{-1/2}$. In practice, at a given point, the liquid is drawn off by the two components (tangential and normal) of the Maxwell stress, producing an ejection. This is the physical origin of electrosprays, with a special case in which the ejection occurs steadily, in the form of a stable, conical charged microjet: the famous Taylor cone-jet mode.

In contrast to the ballistic ejection of liquid from a round orifice, the jet from a Taylor cone is subsequently accelerated along the axial coordinate z by both normal and tangential components of the Maxwell stress. If one feeds a constant liquid flow rate Q from the tube source, a steady regime will be reached when the jet emits exactly the same flow rate Q. Thus, excluding other ejection modes, the process by which a given (sufficiently high) voltage is initially applied to a capillary tube that is feeding a flow of liquid Q, and eventually forms a stable Taylor conical jet, takes some time (see Fig. 8.3B) (Collins, Jones, Harris, & Basaran, 2008; Collins, Sambath, Harris, & Basaran, 2013; Gañán-Calvo et al., 2016; Pillai, Berry, Harvie, & Davidson, 2016). Although that time, comparable to the macroscale capillary-inertial time $t_o = \left(\frac{\rho R_1^3}{\sigma}\right)^{1/2}$, is very large compared to t_e (which is below 1 ms in nearly all cases), it is nevertheless of the order of seconds or below.

When the jet is steadily emitted, since it drags a charge rate from the cone, its presence (space charge effect) modifies the electrostatic potential decay that drives the jet itself. Hence, the applied voltage U should be modified to keep the liquid meniscus in a stable configuration under the existing boundary conditions and geometry of the electrodes and liquid feeding means. On the other hand, the local values of the normal electric field at the liquid surfaces can be so high as to promote local ionization of the surrounding gas, which becomes conductive: the corona discharge phenomenon (Machala, Marode, Laux, & Kruger, 2004). An exhaustive review on water electrosprays under corona discharge is given by Borra (2018). Charged jet, droplets, and ionized gas contribute to the so-called space charge effect that reduces the intensity of the electric field on the meniscus surface under a given applied electric potential. In what follows, corona discharge is not considered, though.

Under the presence of the charged jet and resulting spray, the stability range of applicable voltages is a two-dimensional (2D) region in the plane $\{Q, U\}$ (see Cloupeau & Prunet-Foch, 1989 and Gañán-Calvo et al., 2018, and Fig. 8.4), where the applied voltage U obviously depends on each specific geometry of the electrodes. However, the range of applicable voltages for a given flow rate Q is relatively narrow and does not generally produce significant variations of the jet diameter compared to those caused by the

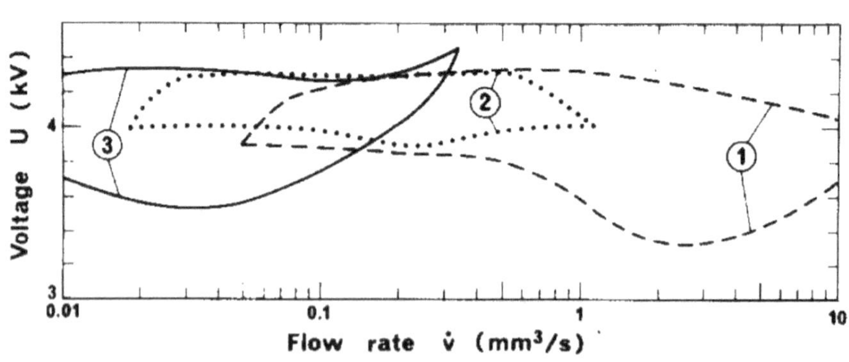

FIG. 8.4 Experimental map of stability of the cone-jet mode for dioxane-formamide mixtures using a cylindrical capillary of 0.5 mm OD. Curves (1)–(3) correspond to conductivities $K = 3.2 \times 10^{-7}$, 3×10^{-6}, and 3.6×10^{-5}, respectively. *From Cloupeau, M., & Prunet-Foch, B. (1989). Electrostatic spraying of liquids in cone-jet mode. Journal of Electrostatics, 22, 135–159.*

change of Q. Consequently, the jet diameter can be considered almost uniquely controlled by the supplied liquid flow rate Q if one operates the electrospray at the average applied voltage to keep the emission stable and steady for each given Q.

8.2 Taylor cone-jets

The (static) charge due to the steady ejection that modifies the electric potential can be expressed assuming the presence of a concentrated line of charges distributed along the axis $\theta = \pi$. Using cylindrical coordinates (r, z) for a slender jet (i.e., such that $z \gg r$), in first approximation that charge distribution produces a potential Φ_J that can be expressed as

$$\Phi_J = A(z)\log(r) \tag{8.5}$$

such that the complete electric potential Φ can be written by superposition as $\Phi = \Phi_T + \Phi_J$. In the following, we assume this simple superposition principle and consider the action of the total potential Φ.

8.2.1 Basic physics and equations

8.2.1.1 The driving electric field (Taylor)

The resulting electric field of Taylor's solution along the axis, for $\theta = \pi$ (i.e., the jet region), is given by

$$E_{z,T} = \frac{\pi}{4}A_T R^{-1/2} = \left(\frac{\pi^2\sigma}{8\varepsilon_o \tan(\alpha_T) Q'^2_{1/2}(\alpha_T)}\right)^{1/2} R^{-1/2}, \tag{8.6}$$

where R can be turned into the cylindrical axial coordinate z. In cylindrical coordinates, the radius of jet surface can be described as $r = a(z)$. In these coordinates, the previously mentioned potential, modified due to the presence of the charged jet, produces the following normal and tangential electric fields on the jet surface:

$$E_n = \frac{A}{a} - (E_{z,T} + A'\log(a))a' \simeq \frac{A}{a} \quad \text{and} \quad E_s = E_{z,T} + A\frac{a'}{a} + A'\log(a), \tag{8.7}$$

respectively, where $a(z)$ is the local jet radius and $A(z)$ is an unknown function, with dimensions of electric potential that should be solved according to the conservation equations and given boundary conditions. Eq. (8.7) is obtained assuming again $a/z \ll 1$ under the slender approximation $a/z \ll 1$. Primes are derivatives with respect to z. In the following, we write these conservation equations and discuss their implications.

8.2.1.2 The ejected charge rate (electric current)

Once a steady Taylor cone-jet regime is established, the charge per unit time (electric current) is transported by two mechanisms:

(1) By the charged moving surface of the jet, with surface speed v_s that can be assumed nearly equal to the average liquid velocity $v = Q/(\pi a^2)$. This is justified by the rapid viscous diffusion of momentum in the radial direction, assuming that $\rho v a^2/(\mu z) \ll 1$, which is generally fulfilled in electrospray, as subsequently discussed. The resulting *convective* electric current is given by

$$I_n = \frac{2Q\varepsilon_o E_n}{a}. \tag{8.8}$$

In first approach, neglecting terms of the order $O(a/z)$ and higher ($a/z \ll 1$), one has $E_n \simeq E_r = A/a$.

(2) By Ohmic *conduction* through the jet bulk that can be written as

$$I_b = \pi a^2 K E_s. \tag{8.9}$$

Thus, the total ejected electric current is written as

$$I = I_n + I_b = \frac{2Q\varepsilon_o E_n}{a} + \pi a^2 K E_s. \tag{8.10}$$

8.2.1.3 The balance of surface stresses

Under the same approximation, the balance of normal stresses on the jet surface can be written as

$$\frac{\sigma}{a} = P + \frac{\varepsilon_o}{2}\left(E_n^2 + (\beta - 1)E_s^2\right), \tag{8.11}$$

where $P = P(z)$ is the pressure in the liquid, and normal viscous stresses are neglected. On the other hand, the tangential stress balance reduces to

$$\tau_s = E_s \varepsilon_o \left(E_n - \beta E_n^i\right) \simeq \varepsilon_o E_s E_n \tag{8.12}$$

under the slender assumption, where τ_s is the tangential viscous stress on the jet surface.

8.2.1.4 The liquid momentum balance

The slender approximation allows the consistent reduction of the momentum equation $D\mathbf{v}/Dt + \nabla P = \mu \nabla^2 \mathbf{v}$ to a one-dimensional (1D) expression as

$$\frac{d}{dz}\left(\frac{\rho v^2}{2} + P\right) = \frac{2\tau_s}{a} + \frac{3\mu}{a^2}\frac{d}{dz}\left(a^2 \frac{dv}{dz}\right), \tag{8.13}$$

where the velocity vector $\mathbf{v}$ can be reduced in first approximation to its average axial component $v = Q/(\pi a^2)$ from the mass conservation.

Substituting the values of the pressure and the surface stress τ_s by those given by Eqs. (8.11), (8.12), the momentum balance along the jet finally reads (Gañán-Calvo, 1999)

$$\frac{d}{dz}\left(\frac{\rho Q^2}{2\pi^2 a^4}+\frac{\sigma}{a}-\frac{\varepsilon_o E_n^2}{2}-(\beta-1)\frac{\varepsilon_o E_s^2}{2}\right)=\frac{2\varepsilon_o E_s E_n}{a}-\frac{3Q\mu}{\pi a^2}\frac{d}{dz}\left(\frac{1}{a}\frac{da}{dz}\right). \quad (8.14)$$

8.2.1.5 Nondimensional equations

Introducing the characteristic values of length $d_o=\left(\frac{\sigma\varepsilon_o^2}{\rho K^2}\right)^{1/3}$, electric field $E_o=\left(\frac{\sigma}{\varepsilon_o d_o}\right)^{1/2}$, flow rate $Q_o=\frac{\sigma\varepsilon_o}{\rho K}$, and electric current $I_o=\left(\frac{\sigma^2\varepsilon_o}{\rho}\right)^{1/2}$, the balance of momentum, the charge transport, and the tangential electric field can be written in nondimensional form as

$$\frac{d}{dx}\left(\frac{\Theta^2}{2\pi^2 f^4}+\frac{1}{f}-\frac{e_n^2}{2}-\frac{(\beta-1)e_s^2}{2}\right)=\frac{2e_s e_n}{f}-\frac{3\eta\Theta}{\pi f^2}\frac{d}{dx}\left(\frac{1}{f}\frac{df}{dx}\right), \quad (8.15)$$

$$\mathcal{I}=\frac{2\Theta e_n}{f}+\pi f^2 e_s, \quad (8.16)$$

$$e_n=a/f, \quad e_s=\frac{\pi}{\sqrt{8\tan(\alpha_T)}}\frac{1}{Q'_{1/2}(\alpha_T)}x^{-1/2}-\frac{d(a\log(f))}{dx}, \quad (8.17)$$

respectively, where

$$f=a/d_o,\ x=z/d_o,\ e_n=E_n/E_o,\ e_s=E_s/E_o,\ \text{and}\ \varphi=A/(d_oE_o) \quad (8.18)$$

are the nondimensional-dependent variables of the problem.

Given the three nondimensional parameters

$$\Theta=Q/Q_o,\quad \eta=\left(\frac{K\mu^3}{\rho\sigma^2\varepsilon_o}\right)^{1/3},\quad \text{and}\quad \beta=\varepsilon_i/\varepsilon_o, \quad (8.19)$$

the system of four equations (8.15–8.17) provide the solution for the four variables f, e_n, e_s, and φ, where $\mathcal{I}=(I_s+I_b)/I_o\equiv I/I_o$ is an eigenvalue of the problem for the condition that the solution matches the Taylor cone. The relationship of the local values of the classical Weber (We) and Ohnesorge (Oh) numbers with the parameters Θ and η is now straightforward:

$$\mathrm{We}=\pi^{-2}\Theta^2 f^{-3}, \qquad \mathrm{Oh}=\eta f^{-1/2}, \quad (8.20)$$

which, naturally, change along the axial jet coordinate with the nondimensional radius f.

Finding the solution of systems (8.15)–(8.17) for the whole parametrical space $\{\Theta,\beta,\eta\}$ can be a tricky mathematical issue, and its treatment is not the object of present illustrative analysis. Thus, in the following, we will discuss a comprehensible approach where the use of dimensional analysis, together with a rigorous analysis of the equations, will provide the crucial information to build simple predictive models for the most useful outcome of electrosprays: the *scaling laws* for the ejected electric current and the size of the ejected droplets.

8.2.2 Taylor cone-jets: Scaling laws

In Eq. (8.15), the four terms on the left-hand side are, informally speaking, the axial resultants of the inertia, surface tension, electrostatic "suction," and electric polar stresses, respectively. On the right-hand side of Eq. (8.15), the two terms are the axial resultants of the tangential stress on the surface and the viscous stresses in the liquid.

The variables f, e_n, e_s, and φ, and the axial coordinate x may acquire very large or small values compared to unity, depending on the values of the three parameters of the problem, $\{\Theta, \beta, \eta\}$. These values define a three-dimensional (3D) parametrical space with some well-defined regions, where universal solutions of the system (8.15)–(8.17) and the eigenvalue $\mathcal{I}$ can be found. These universal solutions imply particular scalings of the variables and the eigenvalue with the three parameters of the problem. In the following, we analyze such scalings, which are usually termed *scaling laws*.

First of all, under the assumption that the electric potential Φ_T is asymptotically dominant on the jet, one always have

$$e_s \sim x^{-1/2}. \tag{8.21}$$

Second, the ejected electric current is determined at the point, where the transition from Ohmic conduction-dominance to surface charge convection-dominance of charge transport occurs. We assume that this transition takes place already in the jet, sufficiently far away from the cone-jet transition to have the influence of Taylor axial field. Thus, electric conduction and convection should be of the same order, which demands

$$\mathcal{I} \sim \Theta e_n/f \sim f^2 e_s. \tag{8.22}$$

These are three common fundamental conditions for all Taylor cone-jet scaling laws. The relative weight and dominance of the different drivers and resistances expressed by the momentum equation (8.15) render four main scenarios, which yield the following four scaling laws (Gañán-Calvo, 2004).

8.2.2.1 Dominance of inertia, electrostatic suction, and tangential electric stress: The IE scaling law

This is the most common operating region of Taylor cone-jets, and occurs for $\Theta \gg \beta$ and $\Theta \gg \eta$. Given that $\beta > 1$, this means that Θ should be sufficiently large for this scaling to apply correctly. In this case, one has that

$$\Theta^2/f^4 \sim e_n^2 \sim e_n e_s x/f, \tag{8.23}$$

which together with Eqs. (8.21), (8.22) yields

$$f \sim \Theta^{1/2}, \quad x \sim \Theta, \quad e_n \sim O(1), \quad e_s \sim \Theta^{-1/2}, \quad \mathcal{I} \sim \Theta^{1/2}. \tag{8.24}$$

In dimensional form, this scaling implies that (Gañán-Calvo, 1999; Gañán-Calvo, 2004; Gañán-Calvo et al., 2018):

(1) The electric current should have a universal expression as

$$I = k_I I_o (Q/Q_o)^{1/2} \equiv k_I (\sigma K Q)^{1/2}. \tag{8.25}$$

This law was first proposed by Gañán-Calvo, Barrero, and Pantano (1993).

(2) The ejected droplets, commensurate with the jet radius, should have diameters following a universal law as

$$d_g = k_d d_o (Q/Q_o)^{1/2} \equiv k_d \left(\frac{\rho \varepsilon_o Q^3}{\sigma K}\right)^{1/6}. \tag{8.26}$$

These universal expressions have been amply supported by most published experimental measurements, approximately yielding prefactors $k_I \simeq 2.5$ and $k_d \simeq 1.0$. Fig. 8.5A and B illustrates the accuracy of these scaling laws against several hundreds of measurements and tens of different liquids. Given k_d, the Weber and Ohnesorge numbers for this scaling law read:

$$\mathrm{We} = \frac{3.782^3}{(2\pi)^2}\Theta^{1/2} = 0.685\Theta^{1/2}, \quad \mathrm{Oh} = (3.782/2)^{1/4}\eta\Theta^{1/4} = 1.173\eta\Theta^{1/4}. \tag{8.27}$$

Thus, assuming a maximum We = 20 for Rayleigh breakup, one would expect a nearly monodisperse charged aerosol for flow rates Q up to $0.685^{-2} \times 20^2 Q_o = 852.5 Q_o$. This limit is marked in Fig. 8.5B: observe the bending of the droplet radius trend below the scaling law that assumes proportionality between the jet radius and the droplet radius right at this value of $\Theta_{lim} = 852.5$. Over Θ_{lim}, the growth rate of nonsymmetric breakup modes begins to grow over that of the axisymmetric one, and the breakup becomes ultimately chaotic, which yields polydisperse and smaller droplets.

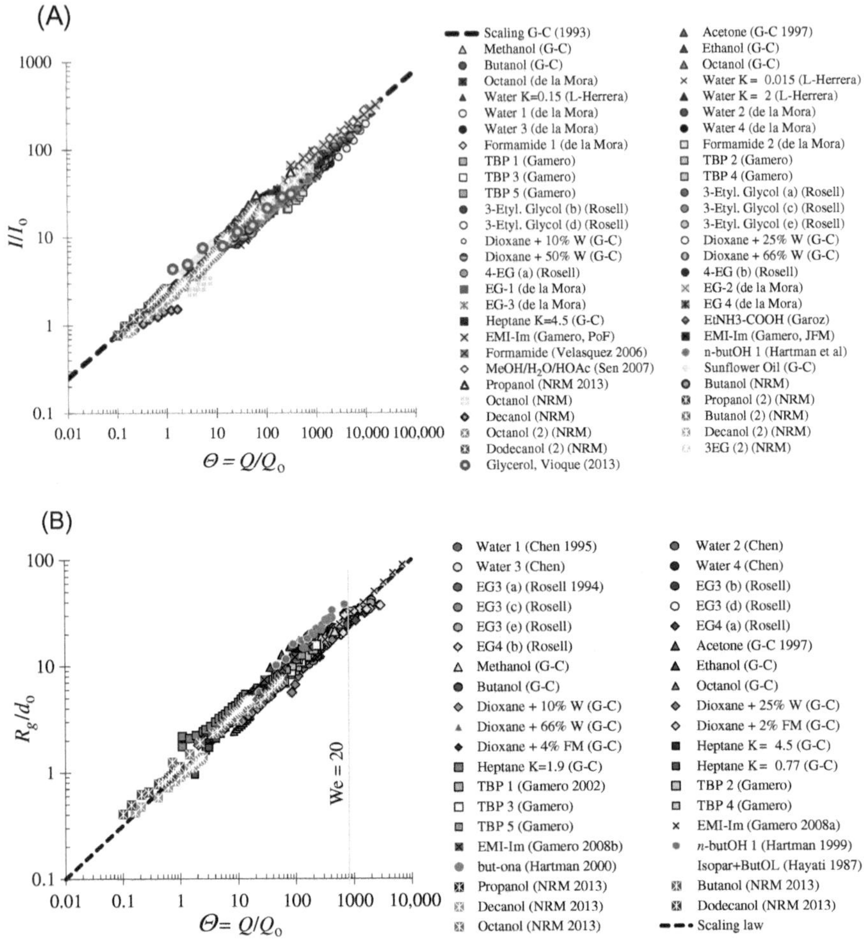

FIG. 8.5 The IE scaling law: (A) electric current $\mathcal{I} = I/I_o$ as a function of the flow rate parameter Θ, measured for 54 different liquids. (B) Measured mean radius of the ejected droplets R_g normalized with d_o as a function of Θ, for 39 different liquids. *Dashed thick black lines* represent Eqs. (8.25), (8.26). *Data taken from the literature (Fernandez de la Mora & Loscertales, 1994; Gamero-Castaño, 2008a, 2008b; Gamero-Castaño & Hruby, 2002; Gañán-Calvo, Dávila, & Barrero, 1997; Gañán-Calvo, Rebollo-Muñoz, & Montanero, 2013; Garoz et al., 2007; Hartman, Brunner, Camelot, Marijnissen, & Scarlett, 1999; Hayati, Bailey, & Tadros, 1987a, 1987b; López-Herrera, Barrero, Boucard, Loscertales, & Márquez, 2004; Rosell-Llompart & de la Mora, 1994; Velásquez-García, Akinwande, & Martínez-Sánchez, 2006).*

8.2.2.2 *Dominance of inertia and polar stress: The IP scaling law*

This scaling law applies for highly polar liquids with $\eta \ll \Theta \ll \beta$, where one has

$$\Theta^2/f^4 \sim \beta e_s^2 \sim e_n e_s x/f, \tag{8.28}$$

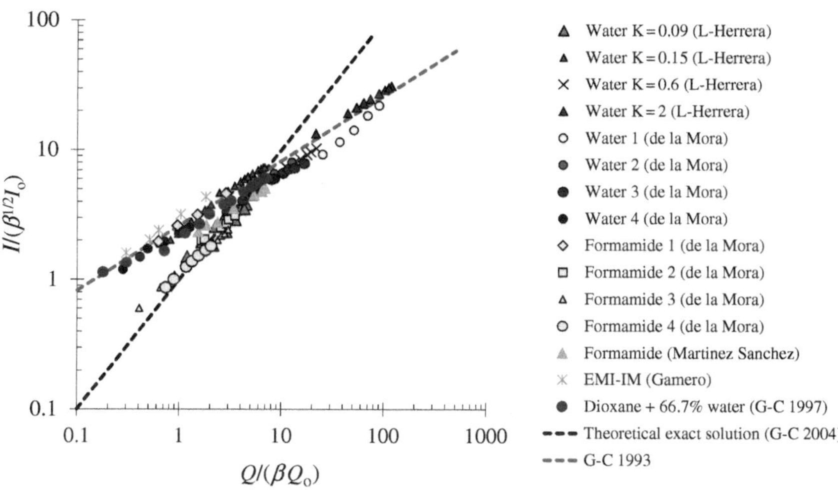

FIG. 8.6 The IP scaling law for the electric current. *Dashed thick black line* represents Eq. (8.30). The *blue dashed line* (*dark gray in print version*) is the IE scaling law (8.25). *Data from different authors (Fernandez de la Mora & Loscertales, 1994; Gamero-Castaño, 2008a; Gañán-Calvo et al., 1997; López-Herrera et al., 2004; Lozano & Martínez Sánchez, 2005).*

which yields

$$f \sim \Theta^{1/2}, \quad x \sim \beta, \quad e_n \sim \left(\beta^{-1}\Theta\right)^{1/2}, \quad e_s \sim \beta^{-1/2}, \quad \mathcal{I} \sim \beta^{-1/2}\Theta. \tag{8.29}$$

This scaling law is approximately fulfilled in some measurements with water and formamide, see Fig. 8.6. In dimensional form, it yields the following expressions for the emitted current and droplet diameter:

$$I = k_{I,IP}\left(\frac{\rho Q^2 K^2}{\beta \varepsilon_o}\right)^{1/2}, \quad d_g = k_{d,IP}\left(\frac{\rho \varepsilon_o Q^3}{\sigma K}\right)^{1/6}, \tag{8.30}$$

where the prefactors $k_{I,IP}$ and $k_{d,IP}$ are in general different from previous k_I and k_d in Eqs. (8.25), (8.26). Since the droplet diameter has the same scaling as the IE law, Eq. (8.26), one expects $k_{d,IP} = k_d$. However, from experimental measurements (see Fig. 8.6), one has $k_{d,IP} \simeq 1$ (Gañán-Calvo, 2004).

Alternatively, in cases where $\beta \ll \Theta \ll \beta^4\eta^3$, one may have

$$\Theta^2/f^4 \sim e_n^2 \sim \beta e_s^2, \tag{8.31}$$

which yields

$$f \sim \beta^{1/6}\Theta^{1/3}, \quad x \sim \beta^{5/3}\Theta^{-2/3}, \quad e_n \sim \beta^{-1/3}\Theta^{1/3}, \quad e_s \sim \beta^{-5/6}\Theta^{1/3}, \\ \mathcal{I} \sim \beta^{-1/2}\Theta. \tag{8.32}$$

This regime is often observed when the so-called "silver bullet" mode (Chen, Pui, & Kaufman, 1995; López-Herrera et al., 2004) occurs for certain voltage and flow rate ranges, which yield

$$I = k'_{I,IP}\left(\frac{\rho Q^2 K^2}{\beta \varepsilon_o}\right)^{1/2}, \quad d_g = k'_{d,IP}\left(\frac{\beta^{1/2}\varepsilon_0 Q}{K}\right)^{1/3}, \tag{8.33}$$

where, again, prefactors $k'_{I,IP}$ and $k'_{d,IP}$ should be experimentally obtained.

8.2.2.3 *Dominance of viscous forces, electrostatic suction, and tangential electric stress: The VE scaling law*

When viscous forces dominate, the jet does not normally break up into droplets but it stretches nearly indefinitely, whipping and winding up toward the electrode where it collects. Thus, this regime does not belong to the *electrospray* realm: in reality, it is termed *electrospinning* (Reneker & Yarin, 2008), and usually occurs when $\beta^4\eta^{-3} \ll \Theta \ll \eta$. The conditions are

$$\eta\Theta/(xf^2) \sim e_n^2 \sim \beta e_n e_s x/f, \tag{8.34}$$

which yield

$$f \sim (\eta\Theta^3)^{1/8}, \quad x \sim (\eta\Theta)^{1/2}, \quad e_n \sim (\eta\Theta^{-1})^{1/8}, \quad e_s \sim (\beta^{-1}\Theta^{-1})^{1/4}, \quad \mathcal{I} \sim \Theta^{1/2}. \tag{8.35}$$

This yields

$$I = k_{I,VE}(\sigma K Q)^{1/2}. \tag{8.36}$$

Given that the scaling for the electric current is identical to that of the IE scaling, it is nearly indistinguishable in experiments, and one has $k_{I,\ VE} \simeq k_I$. However, the nearly indefinite stretching of the liquid filament produces a drastic reduction of the diameter of the fiber along the axial length. Thus, the scaling of the jet radius

$$a = \chi(z)\left(\frac{\eta^9 \varepsilon_0^4 Q^9}{\rho^2 K^4 \sigma^7}\right)^{1/24} \tag{8.37}$$

implies the calculation of a nondimensional function $\chi(z)$ whose values could be rather small for large values of the axial coordinate z. This function $\chi(z)$ could be analytically determined, although the development of nonsymmetric (whipping) instabilities of the fiber will make the calculation invalid for sufficiently large z-values.

8.2.2.4 Dominance of viscous forces, polar forces, and tangential electric stress: The VP scaling law

This scaling may theoretically occur when $\eta \gg \beta^3$ and both $\Theta \gg \eta^{-6}\beta^7$ and $\Theta \gg \eta^{-9}\beta^{11}$ hold. In these conditions, one has

$$\eta\Theta/(xf^2) \sim \beta e_s^2 \sim \beta e_n e_s x/f, \tag{8.38}$$

which yields

$$\begin{aligned} f &\sim (\beta^{-1}\eta\Theta)^{1/2}, \quad x \sim \beta^2\eta^{-1}, \quad e_n \sim (\beta^{-5}\eta^4\Theta)^{1/2}, \\ e_s &\sim \beta^{-1}\eta^{1/2}, \quad \mathcal{I} \sim \beta^{-2}\eta^{3/2}\Theta. \end{aligned} \tag{8.39}$$

In dimensional form, this scaling reads

$$I = k_{I,VP}\left(\frac{\mu^3K^3Q^2}{\beta^4\sigma^2\varepsilon_o^2}\right)^{1/2}, \tag{8.40}$$

where prefactor $k_{I,\ VP}$ is experimentally determined. Note that the radius of the ejected liquid fiber along the axial coordinate z,

$$a(z) = \zeta(z)\frac{1}{\beta}\left(\frac{\mu Q}{\sigma}\right)^{1/2} \tag{8.41}$$

is dependent on the liquid viscosity, surface tension, and the imposed flow rate with the same scaling as in viscous flow focusing (Gañán-Calvo, Pérez-Saborid, López-Herrera, & Gordillo, 2004), but inversely proportional to the liquid polarity. Again, solving the problem entails solving a nondimensional function $\zeta(z)$ of order unity for $z \sim \beta^2\eta^{-1}d_o$, which is out of the scope of this work.

8.2.2.5 Other marginal scalings

At the limits of a stable Taylor cone-jet configuration, when the liquid flow rate decreases and the jet diameter is reduced, the surface tension increases to the point of becoming a significant resistance to the ejection. Hence, in this marginal region, surface tension dominates, and the following scenarios and corresponding scaling laws appear:

1. *Inertia ~ surface tension; driver: electrostatic suction.* This scenario yields

$$f \sim \Theta^{2/3}, \quad x \sim \Theta^{8/3}e_n \sim \Theta^{-1/3}, \quad e_s \sim \Theta^{-4/3}, \quad \mathcal{I} \sim 1. \tag{8.42}$$

2. *Drivers: tangential force and electrostatic suction.* This other scenario yields

$$f \sim x \sim \Theta^{1/3}, \quad e_n \sim e_s \sim \Theta^{-1/6}, \quad \mathcal{I} \sim \Theta^{1/2}. \tag{8.43}$$

3. *Drivers: electrostatic suction and polar force.* In this case, one has

$$f \sim \beta^{1/6}\Theta^{1/3}, \quad x \sim \beta^{7/6}\Theta^{1/3}, \quad e_n \sim \beta^{-1/12}\Theta^{-1/6}, \\ e_s \sim \beta^{-7/12}\Theta^{-1/6}, \quad \mathcal{I} \sim \beta^{-1/4}\Theta^{1/2}. \tag{8.44}$$

4. *Viscous force ~ surface tension; driver: surface stress.* This yields

$$f \sim \Theta^{1/3}, \quad x \sim \eta\Theta^{2/3}, \quad e_n \sim e_s \sim \eta^{-1/2}\Theta^{-1/3}, \quad \mathcal{I} \sim \eta^{-1/2}\Theta^{1/3}. \tag{8.45}$$

5. *Viscous force ~ surface tension; driver: electrostatic suction.* This other viscous scenario yields

$$f \sim \eta^{1/8}\Theta^{3/8}, \quad x \sim \eta^{7/8}\Theta^{5/8}, \quad e_n \sim \eta^{-1/16}\Theta^{-3/16}, \\ e_s \sim \eta^{-7/16}\Theta^{-5/16}, \quad \mathcal{I} \sim \eta^{-3/16}\Theta^{7/16}. \tag{8.46}$$

All of these scenarios are hampered by the limitations inherent in the ultimate enabling mechanism of the electrospray phenomenon, which becomes more important as the characteristic hydrodynamic times diminish (e.g., when the issued flow rates decrease): the charge relaxation on liquid surfaces exposed to intense electric fields. This charge relaxation presupposes a nontrivial electrokinetic process that needs some revision to provide a relatively complete understanding of the physics of electrosprays.

8.3 Electrokinetics of electrospraying

The analyses undertaken so far have obviated the electrokinetic nature of the electric charge because of its complexity. Certainly, a detailed description of the distribution of the different charged species that give net charge to the fluid under an electric field is not essential to describe accurately the physics. That is because the averaging behind the electrostatic Maxwell equations suffices to explain (and describe) the physics of the cone-jet structure and dynamic as demonstrated by the aforementioned works. However, the different behavior of the various ionic species under an electric field is behind the differences observed in electrospray operating in direct polarity (needle voltage over the one of the downstream electrode) or inverse polarity (needle voltage below that of the downstream electrode) (Martin, Perea, Garcia-Ybarra, & Castillo, 2012). Also, it is well known how the polarity affects the performance and operating conditions in terms of applied voltage and flow rate thresholds of electrospray ionization mass spectrometers (ESI-MS), where the polarity is changed accordingly to the analyte of interest (negatively charged ions such as inorganic anions and deprotonated organic acids are analyzed in the negative ion mode) (Cech & Enke, 2001).

8.3.1 Bulk equations

The distribution of charged species alters the electric field by creating a net volumetric charge, ρ_e,

$$\nabla \cdot (\varepsilon \mathbf{E}) = \rho_e = \sum_k e z_k n_k, \tag{8.47}$$

where ϵ is the electrical permittivity, e is the electron charge, n_k is the volumetric concentration of kth species, and z_k is its valence.

The fluid velocity, $\mathbf{u}$, is linked to the electric field through the Maxwell stress tensor τ_M in the momentum equation

$$\rho D_t \mathbf{u} = -\nabla P + \mu \nabla^2 \mathbf{u} + \nabla \cdot \tau_M \quad \text{with} \quad \tau_M = \varepsilon[\mathbf{EE} - (\mathbf{E} \cdot \mathbf{E})\mathbf{I}/2], \tag{8.48}$$

where P is the liquid pressure, $\mathbf{I}$ is the unit stress tensor, and D_t denotes the material derivative following the fluid particle.

Finally, the temporal evolution of n_k responds to the usual transport mechanism of convection and diffusion. In addition, the electrical migration of the ionic species ($z_k \neq 0$) under electric fields has to be considered,

$$\partial_t n_k + \nabla \cdot (\mathbf{u} n_k) = \nabla \cdot [-\omega_k e z_k n_k \mathbf{E} + \omega_k k_h T \nabla n_k] + r_k. \tag{8.49}$$

Note that this mechanism is absent for neutral species since $z_k = 0$. The transport by diffusion and electrical migration is not independent since the surrounding solvent media affects equally both transport mechanisms; therefore, the mobility ω_k is involved in both terms. Above, k_b is the Boltzmann constant and T is the temperature. Besides, the creation/destruction of species by reaction, r_k, could be present (Saville, 1997).

The interdependence among the ionic and neutral species can be quite intricate due to the recombination/dissociation chemical reactions that may appear (Saville, 1997). In the simplest case of a 1-1 electrolyte, three species would be present; neutral (n_o), cation (n_1), and anion (n_2). The formation of the ions occurs by dissociation of the neutral species at a rate proportional to its concentration as $k_+ n_o$, while the recombination rates of ions into the neutral species would be $k_- n_1 n_2$, being the rate of production of ions (Saville, 1997),

$$-r_o = r_1 = r_2 = k_+ n_o - k_- n_1 n_2, \tag{8.50}$$

where k_+ and k_- are rate constants dependent on the ions, the corresponding neutral species, and the solvent. In case of strong electrolytes, the neutral species is fully dissociated and the reaction term is absent. At equilibrium, the concentration of the neutral species could be, therefore, determined directly from the ionic concentrations,

$$k_+ n_o - k_- n_1 n_2 = 0 \quad \Rightarrow n_o = \frac{k_-}{k_+} n_1 n_2. \tag{8.51}$$

For further insight on the subject, the reader is referred to the seminal review of Saville (1997).

8.3.2 The surface charge layer: Debye's length

The applied electric field drives ions from the liquid bulk according to their polarity until they reach the limits of the liquid domain (either free or solid surfaces). Consequently, they accumulate at those surfaces by forming a diffuse layer, namely the Debye layer of a width λ_D, with maximum concentration at the exposed surface. The characteristic width of that layer where the electrical drive is balanced by thermal diffusion, is given by

$$\lambda_D = \sqrt{\frac{\varepsilon k_B T}{e^2 \sum (z_k n_k)}}. \tag{8.52}$$

The electrokinetic nature of the surface charge density is not as clearly stated as the volume charge density. Schnitzer and Yariv (2015) propose that the surface charge density, σ_e, results from the unbalance of surface concentrations of positive and negative species, $\sigma_e = \sum e z_k n_{sk}$ as in the case of the volume charge density (see Eq. 8.47). These surface concentrations, n_{sk}, would be governed by continuity equations which take into account surface dilation, surface convection, and diffusion. Under relatively general situations where the liquid domain presents an axisymmetric geometry (e.g., the Taylor cone-jet), that continuity equations can be written as

$$\partial_t n_{sk} + \partial_s (n_{sk} v_t) + \frac{n_{sk} v_t}{f} \partial_s f + n_{sk} v_n \nabla \cdot \mathbf{n}$$
$$= \omega_k e z_k n_k \mathbf{E} \cdot \mathbf{n} - \omega_k k_B T (\mathbf{n} \cdot \nabla n_k) - \frac{\partial_s (\omega_k^s e z_k n_{sk} f E_t)}{f} + \frac{\partial_s (\omega_k^s k_B T \partial_s n_{sk})}{f}, \tag{8.53}$$

where a surface charge mobility ω_k^s has been included for consistency. In the earlier expression, v_t and v_n denote tangential and normal velocities and f is the radial position of the interface; ∂_t and ∂_s denote partial time and space derivatives, the latter along the surface tangential direction. The relation between bulk and surface concentrations is $n_{sk} = \Gamma_k n_k$, where Γ_k would be a surface adsorption coefficient with a difficult experimental quantification. In contrast, the approach of Mori and Young (2018) does not have any built-in surface charges; the surface charge density σ_e emerges naturally from the bulk charge as a spatial integration across the Debye layer

$$\sigma_e = \sum_k e z_k \int_{-\infty}^{0} n_k dz \simeq \sum_k e z_k \int_{-\lambda_D}^{0} n_k dz. \tag{8.54}$$

Both models would differ when applied to a Taylor cone-jet, an object of subsequent studies. In summary, all the above would provide a complete platform for the detailed description of the electrokinetics and electrochemistry processes entailed in the macroscopic electrospraying phenomenon. The particular case of interest here, the steady Taylor cone-jet mode, implies in general that hydrodynamic times are large compared to charge relaxation to have free charges relaxed at the liquid surfaces exposed to the applied electric field. This ensures that the resultant Maxwell stresses on these surfaces are sufficiently intense to maintain the enormous acceleration associated with the cone-jet configuration. The flow stability can be compromised when the ejected flow rate becomes too small, and the size of the jet is reduced to the point to have hydrodynamic (residence) times at the cone-jet transition region comparable to the charge relaxation time t_e. On the other hand, the thickness of the charge layer (Debye's length) may compromise the bulk charge neutrality situation when the local volumes involved are small. Both situations are associated with the global stability limits of cone-jet electrospray, but their quantitative analysis is the subject of ongoing work.

Once the basic physics and the electrokinetics of the cone-jet mode have been discussed in some detail, in the following we describe a few selected high-value applications of this highly controlled liquid fragmentation mechanism at the micro and nanoscale in analytical chemistry and biochemistry.

8.4 Applications in detection of virus-like particles

As described in the previous section, the electrokinetics of the electrospray are fundamental in determining the charged species ejected in the droplets. Similarly, the liquid's molecular structure and dissociation rate set its electrokinetic behavior, which in turn is related to both, its relative permittivity and electrical conductivity. At the length scale of a typical cone-jet, the electrokinetics may strongly influence the transport of ions within water droplets produced by an electrospray. Indeed, for biological applications of electrosprays, water is generally the solvent of choice because of its well-known biocompatibility and its relatively high electrical conductivity, needed to generate a stable cone-jet. Thus, in the following section, relevant examples that employ water for the generation of bioaerosols with electrosprays in the cone-jet mode are discussed.

8.4.1 Relevance of the cone-jet mode

The periodic breakup of the jet emitted by electrosprays operating in the cone-jet mode leads to the formation of droplets of uniform size. The diameter of the initial droplet (i.e., the mother droplet) may be accurately controlled by adequately adjusting process parameters such as liquid flow rate and applied voltage (Gañán-Calvo, 1999; Gañán-Calvo, 2004; Gañán-Calvo et al., 2016). Indeed, the steady production of small monodispersed droplets, with sizes ranging from the submicrometer 1 < μm to the nanometer length scales, is one of the main features of the cone-jet mode. It is precisely such capability to steadily generate tiny droplets that turns the electrospray into a highly attractive tool to deliver analytes for applications in airborne particles' detection platforms.

Airborne particles, also called aerosols, are either liquid droplets or solids with diameters typically smaller than 100 μm suspended in air. Among those, biological particles such as proteins, bacteria, and viruses have sparked attention due to their threat to human health, especially when their size is in the submicrometer range. For instance, viruses have characteristic sizes of the order of 25–400 nm (Madigan, Bender, Buckley, Sattley, & Stahl, 2015) and in natural systems they tend to couple with larger particles and aggregate leading to a virus size distribution (Hogan, Kettleson, Ramaswami, Chen, & Biswas, 2006). With a broad size range, those aerosols can be inhaled and pass through the human respiratory tract and while larger particles may deposit onto the nasal cavity and the tracheobronchial region, smaller particles may penetrate the gas-exchange regions of the lungs (alveolar region) and settle there. For rapid identification of these biological particles, an online real-time detection approach is desired to minimize situations of risk exposure and to implement appropriate control protocols. Delay in identification of such particles may have critical consequences for human health and the environment.

Aerosol scientists have developed instruments that allow accurate classification and sizing of micro- and nanoparticles online in real time. The instrumentation includes scanning mobility particle sizers (SMPSs) and differential mobility analyzers (DMAs) (Kaufman, 1998; Knutson & Whitby, 1975; Kousaka, Okuyama, Adachi, & Mimura, 1986). The operating principle of those instruments relies on the electrical mobility of the particles, Z_p, which is dictated by their size and their charging state:

$$Z_p = peC_c/3\pi\mu D_p, \tag{8.55}$$

where D_p is the mobility diameter of the particles, p is the number of elementary charges, e is the electron elementary charge, μ is the gas viscosity, and C_c is the Cunningham slip correction factor $(C_c(D_p) = 2.514\lambda_l/D_p + 0.8\lambda_l/D_p \exp(-0.55\lambda_l/D_p)$, where λ_l is the gas mean free path). An SMPS uses a bipolar charger to establish a known equilibrium charge distribution. Such bipolar chargers are also called neutralizers, since the overall eventual charge level of all particles is nearly zero. In a standard electrical mobility analysis, an aerosol is first introduced to an SMPS and subsequently to the DMA. Such analyzer consists of a channel between two cylindrical electrodes arranged concentrically. The aerosol flow is introduced upstream, adjacent to one of those electrodes, generally the external one; while a sheath flow fills the rest of the channel. As both gas currents transport the particles through the channel, a voltage difference between the electrodes drives particles of appropriate polarity across the channel. In DMAs, a small flow that exits through a port located downstream in the counter electrode (generally the inner one) extracts particles that have mobilities in a narrow range. Aerosol particles are thus classified based on a balance of drag and electrical forces as they flow downstream, toward the exit port. By gradually varying the applied voltage, a range of particles can be classified based on their electrical mobility, thus obtaining a particle size distribution. The DMA is also known as ion mobility spectrometer (IMS), electrical aerosol analyzer (EAA), or scanning electrical mobility spectrometer. Generally, in the standard commercial configuration, the SMPS and the DMA are coupled together into a single device and often referred to as SMPS. Hereafter, the term SMPS implies the use of a DMA unless otherwise stated. Since the different electrical mobilities are measured sequentially in an SMPS, the measurement of a size distribution requires only between 2 and 6 min, depending on the SMPS model, its settings, and required accuracy. For details on the functioning principle of the SMPS, the interested reader may refer to Knutson and Whitby (1975) and Flagan (2008) and references therein.

The versatility of electrical mobility-based analyses has prompted researchers to couple aerosol instrumentation with electrosprays, operating in the cone-jet mode, to investigate the size of proteins and virus particles and their aggregation state; a technique called electrospray size spectrometry analysis (ESSA) or charge-reduced electrospray size spectrometry (Bacher et al., 2001; Hogan et al., 2006; Kaufman, 1998; Modesto-Lopez, Kettleson, & Biswas, 2011) (see Fig. 8.7). In such methods, inherently charged droplets, containing biological particles, are generated with the electrospray from a sampling liquid and subsequently classified with an SMPS based on the particles electrical mobility. The advantage of the electrospray is twofold: (i) because the as-produced droplet is relatively small, it reduces the error associated with a large amount of undesired material (background) dissolved in the sampling liquid, which may mask the analyte particle in the detection instrument; and (ii) by

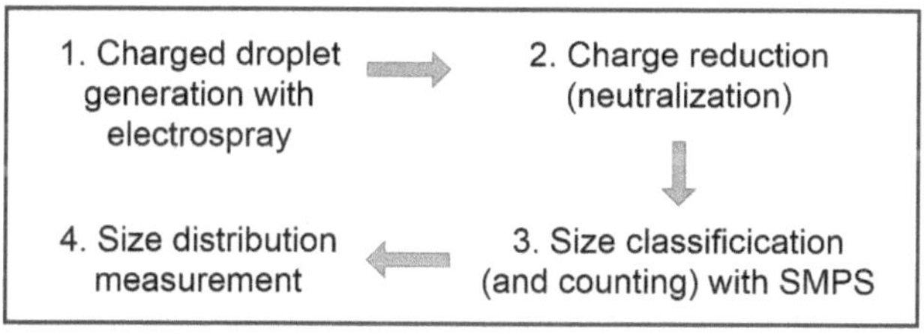

FIG. 8.7 General ESSA procedure.

controlling the concentration of analytes in the sampling liquid (for instance, through proper dilution) the condition of one particle per droplet may be readily achieved, thus facilitating its analysis and detection. In this section, we discuss the most relevant studies that employ electrosprays as generators of analytes for online real-time analyses of biological particles.

8.4.2 Electrospray-enabled electrical mobility measurements

Electrosprays are known as generators of highly charged droplets and particles (Gañán-Calvo, 2004; Gañán-Calvo et al., 2018; Modesto-López & Biswas, 2010; Modesto-López, Thimsen, Collins, Blankenship, & Biswas, 2010; Rosell-Llompart, Grifoll, & Loscertales, 2018), which constitute the analytes. Therefore, prior to performing analyses with aerosol instrumentation, the highly charging state of the electrosprayed particles must be reduced (i.e., neutralized) to a known equilibrium condition (Knutson & Whitby, 1975; Wiedensohler, 1988). This charge reduction of electrospray-generated analytes prevents particle loss due to electrostatic attraction and reduces misclassification by the instrument. Until the last decade, it was common to employ radioactive sources to neutralize electrosprayed particles and brought them to the so-called bipolar equilibrium condition, consisting of a symmetrical distribution of positively and negatively charged particles. The radioactive sources typically employed by aerosol scientists were Po-210 (alpha particles with a maximum α energy of 5.3 MeV) and Kr-85 (beta particles with a maximum β energy of 0.695 MeV) (Forsyth, Liu, & Romay, 1998). The bipolar equilibrium condition was achieved through random thermal collisions between ions emitted from the radioactive source and the highly charged particles. The charge reduction takes place in a so-called neutralization chamber as depicted in Fig. 8.8, downstream of the electrospray generation point. Charge reduction prevents droplets reaching their charge limit, above which they undergo fissions by so-called Coulombic explosions (Gomez & Tang, 1994).

Relevant biological particles reported in the literature include bacteriophages MS2, T2, T4, virus-like particles (VLP) from nonenveloped murine polyomavirus (MPV), and proteins such as ovalbumin (OVA). The commonly measured bioparticles and their characteristic size are summarized in Table 8.1.

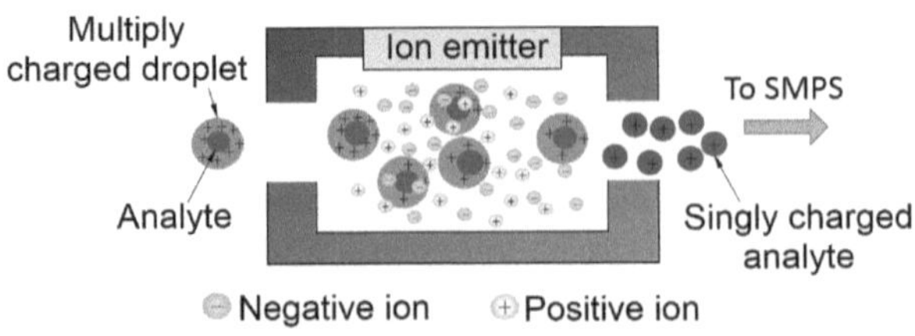

FIG. 8.8 Neutralization chamber.

TABLE 8.1 Characteristic sizes of measured bioparticles.

Particle	Size
Bacteriophage MS2	Head: 27.5 nm (Golmohammadi, Valegård, Fridborg, & Liljas, 1993)
Bacteriophages T2 and T4	Head: 86 × 119.5 nm; tail: 21 × 100 nm (Mesyanzhinov, 2004; Mesyanzhinov et al., 2004)
Bacteriophage λ	Head: 60 nm; tail: 15 × 135 nm (Casjens & Hendrix, 2015; Hendrix & Duda, 1992)
VLPs from nonenveloped MPV	~40 nm (Pease et al., 2009)
OVA	~6 nm (Erickson, 2009)
Escherichia coli	Diameter: 600–800 nm, length: 800–1200 nm (Jung, Lee, & Kim, 2009b)
Liposomes	50–150 nm (Chattopadhyay, Modesto-Lopez, Venkataraman, & Biswas, 2010)

The work of Hogan et al. (2006) reported on the use of an electrospray in the cone-jet mode to generate droplets of 170 nm from different sampling liquids, which contained bacteriophages MS2, T2, or T4 (Fig. 8.9). Using a Po-210 source to neutralize the droplets coupled with a commercial SMPS, they obtained a mobility diameter of bacteriophage MS2 of 24.13 nm (Fig. 8.10), which remained viable after the electrospray process. Conversely, Hogan and coworkers point out that bacteriophages T2 and T4 have lengths larger than the characteristic diameters of the electrospray jet and droplets generated under their experimental conditions, As a result, they report fragmentation of the bacteriophages at the virus capsid head-tail noncovalent interface during either jet formation or jet breakup processes. They found that the sizes of bacteriophage T4's capsid heads were larger than those of bacteriophage T2, with a mean peak mobility diameter of 88.32 nm for T4 and 87.03 nm for T2. Nevertheless, they were unable to detect tail and tail-fiber fragments by size spectrometric analyses. Hogan

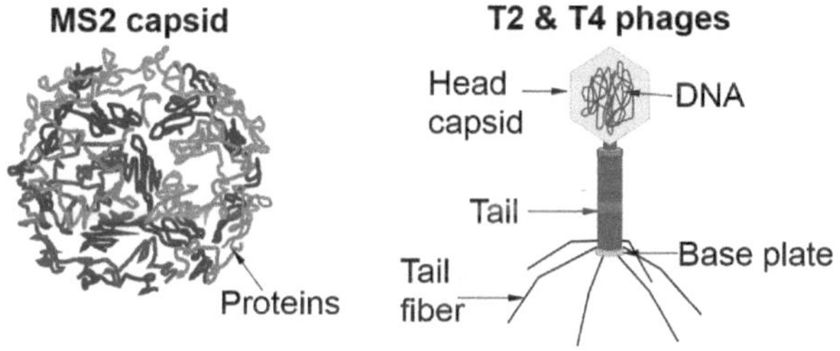

FIG. 8.9 Structures of MS2, T2, and T4 bacteriophages.

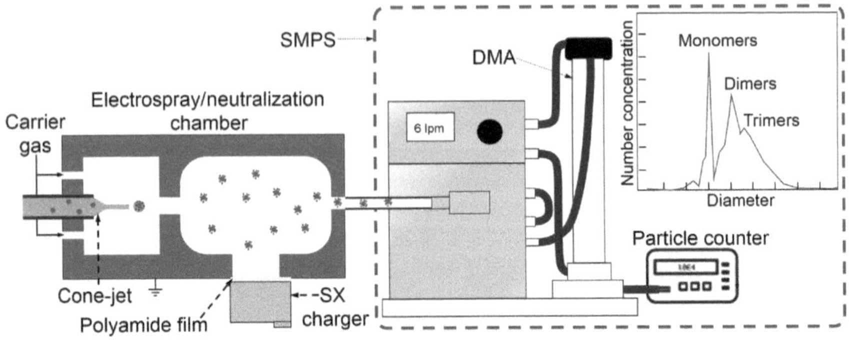

FIG. 8.10 Electrospray coupled with a SXC system. *Adapted from Modesto-Lopez, L. B., Kettleson, E. M., & Biswas, P. (2011). Soft X-ray charger (SXC) system for use with electrospray for mobility measurement of bioaerosols. Journal of Electrostatics, 69(4), 357–364.*

and collaborators argued that the contractile tails of bacteriophage T2 and T4 virions mechanically deform to a varying degree while confined within the smaller jet and droplets. Jung, Lee, and Kim (2009a) used an in-house built, charge-reduced electrospray system to generate nonagglomerated bacteriophage aerosols consisting of MS2 particles. The authors operated the electrospray in the cone-jet combined with two Po-210 sources aiming at neutralizing the electrosprayed droplets and thus prevent their breakup through Coulombic explosions. They also measured the particle size using an SMPS. Furthermore, the authors were able to operate the cone-jet mode for a range of liquid flow rates, 5–25 μL/h, and voltages, 8–10 kV, and report an initial droplet size of 100–200 nm. It is noteworthy to mention that Jung et al. (2009a) obtained a highly monodisperse size distribution of MS2 particles, with a mean mobility diameter of 23.8 nm, for the range of liquid flow rates and voltages tested, and were able to steadily sustain the cone-jet mode during the entire system operation (~1 h). Their measurements also agreed with the diameter obtained by Hogan et al. (2006) for MS2. Later on, Jung et al. (2010) used their electrospray-SMPS system to measure the size of *Escherichia coli* and reported a geometric mean mobility diameter of 779 nm. They concluded that nonagglomerated particles emitted via the cone-jet mode of electrospray contributed to improve quantitative analyses of bacteria viability.

Pease et al. (2009), similarly to the work of Hogan et al. (2006), used a commercial electrospray system operated in the cone-jet mode and a Po-210 source coupled with an SMPS system to quantitatively characterize virus-like particles derived from the nonenveloped murine polyomavirus (MPV). Their results, which were supported by transmission electron microscopy (TEM) images, evidenced the applicability of charge-reduced electrospray in size spectrometry. Pease and coworkers concluded that the electrospray-DMA system is a rapid and cost-effective technique to obtain statistically significant size distributions. They noted that analyses with the electrospray-DMA system were completed within an hour range, whereas each histogram from TEM images required at least a day to prepare. Furthermore, sampling characteristics of each technique differ significantly, since the electrospray-DMA system allowed to analyze of the order of 10^5 particles in comparison to the 200 of TEM. Therefore, the combination of speed and larger sample size of the electrospray-DMA system improved the ability to capture multimodal distributions and discern subtle shifts in size between VLPs. The rapid availability of results is key in situations where the human respiratory system is compromised, and has motivated studies of drug delivery via inhalation. For instance, aiming at elucidating the mechanism of controlled drug delivery into the respiratory system, Chattopadhyay et al. (2010) implemented a commercial electrospray in the cone-jet mode to aerosolize liposomes. Subsequently, they neutralized the liposomes on-line with a Po-210 source and sized them with an SMPS. Liposomes are self-assembled, hollow structures made of phospholipids, which have been explored as drug carriers for controlled release of insulin and various anticancer drugs to the lungs. Small-sized liposomes dissolve relatively rapid in the alveolar fluid, and may accelerate the internalization of the drug by epithelial cells. In their study, Chattopadhyay and colleagues explored the atomization of nanometer-sized liposomes via electrospray in the cone-jet mode as a tool for drug delivery. Their sampling liquids, contained round-shaped vesicles with sizes ranging from 50 nm up to 250 nm, as demonstrated by TEM analyses. However, they found that the liposomes underwent deformation as they passed through the cone-jet of the electrospray and their morphology changed to a rod-like shape. Such variation is in line with the mechanical deformation of T2 and T4 particles reported by Hogan et al. (2005) as the bacteriophages were confined within the jet. These works prove the electrospray, coupled with bipolar ion emitters, as reliable tool for size spectrometry analyses.

Nevertheless, because of safety regulations that prevent the use of radioactive isotopes, researchers have sought for alternative ion sources, including soft X-rays (SX) and corona discharge. Modesto-Lopez et al. (2011) developed, for the first time, an in-house electrospray-soft X-ray charger (SXC) system (Fig. 8.10) for use in electrospray size spectrometry of bioaerosols, thus providing an alternative to systems using radioactive neutralizers. They also compared the efficacy of the SXC as neutralizer of electrosprayed particles to bipolar ion emitters such as Po-210 and Kr-85 sources. In their approach, Modesto-López and coworkers used a tapered stainless-steel capillary of 160 μm (internal diameter) to generate droplets with size of the order of 2.5 μm, which contained the analyte particles. They used a commercial SXC (Hamamatsu Photonics) operating with an input potential of 12 V, which generates X-rays with an intensity of

3.5–9.5 keV and wavelength of 0.12–0.41 nm from a circular beryllium window of approximately 1.5 cm in diameter. The SXC has the capability to generate ion number densities ($10^{14} m^{-3}$) one and two orders of magnitude larger than the Po-210 and Kr-85 sources, respectively. Similarly, its ion production rate ($10^{16} m^{-3} s^{-1}$) is significantly greater than both sources (Po-210: $10^{15} m^{-3} s^{-1}$; Kr-85: $10^{13} m^{-3} s^{-1}$). Modesto-Lopez et al. (2011) measured the mobility diameter of 60-nm polystyrene latex (PSL) particles (as a reference), OVA protein, and MS2 particles with their electrospray-SXC system. For the PSL particles, they obtained a very narrow peak, as expected, corresponding to a monodisperse particle size distribution, thus indicating the correct functioning of their system. For the MS2 particles, they reported a mobility diameter of 24.1 nm, which agrees with the data reported by Hogan et al. (2006) and Jung et al. (2009a). In the case of OVA, Modesto-Lopez and coworkers were able to detect monomers, dimers, and trimers with sizes of 6.38, 7.91, and 8.82 nm, respectively, in a similar fashion as the results of Kaufman (1998) for other albumin complexes. Furthermore, the use of a Po-210 source for neutralizing electrosprayed particles resulted in a particle loss of 42%, 12.7%, and 23% for PSL, MS2, and OVA particles, respectively, compared to the SXC. Particle loss was significantly higher when a Kr-85 source was used. They demonstrated that photoionization with an SXC was equally effective in neutralizing multiply charged particles, in the nanometer-size range, as the commonly used Po-210 diffusion neutralizer. In addition, the SXC has an external controller, which allows for switching on/off the device upon demand, thus minimizing risk exposure. Later on, Clouet-Foraison et al. (2017) applied a commercial technology as that developed by Modesto-Lopez et al. (2011) to quantify nonhigh-density-lipoprotein particle concentration using an electrospray-SXC system coupled with an SMPS.

8.5 Concluding remarks

A comprehensive review of the physics of liquid ejection and droplet formation with electrosprays in the cone-jet mode has been provided. In addition, scaling laws determining the characteristic droplet size ejected in such mode within different scenarios were discussed in details. Furthermore, a comprehensive discussion of the electrokinetics of electrosprays and their implication on the evolution of charged species was also included. Finally, high-added value applications of electrosprays, related to the field of analytical chemistry were presented. In particular, the capability of the electrospray as enabling tool to generate micro- and nanodroplets containing biological analytes and its coupling with aerosol-based instrumentation for online real-time analyses of viruses and proteins were reviewed.

References

Bacher, G., Szymanski, W. W., Kaufman, S. L., Zllner, P., Blaas, D., & Allmaier, G. (2001). Charge-reduced nano electrospray ionization combined with differential mobility analysis of peptides, proteins, glycoproteins, noncovalent protein complexes and viruses. *Journal of Mass Spectrometry*, *36*(9), 1038–1052.

Borra, J.-P. (2018). Review on water electro-sprays and applications of charged drops with focus on the corona-assisted cone-jet mode for high efficiency air filtration by wet electro-scrubbing of aerosols. *Journal of Aerosol Science*, *125*, 208–236.

Casjens, S. R., & Hendrix, R. W. (2015). Bacteriophage lambda: Early pioneer and still relevant. *Virology*, *479–480*, 310–330.

Cech, N. B., & Enke, C. G. (2001). Practical implications of some recent studies in electrospray ionization fundamentals. *Mass Spectrometry Reviews*, *20*(6), 362–387.

Chattopadhyay, S., Modesto-Lopez, L. B., Venkataraman, C., & Biswas, P. (2010). Size distribution and morphology of liposome aerosols generated by two methodologies. *Aerosol Science and Technology*, *44*(11), 972–982.

Chen, D.-R., Pui, D. Y. H., & Kaufman, S. L. (1995). Electrospraying of conducting liquids for monodisperse aerosol generation in the 4 nm to 1.8 μm diameter range. *Journal of Aerosol Science*, *26*, 963–977.

Clouet-Foraison, N., Gaie-Levrel, F., Coquelin, L., Ebrard, G., Gillery, P., & Delatour, V. (2017). Absolute quantification of bionanoparticles by electrospray differential mobility analysis: An application to lipoprotein particle concentration measurements. *Analytical Chemistry*, *89*(4), 2242–2249.

Cloupeau, M., & Prunet-Foch, B. (1989). Electrostatic spraying of liquids in cone-jet mode. *Journal of Electrostatics*, *22*, 135–159.

Collins, R. T., Jones, J. J., Harris, M. T., & Basaran, O. A. (2008). Electrohydrodynamic tip streaming and emission of charged drops from liquid cones. *Nature Physics*, *4*, 149–154.

Collins, R. T., Sambath, K., Harris, M. T., & Basaran, O. A. (2013). Universal scaling laws for the disintegration of electrified drops. *PNAS*, *110*, 4905–4910.

Erickson, H. P. (2009). Size and shape of protein molecules at the nanometer level determined by sedimentation, gel filtration, and electron microscopy. *Biological Procedures Online*, *11*(1), 32–51.

Fenn, J. B., Mann, M., Meng, C. K., Wong, S. F., & Whitehouse, C. M. (1989). Electrospray ionization for mass spectrometry of large biomolecules. *Science*, *246*, 64–71.

Fernández de la Mora, J. (2007). The fluid dynamics of Taylor cones. *Annual Review of Fluid Mechanics*, *39*, 217–243.

Fernandez de la Mora, J., & Loscertales, I. G. (1994). The current transmitted through an electrified conical meniscus. *Journal of Fluid Mechanics*, *260*, 155–184.

Flagan, R. C. (2008, March). Differential mobility analysis of aerosols: A tutorial. *Kona Powder and Particle Journal*, *26*, 254–268.

Forsyth, B., Liu, B. Y. H., & Romay, F. J. (1998). Particle charge distribution measurement for commonly generated laboratory aerosols. *Aerosol Science and Technology*, *28*(6), 489–501.

Gamero-Castaño, M. (2008a). Characterization of the electrosprays of 1-ethyl-3-methylimidazolium bi(trifluoromethylsulfonyl)imide in vacuum. *Physics of Fluids*, *20*. 0321031-11.

Gamero-Castaño, M. (2008b). The structure of electrospray beams in vacuum. *Journal of Fluid Mechanics*, *604*, 339–368.

Gamero-Castaño, M., & Hruby, V. (2002). Electric measurements of charged sprays emitted by cone-jets. *Journal of Fluid Mechanics*, *459*, 245–276.

Gañán-Calvo, A. M. (1998). Generation of steady liquid microthreads and micron-sized monodisperse sprays in gas streams. *Physical Review Letters*, *80*, 285–288.

Gañán-Calvo, A. M. (1999). The surface charge in electrospraying: Its nature and its universal scaling laws. *Journal of Aerosol Science, 30*(7), 863–872.
Gañán-Calvo, A. M. (2004). On the general scaling theory for electrospraying. *Journal of Fluid Mechanics, 507*, 203–212.
Gañán-Calvo, A. M., Barrero, A., & Pantano, C. (1993). The electrohydrodynamics of electrified conical menisci. *Journal of Aerosol Science, 24*, S19–S20.
Gañán-Calvo, A. M., Dávila, J., & Barrero, A. (1997). Current and droplet size in the electrospraying of liquids. Scaling laws. *Journal of Aerosol Science, 28*, 249–275.
Gañán-Calvo, A. M., López-Herrera, J. M., Rebollo-Muñoz, N., & Montanero, J. M. (2016). The onset of electrospray: The universal scaling laws of the first ejection. *Science Reports, 6*(1), 32357.
Gañán-Calvo, A. M., & Montanero, J. M. (2009). Revision of capillary cone-jet physics: Electrospray and flow focusing. *Physical Review E, 79*, 066305.
Gañán-Calvo, A. M., Pérez-Saborid, M., López-Herrera, J. M., & Gordillo, J. M. (2004). Steady high viscosity liquid micro-jet production and fiber spinning using co-flowing gas conformation. *European Physical Journal B: Condensed Matter and Complex Systems, 39*, 131–137.
Gañán-Calvo, A. M., Rebollo-Muñoz, N., & Montanero, J. M. (2013). Physical symmetries and scaling laws for the minimum or natural rate of flow and droplet size ejected by Taylor cone-jets. *New Journal of Physics, 15*, 033035.
Gañán-Calvo, A. M., López-Herrera, J. M., Herrada, M. A., Ramos, A., & Montanero, J. M. (2018). Review on the physics of electrospray: From electrokinetics to the operating conditions of single and coaxial Taylor cone-jets, and AC electrospray. *Journal of Aerosol Science, 125*, 32–56.
Garoz, D., Bueno, C., Larriba, C., Castro, S., Romero-Sanz, I., Mora, J. D. L., … Saito, G. (2007). Taylor cones of ionic liquids from capillary tubes as sources of pure ions: The role of surface tension and electrical conductivity. *Journal of Applied Physics, 102*(6), 064913.
Golmohammadi, R., Valegård, K., Fridborg, K., & Liljas, L. (1993). The refined structure of bacteriophage MS2 at 2 · 8 Å resolution. *Journal of Molecular Biology, 234*(3), 620–639.
Gomez, A., & Tang, K. (1994). Charge and fission of droplets in electrostatic sprays. *Physics of Fluids, 6*(1), 404–414.
Hartman, R. P. A., Brunner, D. J., Camelot, D. M. A., Marijnissen, J. C. M., & Scarlett, B. (1999). Electrohydrodynamic atomization in the cone-jet mode physical modeling of the liquid cone and jet. *Journal of Aerosol Science, 30*, 823–849.
Hayati, I., Bailey, A. I., & Tadros, T. F. (1987a). Investigations into the mechanism of electrohydrodynamic spraying of liquids: I effect of electric field and the environment on pendant drops and factors affecting the formation of stable jets and atomization. *Journal of Colloid and Interface Science, 117*, 202–221.
Hayati, I., Bailey, A. I., & Tadros, T. F. (1987b). Investigations into the mechanism of electrohydrodynamic spraying of liquids: II mechanism of stable jet formation and electrical forces acting on a liquid cone. *Journal of Colloid and Interface Science, 117*, 222–230.
Hendrix, R. W., & Duda, R. L. (1992). Bacteriophage λ PaPa: Not the mother of all λ phages. *Science, 258*(5085), 1145–1148.
Hogan, C., Kettleson, E., Lee, M.-H., Ramaswami, B., Angenent, L., & Biswas, P. (2005). Sampling methodologies and dosage assessment techniques for submicrometre and ultrafine virus aerosol particles. *Journal of Applied Microbiology, 99*(6), 1422–1434.
Hogan, C. J., Kettleson, E. M., Ramaswami, B., Chen, D. R., & Biswas, P. (2006). Charge reduced electrospray size spectrometry of mega- and gigaDalton complexes: Whole viruses and virus fragments. *Analytical Chemistry, 78*(3), 844–852.
Jaworek, A., & Krupa, A. (1999). Jet and drops formation in electrohydrodynamic spraying of liquids. A systematic approach. *Experiments in Fluids, 27*(1), 43–52.
Jaworek, A., & Sobczyk, A. (2008). Electrospraying route to nanotechnology: An overview. *Journal of Electrostatics, 66*(3-4), 197–219.

Jung, J. H., Lee, J. E., Hwang, G. B., Lee, B. U., Lee, S. B., Jurng, J. S., & Bae, G. N. (2010). Electrospray-assisted ultraviolet aerodynamic particle sizer spectrometer for real-time characterization of bacterial particles. *Analytical Chemistry, 82*(2), 664–671.

Jung, J. H., Lee, J. E., & Kim, S. S. (2009a). Generation of nonagglomerated airborne bacteriophage particles using an electrospray technique. *Analytical Chemistry, 81*(8), 2985–2990.

Jung, J. H., Lee, J. E., & Kim, S. S. (2009b). Thermal effects on bacterial bioaerosols in continuous air flow. *The Science of the Total Environment, 407*(16), 4723–4730.

Kaufman, S. L. (1998). Analysis of biomolecules using electrospray and nanoparticle methods. *Journal of Aerosol Science, 29*(5–6), 537–552.

Knutson, E., & Whitby, K. (1975). Aerosol classification by electric mobility: Apparatus, theory, and applications. *Journal of Aerosol Science, 6*(6), 443–451.

Kousaka, Y., Okuyama, K., Adachi, M., & Mimura, T. (1986). Effect of Brownian diffusion on electrical classification of ultrafine aerosol particles in differential mobility analyzer. *Journal of Chemical Engineering of Japan, 19*(5), 401–407.

López-Herrera, J. M., Barrero, A., Boucard, A., Loscertales, I. G., & Márquez, M. (2004). An experimental study of the electrospraying of water in air at atmospheric pressure. *Journal of the American Society for Mass Spectrometry, 15*, 253–259.

Lozano, P., & Martínez-Sánchez, M. (2005). On the dynamic response of externally wetted ionic liquid ion sources. *Journal of Physics D: Applied Physics, 38*, 2371–2377.

Machala, Z., Marode, E., Laux, C., & Kruger, C. (2004). DC glow discharges in atmospheric pressure air. *Journal of Advanced Oxidation Technologies, 7*(2), 133–137.

Madigan, M. T., Bender, K. S., Buckley, D., Sattley, W. M., & Stahl, D. A. (2015). *Brock biology of microorganisms*. New York: Pearson.

Martin, S., Perea, A., Garcia-Ybarra, P. L., & Castillo, J. L. (2012). Effect of the collector voltage on the stability of the cone-jet mode in electrohydrodynamic spraying. *Journal of Aerosol Science, 46*, 53–63.

Mesyanzhinov, V. V. (2004). Bacteriophage T4: Structure, assembly, and initiation infection studied in three dimensions. *Advances in Virus Research, 63*, 287–352.

Mesyanzhinov, V. V., Leiman, P. G., Kostyuchenko, V. A., Kurochkina, L. P., Miroshnikov, K. A., Sykilinda, N. N., & Shneider, M. M. (2004). Molecular architecture of bacteriophage T4. *Biochemistry, 69*(11), 1190–1202.

Modesto-López, L. B., & Biswas, P. (2010). Role of the effective electrical conductivity of nanosuspensions in the generation of TiO_2 agglomerates with electrospray. *Journal of Aerosol Science, 41*(8), 790–804.

Modesto-Lopez, L. B., Kettleson, E. M., & Biswas, P. (2011). Soft X-ray charger (SXC) system for use with electrospray for mobility measurement of bioaerosols. *Journal of Electrostatics, 69*(4), 357–364.

Modesto-López, L. B., Thimsen, E. J., Collins, A. M., Blankenship, R. E., & Biswas, P. (2010). Electrospray-assisted characterization and deposition of chlorosomes to fabricate a biomimetic light-harvesting device. *Energy & Environmental Science, 3*(2), 216–222.

Montanero, J. M., & Gañán-Calvo, A. M. (2020). Dripping, jetting and tip streaming. *Reports on Progress in Physics, 83*(9), 097001.

Mori, Y., & Young, Y.-N. (2018). From electrodiffusion theory to the electrohydrodynamics of leaky dielectrics through the weak electrolyte limit. *Journal of Fluid Mechanics, 855*, 67–130.

Pantano, C., Gañán-Calvo, A. M., & Barrero, A. (1994). Zeroth-order, electrohydrostatic solution for electrospraying in cone-jet mode. *Journal of Aerosol Science, 25*, 1065–1077.

Pease, L. F., Lipin, D. I., Tsai, D.-H., Zachariah, M. R., Lua, L. H., Tarlov, M. J., & Middelberg, A. P. (2009). Quantitative characterization of virus-like particles by asymmetrical flow field flow fractionation, electrospray differential mobility analysis, and transmission electron microscopy. *Biotechnology and Bioengineering, 102*(3), 845–855.

Pillai, R., Berry, J. D., Harvie, D. J. E., & Davidson, M. R. (2016). Electrokinetics of isolated electrified drops. *Soft Matter, 12*, 3310–3325.

Reneker, D. H., & Yarin, A. L. (2008). Electrospinning jets and polymer nanofibers. *Polymer, 49*, 2387–2425.
Rosell-Llompart, J., & de la Mora, J. F. (1994). Generation of monodisperse droplets 0.3 to 4 micrometre in diameter from electrified cone-jets of highly conducting and viscous liquids. *Journal of Aerosol Science, 25*, 1093–1119.
Rosell-Llompart, J., Grifoll, J., & Loscertales, I. G. (2018). Electrosprays in the cone-jet mode: From Taylor cone formation to spray development. *Journal of Aerosol Science, 125*, 2–31.
Saville, D. A. (1997). Electrohydrodynamics: The Taylor-Melcher leaky dielectric model. *Annual Review of Fluid Mechanics, 29*, 27–64.
Schnitzer, O., & Yariv, E. (2015). The Taylor–Melcher leaky dielectric model as a macroscale electrokinetic description. *Journal of Fluid Mechanics, 773*, 1–33.
Taylor, G. (1964). Disintegration of water drops in electric field. *Proceedings of the Royal Society of London A, 280*, 383–397.
Velásquez-García, L., Akinwande, A., & Martínez-Sánchez, M. (2006). A planar array of micro-fabricated electrospray emitters for thruster applications. *Journal of Microelectromechanical Systems, 15*(5), 1272–1280.
Whitehouse, C. M., Dreyer, R. N., Yamashita, M., & Fenn, J. B. (1985). Electrospray interface for liquid chromatographs and mass spectrometers. *Analytical Chemistry, 57*, 675–679.
Wiedensohler, A. (1988). An approximation of the bipolar charge distribution for particles in the submicron size range. *Journal of Aerosol Science, 19*(3), 387–389.

CHAPTER

9

Ultrasonic nebulizers

Krzysztof Jankowski

Faculty of Chemistry, Department of Analytical Chemistry, Warsaw University of Technology, Warsaw, Poland

9.1 Introduction

With pneumatic nebulizers, only a relatively small fraction of the sample solution is converted to the very fine aerosol that ultimately reaches an excitation/ionization source. Alternatively, sound can be used instead of a gas as the energy source for converting a liquid to an aerosol. The driving force for this generation is an ultrasonic beam generated by a piezoelectric transducer. These devices produce small droplets (below 5 μm) with high sphericity and uniform size distribution compared to those produced by pneumatic nebulizers (Barreras, Amaveda, & Lozano, 2002; Kooij, Astefanei, Corthals, & Bonn, 2019; Kudo, Sekiguchi, Sankoda, Namiki, & Nii, 2017; Rajan & Pandit, 2001).

Ultrasonic nebulizers (USNs) incorporate a piezoelectric plate vibrating at high frequencies (100 kHz–10 MHz) in order to produce aerosol with mean drop diameters that appear to be a function of exciting frequency. In the exemplary measurement, for the frequency of 1.65 MHz, the droplet size distribution function presents a main peak between 3 and 5 μm.

Due to their high efficiency, USNs have the advantage of generating high aerosol loads with a greater fraction of primary aerosol delivered to the excitation/ionization source. Ultrasonic nebulizers improve analyte transport compared with pneumatic devices, but usually require partial or total solvent removal. Aerosol desolvation, using a heated section followed by a condenser, not only reduces the solvent loading in the plasma, but also increases the analyte transport relative to transport without desolvation. The high efficiency of aerosol generation results in improved sensitivity and lower limits of detection, provided that aerosol desolvation is used to reduce the solvent load reaching the plasma. Limits of detection

Analytical Nebulizers
https://doi.org/10.1016/B978-0-323-91181-8.00005-9

achieved with USNs are superior, by up to a factor of 10, to those obtained with a conventional setup (Castillano, Vela, Caruso, & Story, 1992).

Moreover, aerosol formation is primarily independent of the nebulizer gas flow rate. This permits USN interfacing with low power, low gas flow-rate excitation/ionization sources including microwave plasmas and miniaturized discharges for OES and ambient MS (Jankowski, Karmasz, Ramsza, & Reszke, 1997; Pintabona, Astefanei, Corthals, & van Asten, 2019; Song, You, Perdomo, & Evans-Nguyen, 2020).

Unfortunately, USNs, in general, also suffer from some limitations. Most conventional USN systems tend to be large and complicated to use. Moreover, this type of nebulizer is very sensitive to the sample matrix concentration. Both matrix effects and memory effects are well documented in USN including larger washout times due to analyte binding to the heating chamber walls. Spectrometric results produced with USN show relatively poor long-term and short-term stability due to heat generation by the transducer. Finally, USNs are significantly more expensive than the pneumatic designs (Budic, 2001; Masson, Vives, Orignac, & Prunet, 2000).

9.2 Design and fundamentals

9.2.1 Droplet formation mechanism

There are two widely accepted mechanisms of liquid disintegration during USN: cavitation mechanism and capillary wave mechanism. The former occurs when a liquid film is sonicated, microbubbles are formed under the liquid film, and which eventually collapse generating droplets. According to the second, liquid capillary waves are formed by unstable oscillations and the peaks of these waves are subsequently torn away from the bulk liquid producing droplets. However, further research showed that a combination of these two mechanisms is the source of aerosol formation. Nevertheless, these explanations are still subject to controversy (Kooij et al., 2019; Kudo et al., 2017; Rajan & Pandit, 2001).

At high frequencies (MHz level) and high power level (tens of Watts), cavitation is the main mode. At the high frequencies commonly used in modem USNs, typically 1 MHz or greater, the mechanism of aerosol production changes from cavitation to capillary wave formation due to the relatively low power applied. With the capillary wave mechanism, power density in the liquid surface, rather than the operating frequency, becomes of major importance (Kudo et al., 2017).

As a result of the interaction between the ultrasonic waves and the liquid film, a very fine aerosol of a low droplet velocity is obtained. At small liquid flow rates, a regular liquid layer of a few millimeter diameter is formed in the center of the vibrating plate, from which aerosol droplets are ejected perpendicularly to the free liquid film surface. When liquid is delivered onto the surface of the transducer as a flowing stream, a thin liquid film is formed, the thickness of which varies from 80 to 300 μm depending on the liquid flow rate. The maintaining of a regular liquid layer in the center of the plate is of essential importance for the aerosol flow rate stability, since thin film geometries led to smaller droplets and higher nebulization rates compared with that obtained using a set of individual sessile droplets as the source liquid geometry. The liquid film thickness effects no longer exist when a critical value of about 60 μm is reached (Collins et al., 2012; Kooij et al., 2019; Kudo et al., 2017; Rajan & Pandit, 2001).

In contrast, at high liquid sample flow rates (1 mL min^{-1} or more), a pronounced fraction of large droplets can be formed. The observed sample flow rate effects may be due to the ultrasonic cavitation phenomenon, which occurs when the liquid film thickness grows.

9.2.2 Prediction of droplet size

To predict droplet sizes produced with USN, experimental correlations have been proposed (Kooij et al., 2019), which account to the effects of both a liquid properties including flow rate, viscosity, density, and surface tension, and ultrasonic properties like amplitude, frequency, and the area of vibrating surface. Under thin film conditions, the droplet size is a function of the frequency of the ultrasound wave and the droplet diameter decreases by an increase of the working frequency, as shown in the following equation:

$$d_p = 0.34 \cdot \left(\frac{8\pi\sigma}{\rho_{liq}F^2}\right)^{1/3} \tag{9.1}$$

where d_p is the droplet size (cm), σ the surface tension (dyne cm^{-1}), F the forcing sound frequency (Hz), and ρ_{liq} the liquid density (g mL^{-1}).

At high sample flow rates, another empirical equation (Eq. 9.2) to predict the droplet size (d_p) has been proposed (Rajan & Pandit, 2001):

$$d_p = 31.7\left(\frac{\sigma}{\rho_{liq}}\right)^{0.354} \eta^{0.303}\, Q^{0.139} \tag{9.2}$$

where η is a liquid viscosity (cP) and Q is a volumetric liquid flow rate (L min^{-1}).

When applying aerosol desolvation (Kuo, Chan, Lin, Huang, & Chen, 2019), the resulting dry particle diameter, d_{dry}, depends on the volume fraction of the solute, F_v, and the droplet diameter, d_p, according to the following equation:

$$d_{dry} = d_p(F_v)^{1/3} \tag{9.3}$$

Alternatively, the expected dry particle size (Rodes, Smith, Crouse, & Ramachandran, 1990) can be calculated from the following equation:

$$d_{dry} = \left(\frac{K\sigma C_w}{\rho_{sol}\rho_{liq}F^2} \right)^{1/3} \tag{9.4}$$

where K is a constant that equals 9.877×10^{-7}, C_w is the liquid dissolved solid content ($mg\,L^{-1}$), and ρ_{sol} is the composite dry density of dissolved solids ($g\,cm^{-3}$).

9.2.3 Typical designs of USNs and desolvation systems

A variety of USN arrangements have been developed for use in analytical spectrometry, depending on a specific application (Al-Jumaily & Meshkinzar, 2017). A majority of the designs now in use can be defined as "high-flow" nebulizers as they use about 1 mL of analyte per minute. With these nebulizers, improvements in limits of detection up to 10-fold are achieved when they are coupled to a desolvation system. The use of higher liquid flow rates results in both an increase in the droplet size and a decrease in the nebulization efficiency. At sample flow rates of over $100\,\mu L\,min^{-1}$ worsening of plasma stability is observed due to the too-large water loading to the plasma. In turn, ultrasonic micronebulizers are capable of operating at sample flow rates down to $5\,\mu L\,min^{-1}$. Importantly, at small analyte flows it is not necessary to use any desolvation unit (Tarr, Zhu, & Browner, 1993).

The components of a conventional high-flow USN are an ultrasonic transducer, carrier gas inlet, sample inlet for introducing liquid sample across an oscillating piezoelectric transducer, spray chamber, drain, cooling medium inlet and outlet, and desolvation system (Fig. 9.1). Both the excess sample from the spray chamber and the condensed sample solvent from the condenser are removed using an integrated drain pump.

A material characterized by a high acoustic transformation coefficient is necessary for transducer construction. A piezoelectric plate, commonly made of lead zirconate titanate, barium titanate, or quartz, undergoes mechanical deformation when an electric field is applied. Piezoelectric devices made of ceramic material operate effectively over a wide range of frequencies, from 30 kHz to 10 MHz. Lead titanate and lead zirconate transducers use less power than barium titanate transducers and withstand better at higher temperatures.

Unfortunately, continuous-flow USNs are relatively bulky, transducer-position sensitive, and generate considerable heat during operation. A transducer plate of about 20 mm diameter is recommended to operate within 15 degrees of the vertical to enable free outflow of excess liquid to the drain. Transducer orientation is critical for key nebulizer

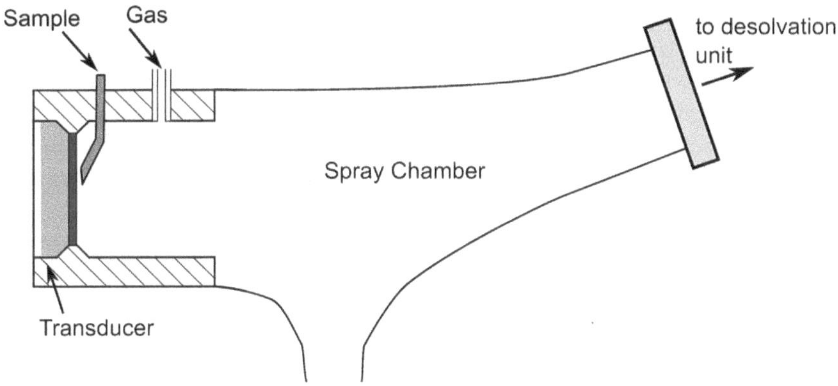

FIG. 9.1 Conventional high-flow USN device. *No permission required.*

operational parameters including wash-out times, aerosol density stability, and the mean droplet diameter, thus affecting the reproducibility of determinations and limits of detection.

A common feature in the design of most USNs is the use of a water-cooled system positioned behind the transducer. To improve nebulization stability and to avoid overheating of the transducer, it is cooled with water at room temperature. More heat is produced when the megahertz frequency is applied. Overheating the transducer results in its diminished lifetime or complete loss of piezoelectric properties leading to entire nebulizer failure (Castillano et al., 1992; Jankowski et al., 1997; Tarr et al., 1993).

Another major consideration of USN design is the protection of the transducer working surface. Three main protective strategies are employed: gas or liquid cooling, use of protective film or bonded chemically resistant plate, acoustically matched to the resonant frequency. For a liquid cooling system, the use of deionized, degassed water circulating by means of a closed circulator bath is preferred. Gas cooling is less efficient due to the low acoustic impedance of air or nitrogen. Alternatively, ultrasonic waves are propagated from a transducer through the cooling medium to the separate quartz plate. Here, the aerosol temperature in the spray chamber does not exceed 30°C, and thus the increase in solvent loading to the plasma caused by vaporization does not seem to be significant.

In a number of USN systems, the desolvation unit is utilized to obtain a dry aerosol. By various heating devices, the primary aerosol is converted into a tertiary aerosol. Next, solvent overloading is minimized by excess vapor removal using cryocondensers or membrane interfaces to desolvate both aqueous and volatile organic solvents (Castillano et al., 1992; Tarr et al., 1993; Grotti, Lagomarsino, & Frache, 2004). In general, desolvation systems can be divided in two groups: those which employ a solvent load reduction in a thermostated spray chamber and those which employ a heating step for improving solvent vaporization followed by a vapor removal step. In the first case only solvent load reduction is possible, whereas when using two-step desolvation systems the analyte transport rate is also increased.

Aerosol heating can take place in the spray chamber or in a glass extension tube placed just at the exit of the spray chamber. Convective heating is a common solution. Yet, aerosol heating can also be done by absorption of radiation, such as infrared or microwave radiation (Anderlini & Beauchemin, 2018; Grindlay, Maestre, Mora, Hernandis, & Gras, 2005; Grotti et al., 2004; Tarr et al., 1993). Aerosol radiative heating is quicker than resistive heating. Microwave radiation seems to be a more uniform way of aerosol desolvation than conductive heating due to avoidance of the problems associated with overheating of the tube walls, since they would remain relatively cold as compared to the aerosol.

Vapor removal is usually carried out by rapid condensation on the cold surfaces of cryocondensers. Alternatively or sometimes in series, humid transport gas passed through a membrane desolvator and solvent vapor is removed using a semipermeable membrane. Membrane desolvator provides enhanced desolvation by selective permeation and counter gas purging (Fig. 9.2). However, for membrane desolvators, vapor removal rate is not as high as that of condensation systems (Brenner, Zhu, & Zander, 1996).

Aerosol desolvation is a common method for overcoming spectral and nonspectral interferences caused by the presence of solvent in the excitation/ionization source and for improving the analyte transport rate when introducing liquid samples. Interferences are reduced for ICP-OES and ICP-MS, lowering analyte limits of detection. Yet, the distance covered

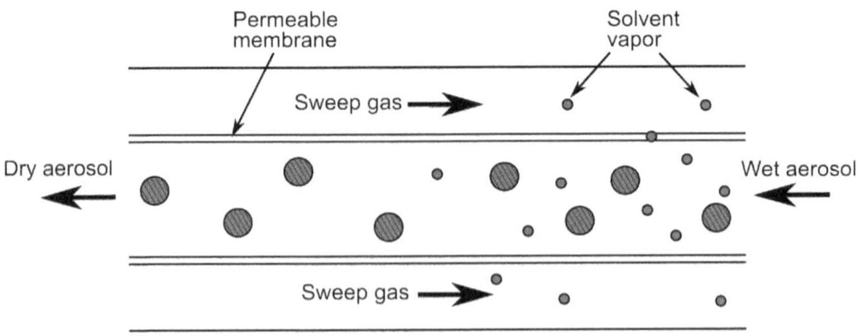

FIG. 9.2 A principle of membrane desolvator. *No permission required.*

by the aerosol to reach the discharge tube considerably lengthens, which results in an increase in memory effects. Conversely, USN without desolvation unit exhibits less memory effects and allows for rapid sample changes (Anderlini & Beauchemin, 2018; Brenner et al., 1996; Budic, 2001; Masson et al., 2000).

More recently, a new miniature ultrasonic nebulization approach for liquid microsample delivery based on surface acoustic waves (SAWs) has been developed. Operating at megahertz level frequencies, SAWs are nanometer-order amplitude sound waves that propagate along the surface of a chip-scale piezoelectric substrate patterned with an interdigitated transducer (ITD) (Fig. 9.3). The SAW nebulizer (SAWN) planar chip is positioned several centimeters below the orifice and centered on the MS inlet. The SAW device on a lithium niobate piezoelectric transducer requires only a low operating power of 1–5 W and, therefore, provides a viable and efficient ionization source for the real-time analysis of a wide range of compounds. Unlike conventional ultrasound where the sound waves propagate through the entire bulk of a medium, most of the acoustic energy associated with the SAW is induced on the horizontal free surface of a liquid film. The SAW is therefore an extremely efficient energy transfer mechanism from the piezoelectric substrate to the liquid with linear relationship between the droplet diameter and the mechanical power coupled into the liquid. SAW nebulizers and other micronebulizers are built without a desolvation unit, unlike most high-flow USN

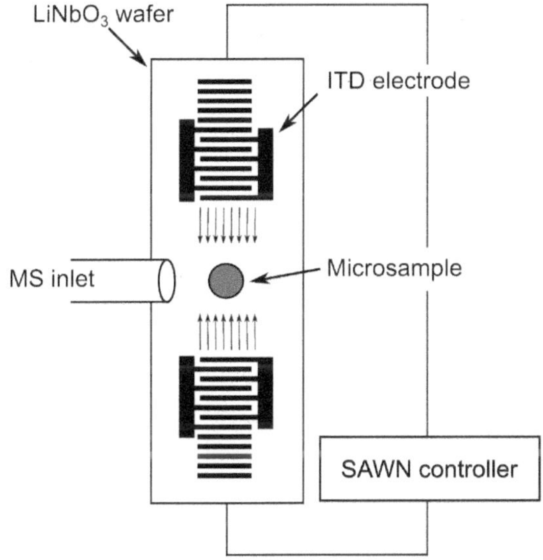

FIG. 9.3 Schematic presentation of SAW nebulizer principle. *No permission required.*

systems (Collins et al., 2012; Heron, Wilson, Shaffer, Goodlett, & Cooper, 2010).

SAW-based nebulizers have considerable advantages over other USNs, in that they use significantly less power (a few Watts) and generate droplets within the size range of 1–5 μm and of a greater initial velocity. Furthermore, the typically high frequencies and low powers required for SAW nebulization do not result in cavitation-induced molecular damage, showing almost negligible denaturation of large biomolecules such as DNA, peptides, or proteins (Heron et al., 2010), which is particularly crucial for sample delivery in ambient mass spectrometry.

The next brand-new, advanced USNs based on a vibrating perforated mesh or array of micro-orifices offer new perspectives for localized and continuous sample delivery. These nebulizers use the same piezoelectric element as conventional USNs but are smaller and provide a more uniform droplet size defined by the dimension of the orifices (Fig. 9.4). Yet, vibrating mesh nebulizers suffer from clogging of the mesh orifices (Hayashi, Furuzawa, Tanaka, & Hiraide, 2004; Kuo et al., 2019).

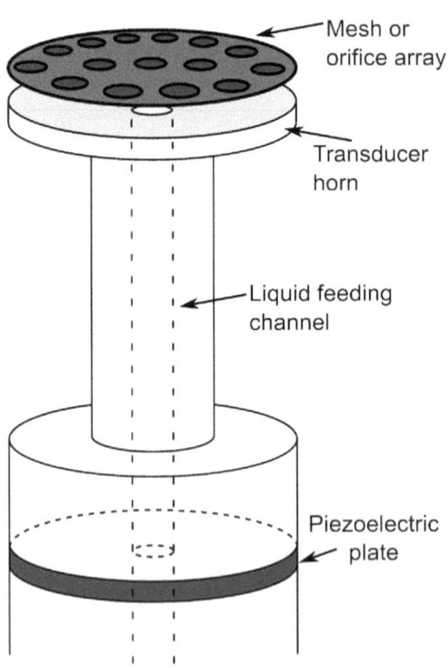

FIG. 9.4 A mesh USN design. *No permission required.*

9.2.4 Additional effects of high aerosol loading

Generating a primary aerosol with a very small mean drop size is an alternative to conventional nebulization, as such fine aerosol drops should pass to the plasma with high efficiency. This results in higher ionization/excitation rates, higher sensitivity, and reduced matrix effects. In spite of the aforementioned features and advantages of the USNs, direct aerosol introduction into low-power discharges generally results in deleterious effects on the plasma characteristics including reduction of atomization efficiency and lowering of plasma temperature, because the robustness of the discharge is insufficient to accept high solvent loading. These phenomena arise because significant amounts of plasma energy are used for vaporization of the solvent. For low-pressure ICP and microwave or glow discharges, the limited discharge capacity for large gas flows and high vapor loads has to be taken into account and aerosol desolvation is required. As a result, atomic and ionic lines with energies above the bond energies of water are markedly enhanced. Ion—atom intensity ratios and excitation and ionization temperatures are also affected by aerosol desolvation (Borkowska-Burnecka, Leśniewicz, & Żyrnicki, 2006; Budic, 2001; Masson et al., 2000).

Yet, the analytical performance can be degraded by the higher salt loading from the sample matrix. Using USN in comparison to the PN, the water transport to the plasma is several times lower while the matrix transport is even 10 times higher. It may lead to lowering in plasma robustness and promotes matrix effects of different origin.

9.3 Applications

9.3.1 Aqueous solution nebulization for plasma OES and MS

The availability of stable, reliable USNs for plasma optical emission and mass spectrometry has led to numerous applications (Table 9.1).

With a USN, a major increase in analyte transport compared with a PN causes the enhancement of line intensities and a decrease in limits of detection. Also, increased sensitivity is observed as the signal-to-background ratio is higher for USNs compared to pneumatic nebulizers. The robustness of the plasma in the axial and radial mode for aerosol injection using USN is significantly lower in comparison with conventional nebulization. The differences in analytical figures of merit obtained at the individual nebulizers are related to various amounts of solvent loaded to the plasma. Water aerosol selectively reduces the intensities of atomic and ionic lines. Generally, matrix effects were

TABLE 9.1 Selected applications of USNs in analytical spectrometry.

Nebulizer	Desolvation	Spectrometric technique	Application	Reference
Commercial high-flow USN	Electrothermal desolvation system	ICP-OES	REEs in geological materials	Brenner and Dorfman (1993)
Commercial high-flow USN	Electrothermal +membrane desolvation system	ICP-OES	Elements in wine	de Andrade et al. (2020)
Commercial high-flow USN	Electrothermal heating + Peltier cooling system	ICP-OES	Elements in rye grass	Hoenig, Doc; ekalová, and Baeten (1998)
Commercial high-flow USN	Electrothermal +membrane desolvation system	ICP-MS	Bromine in water	Kahen, Jorabchi, and Montaser (2006)
Commercial high-flow USN	Electrothermal +membrane desolvation system	ICP-MS (sector field MS)	Elements in snow	Bensimon, Bourquin, and Parriaux (2000)
Commercial high-flow USN	Electrothermal +membrane desolvation system	ICP-OES	Elements in petrochemicals	Botto and Zhu (1994)
Commercial high-flow USN	Electrothermal desolvation system	ICP-MS	Elements in crude oil	Duyck, Miekeley, Porto da Silveira, and Szatmari (2002)
Purpose-built USN, 1.7 MHz	No desolvation	LIBS	Metals in water	Zhong, Zheng, Lu, Cheng, and Xiu (2015)
Commercial geyser-type USN, 2.4 MHz	Electrothermal +membrane desolvation system	LIBS	Elements in drinking and river water	Aras, Yeşiller, Ateş, and Yalçin (2012)

TABLE 9.1 Selected applications of USNs in analytical spectrometry—cont'd

Nebulizer	Desolvation	Spectrometric technique	Application	Reference
Commercial high-flow USN	Electrothermal desolvation system	FF-AAS	Ag, Cd, Pb, Zn in biological and sediment CRMs	Ribeiro, Gallindo Borges, Vieira, and Curtius (2007)
Purpose-built USN, 2.5 MHz	No desolvation	CD-IMS	Propoxur in wastewater	Tabibi and Jafari (2020)
Commercial high-flow USN	Electrothermal desolvation system	HPLC-MIP-MS	As species in human urine	Chatterjee, Shibata, Tao, Tanaka, and Morita (2004)
Commercial low-flow USN	No desolvation	ICP-OES	As, Sb, Se in drinking water	Tyburska et al. (2010)
Commercial SAWN	No desolvation	Ambient MS	Organic explosives	Pintabona et al. (2019)
Purpose-built SAWN, 30 MHz	No desolvation	Ambient MS	Pharmaceuticals in physiological fluids	Ho et al. (2011)
Commercial geyser-type USN, 2.4 MHz	No desolvation	AMS	Elements in airborne aerosol	O'Brien et al. (2019)

No permission required.

higher at the USN than at the PN. For ICP OES, use of a dry aerosol increases interferences, while ICP-MS prefers dry aerosol to reduce interferences of oxide ions. To show an example, the decrease in CeO^+/Ce^+ ratio from 3% at pneumatic nebulization to 0.7% at USN with desolvation unit and even 0.05% at USN with a membrane desolvator in series was reported.

The limits of detection for a number of elements measured in ICP-OES coupled with different nebulizers are given in Table 9.2.

The ultrasonic nebulizer gives LODs that are approximately an order of magnitude better than with a pneumatic nebulizers. Thus, the USN-ICP-

TABLE 9.2 Comparison of limit of detection values (ng mL^{-1}) by ICP-OES and ICP-MS with PN and USN.

Element	PN-ICP -OES	USN-ICP -OES	PN-ICP -MS	USN-ICP -MS
Al	9	0.5	0.009	0.001
Cr	4	0.2	0.05	0.001
Cu	2	0.3	0.002	0.0004
Ni	10	0.4	0.004	0.0005
Pb	27	0.1	0.001	0.0001
Zn	2	0.2	0.005	0.0005

No permission required.

OES technique is capable of determining very low rare earth element concentrations in chemically resistant geological materials, permitting small amounts of sample to be analyzed (Brenner & Dorfman, 1993).

The method developed, using ICP-OES coupled to an ultrasonic nebulizer, proved suitable for routine analysis of minerals in wine with results for LOD and RSD comparable to the use of ICP-MS. Signal suppression was observed for the highest water loading. Conversely, signal enhancement occurred for the lowest water loading (de Andrade et al., 2020).

The determination of 13 trace elements in common environmental matrices by USN-ICP-OES was successful, except for selenium and arsenic, when argon was used as internal standard (Hoenig et al., 1998). Desolvation-induced nonlinearity in the calibration curve for bromine determination by USN-ICP-MS was observed and reduced by the addition of excess sodium chloride to ammonium bromide standard solutions (Kahen et al., 2006). The USN sample introduction technique was employed to improve the sensitivity in microwave plasmas including a microwave plasma torch (Yang, Zhang, Yu, & Jin, 2000).

9.3.2 Nebulization of organic liquids

Volatile organics overload the plasma causing plasma deterioration and generate high reflected power that could cause plasma extinguishing. In addition, emission line broadening effects are observed with ICP-OES detection. In ICP-MS analysis, the problems are even more severe since the incomplete carbon combustion leads to high carbon deposition in the vicinity of cone orifices and inside ion optics section that drastically reduces ion extraction efficiency. In order to reduce this effect, nebulization with an USN coupled with a cooled condenser was tested for gasoline, where a low dilution factor of five was found acceptable for the ICP due to lower solvent load. Ultrasonic nebulization permitted further improvements in organic solvent introduction, because of the combination of two effects: a more efficient nebulization, together with desolvation of organic aerosol (Botto & Zhu, 1994).

Removal of the organic matrix by the USN-membrane desolvator system was sufficient not only for good ICP operability but for the nearly complete elimination of calibration bias between the various solvent systems. It allowed accurate determination of elements in a variety of organic liquids using a single calibration procedure. Universal calibration showed independence of sample matrix type and proved to be valid for ICP-OES analysis of multiple organic solvents (Botto & Zhu, 1996). Using a 140 mm-long membrane desolvator, it was possible to establish a single calibration for direct trace element analysis of gasoline, refinery naphtha streams, and volatile organic solvents by ICP-OES. Despite expected changes in sample composition and volatility, quantitation can be performed using a two-point calibration and at most a single internal standard element.

Sensitivity and operability of ICP-MS for toluene were greatly enhanced with the USN. The 'theoretical quantity' of toluene vapor loaded into the plasma (Q_{VSPL}) was calculated in $mg\,s^{-1}$ according to the ideal gas law equation (Eq. 9.5) (Duyck et al., 2002):

$$Q_{VSPL} = \frac{Q_g p_{sat} M}{RT_C} \tag{9.5}$$

where Q_g is the nebulizer gas flow rate ($L\,min^{-1}$), T_C is the cooling temperature in K, p_{sat} is the saturated vapor pressure (in mm Hg) for each T_C, R is the gas constant, and M is the molecular mass of the solvent ($g\,mol^{-1}$).

As a minimum value of Q_{VSPL} was obtained in the range of −10°C to −20° C, the USN cooling system was set at its operating temperature T_C of −14°C. Operability of the ICP for toluene was further improved with the use of USN in combination with oxygen addition into the outer argon streams. For ICP-MS, carbon build up at the MS interface was minimized by optimization of the argon-to-oxygen ratio in the plasma and by the aerosol desolvation. For crude oil samples and their asphaltenic fractions, the standard addition technique combined with internal standardization was successfully applied for trace element determination (Duyck et al., 2002).

9.3.3 Other spectroscopies: LIBS, AAS, and GD

Ultrasonic nebulizers have found broad application in various spectroscopic techniques. In laser-induced breakdown spectroscopy (LIBS), the use of USN improves ablation efficiency by introducing a sample as a dense cloud of droplets. A purpose-built USN, operating at 1.7 MHz without desolvation, was tested for determining metals in aqueous solutions (Zhong et al., 2015).

In contrast, a 2.4 MHz USN combined with a tandem heater-condenser-membrane dryer unit was utilized for fast determination of metals in drinking and river water. The use of membrane dryer after the desolvation unit provided two to three times enhancements in LOD values (Aras et al., 2012).

The coupling of modern USN systems to flame furnace atomic absorption spectrometry (FF AAS) has proved to be an extremely sensitive, high precision, and robust technique for the determination of trace elements in certified reference biological and sediment materials (Ribeiro et al., 2007).

Atmospheric pressure glow microdischarge (μAPGD) generated between a He microjet anode and a flowing liquid cathode was coupled with USN. It boosted intensities of Ca, Cd, In, K, Li, Mg, Mn, Na, and Sr lines and improved the sensitivity of μAPGD-OES by approximately one order of magnitude (Greda, Jamroz, & Pohl, 2016).

An ultrasonic nebulization multiple channel electrospray ionization source, which generates multiply charged peptides and proteins ions, was developed. In this arrangement, electrospray from an MC-ESI source produced the charged methanol droplets and a USN produces aerosols from the sample solution. The sample aerosols were then purged to interact with the charged methanol droplets. Then, electrospray ionization of peptides and proteins occurred from the newly formed droplets prior to their MS detection (Shiea, Chang, Lin, & Jiang, 2001).

Ultrasonic nebulization can be used for transferring of thermally unstable compounds at ambient temperature, such as Propoxur insecticide, prior their determination by corona discharge ionization ion mobility spectrometry, thus avoiding the thermal decomposition of the analytes (Tabibi & Jafari, 2020).

A new method combining ultrasonic nebulization extraction and low-pressure photoionization (UNE-LPPI) has been developed and employed for the rapid mass spectrometric analysis of various analytes in sample matrices. Feasibility of the technique for detecting a variety of semi-/nonvolatile compounds, such as polycyclic aromatic hydrocarbons, amino acids, dipeptides, drugs, nucleic acids, alkaloids, and steroids was examined. Then, the quantification capability of UNE-LPPI was demonstrated both using soil sample spiked with different amounts of phenanthrene and pyrene, and determining several compounds in tablets and biological tissues (Liu et al., 2015).

9.3.4 Interface for hyphenated techniques

An USN coupled with a tubular PTFE membrane desolvator was used to interface liquid chromatography with helium microwave-induced plasma atomic emission detector. Membrane desolvator conditions such as temperature, solvent composition, and counter current gas flow were optimized. Interestingly, maximum transport efficiency of semivolatile organic compound was obtained at relatively low desolvator temperatures (Das & Carnahan, 2001). The addition of dilute solutions of inorganic chemicals such as NaOH or sulfuric acid enhanced analyte transport, likely by encapsulating the analyte within desolvated $NaOH/H_2SO_4$ particles to avoid analyte volatilization during transport through the heated membrane desolvator.

For a HPLC-MIP-MS system, interchanging of the PN interface with an USN caused 3- to 6-fold increase in the MS signals for the anionic and 6- to 12-fold for cationic arsenic compounds as compared to PN. The high transport efficiency improved the sensitivity and limits of detection for arsenic species (Chatterjee et al., 2004). The desolvation process further enhanced the analyte transport to the plasma relative to transport without desolvation and improved the limits of detection. Also, SAWN could serve for interfacing HPLC and MS (Tveen-Jensen et al., 2015).

A USN interface for capillary electrophoresis and ICP-MS was examined for determining the metal content of metalloproteins in biological samples. This arrangement offered several advantages over the PN interface including better separation resolution, improved sensitivity, and better immunity to mechanical parameters (Lu & Barnes, 1996).

9.3.5 Direct chemical vapor generation techniques

Direct vapor generation can be accomplished in the continuous mode, low-flow USN. The DVG nebulizer incorporates 2–4 solution capillaries with independent liquid feeding and a common gas inlet. An extremely short reaction time between sample, acid, and reductant/oxidant and a rapid separation of the reaction products is obtained by nebulizing the reagents at the quartz vibrating plate, generating aerosol at the entrance to the spray chamber. The water-cooled USN was operated at 1.65 MHz with a forward power of 80 W. The applicability and reliability of the ultrasonic nebulization vapor generation technique with ICP or MIP detection for trace element determination have been demonstrated (Ślachciński, 2014).

Tyburska et al. (2010) integrated a direct hydride generator system in which the spray chamber and USN were modified for hydride generation. The efficient production of volatile hydride-forming elements resulted in the significant increase of analyte intensities and improved limits of detection compared to conventional pneumatic nebulization.

The multimode sample introduction system based on USN provides convenient means of determining elements, either by nebulization or vapor generation, or simultaneously in the same device (Ślachciński, 2014).

9.3.6 Solution nebulization for ambient MS

Surface acoustic wave nebulization (SAWN) has been proposed as an efficient tool for simultaneous transferring and ionizing analytes from a microsample solution into the gas phase for subsequent MS detection under ambient conditions. With a power consumption of only a few watts, the nebulization results in the formation of small droplets that, when introduced into a mass spectrometer, subsequently evaporate and facilitate analyte ionization. It has been shown that ions produced with SAWN possess a lower internal energy compared to electrospray ionization (ESI).SAW ionization can be considered a soft ionization technique that results in little to no ion fragmentation, as can be observed by the unmodified chemical structure of analytes. Despite the fact that SAWN is considered to be a low-energy ionization technique, excellent sensitivity was obtained for organic explosives using a piezoelectric chip with a dual interdigitated transducer (Pintabona et al., 2019).

When used in mass spectrometry, SAWN can be performed either in a pulsed mode, thus conveniently coping with discontinuous MS analyzers, or in a continuous mode like ESI. The continuous SAWN produced an aerosol, which was sampled on a microsecond timescale by a linear ion trap mass spectrometer, and produced multiply charged peptide precursor ions. The signals observed on the mass spectra were shifted to higher m/z compared to those obtained for the same sample analyzed by ESI (Heron et al., 2010).

Within the field of proteomics, there is considerable interest in the design of miniaturized lab-on-a-chip devices in order to process proteins to peptides and carry out the accurate determinations of target analytes at the parts per billion level in low volume samples of physiological fluids and tissues. Using SAW-MS, drugs or heavy metals present in nanomole per liter concentrations can be detected in both positive and negative ion modes from untreated physiological fluids or tap water samples (Ho et al., 2011).

While many versatile ionization sources have been developed for ambient MS, challenges lie in achieving efficient sample transport and ionization. The use of corona discharge (CD) coupled to a SAWN device enhanced sampling rate for both polar and nonpolar analytes by up to four orders of magnitude. The generation of multiply charged peptide ion signals comparable to that of conventional ESI was demonstrated for SAWN-CD. Similar signals were observed for the arrangement consisted of a conventional USN and the CD. Both SAWN-CD and USN-CD may potentially serve as a flexible ionization tool because they can ionize both nonpolar and polar molecules including peptides (Song et al., 2020).

As is well known, ambient MS suffers from low sensitivity for detecting analytes present inside the sample. On-line ultrasonic extraction and nebulization combined with carbon fiber ionization (UEN/CFI-MS) was developed for real-time extraction and nebulization of synthetic drugs present inside the sample prior to screening analysis by AMS. A USN could produce a high-frequency vibration to realize on-line extraction and nebulization of the sample. Then, the produced aerosol could be immediately ionized by carbon fiber ionization mass spectrometry. Compared with existing ambient ionization methods, UEN/CFI-MS exhibits excellent capability to extract analytes inside sample within seconds. Owing to the combination of carbon fiber ionization, UEN/CFI-MS can perform sensitive analysis of analytes of different polarity and shows few limitations for the polarity of potential extractants (Wang et al., 2020).

9.4 Conclusions

Both high-flow and low-flow USNs operated with or without aerosol desolvation proved to be a viable alternative to PN, and they are currently used in routine analysis in the parts per trillion levels. USNs can operate with or without an aerosol desolvation stage depending on the target application. Dense wet aerosols are useful for direct vapor generation techniques, sampling for ambient MS, and in the case of interfacing in hyphenated techniques. Surface acoustic wave nebulization is a technique to generate ions compatible with ambient MS of minute samples providing sensitive detection at the femtomole level. SAW nebulization demonstrates considerable promise as it has shown almost negligible denaturation of proteins or biomolecules during liquid sample disintegration. SAWN can be a viable, low-energy alternative to other soft ionization techniques for the MS analysis of proteomic samples and may permit even softer energy deposition than ESI.

Acknowledgments

This work was supported by Warsaw University of Technology.

References

Al-Jumaily, A. M., & Meshkinzar, A. (2017). On the development of focused ultrasound liquid atomizers. *Advances in Acoustics and Vibration, 2017*. https://doi.org/10.1155/2017/7861726.

Anderlini, T. K., & Beauchemin, D. (2018). Improvement of sample introduction to inductively coupled plasma optical emission spectrometry using an ultrasonic nebulizer with an infrared heated pre-evaporation tube. *Journal of Analytical Atomic Spectrometry*, *33*(1), 127–134. https://doi.org/10.1039/c7ja00340d.

Aras, N., Yeşiller, S. U., Ateş, D. A., & Yalçin, S. (2012). Ultrasonic nebulization-sample introduction system for quantitative analysis of liquid samples by laser-induced breakdown spectroscopy. *Spectrochimica Acta Part B: Atomic Spectroscopy, 74–75*, 87–94. https://doi.org/10.1016/j.sab.2012.06.017.

Barreras, F., Amaveda, H., & Lozano, A. (2002). Transient high-frequency ultrasonic water atomization. *Experiments in Fluids, 33*(3), 405–413. https://doi.org/10.1007/s00348-002-0456-1.

Bensimon, N., Bourquin, J., & Parriaux, A. (2000). Determination of ultra-trace elements in snow samples by inductively coupled plasma source sector field mass spectrometry using ultrasonic nebulization. *Journal of Analytical Atomic Spectrometry, 15*, 731–734. https://doi.org/10.1039/b000460j.

Borkowska-Burnecka, J., Leśniewicz, A., & Żyrnicki, W. (2006). Comparison of pneumatic and ultrasonic nebulizations in inductively coupled plasma atomic emission spectrometry—Matrix effects and plasma parameters. *Spectrochimica Acta Part B: Atomic Spectroscopy, 61*(5), 579–587. https://doi.org/10.1016/j.sab.2006.04.005.

Botto, R. I., & Zhu, J. J. (1994). Use of an ultrasonic nebulizer with membrane desolvation for analysis of volatile solvents by inductively coupled plasma atomic emission spectrometry. *Journal of Analytical Atomic Spectrometry, 9*(9), 905–912. https://doi.org/10.1039/ja9940900905.

Botto, R. I., & Zhu, J. J. (1996). Universal calibration for analysis of organic solutions by inductively coupled plasma atomic emission spectrometry. *Journal of Analytical Atomic Spectrometry, 11*(9), 675–681. https://doi.org/10.1039/ja9961100675.

Brenner, I. B., & Dorfman, E. (1993). Application of ultrasonic nebulization for the determination of rare earth elements in phosphates and related sedimentary rocks using inductively coupled plasma atomic emission spectrometry with comments on dissolution procedures. *Journal of Analytical Atomic Spectrometry, 8*(6), 833–838. https://doi.org/10.1039/ja9930800833.

Brenner, I. B., Zhu, J., & Zander, A. (1996). Evaluation of an ultrasonic nebulizer-membrane separation interface (USN-MEMSEP) with ICP-AES for the determination of trace elements by solvent extraction. *Fresenius' Journal of Analytical Chemistry, 355*(7–8), 774–777. https://doi.org/10.1007/s0021663550774.

Budic, B. (2001). Effects of desolvation on line emission intensities and analyte transport in inductively coupled plasma atomic emission spectrometry using ultrasonic nebulization. *Journal of Analytical Atomic Spectrometry, 16*(2), 129–134. https://doi.org/10.1039/b004050i.

Castillano, T. M., Vela, N. P., Caruso, J. A., & Story, W. C. (1992). Evaluation of an ultrasonic nebulizer for sample introduction in inductively coupled plasma atomic emission spectrometry. *Journal of Analytical Atomic Spectrometry, 7*(6), 807–811. https://doi.org/10.1039/ja9920700807.

Chatterjee, A., Shibata, Y., Tao, H., Tanaka, A., & Morita, M. (2004). High-performance liquid chromatography-ultrasonic nebulizer high-power nitrogen microwave-induced plasma mass spectrometry, real-time on-line coupling for selenium speciation analysis. *Journal of Chromatography A, 1042*(1–2), 99–106. https://doi.org/10.1016/j.chroma.2004.05.001.

Collins, D. J., Manor, O., Winkler, A., Schmidt, H., Friend, J. R., & Yeo, L. Y. (2012). Atomization off thin water films generated by high-frequency substrate wave vibrations. *Physical Review E—Statistical, Nonlinear, and Soft Matter Physics, 86*(5). https://doi.org/10.1103/PhysRevE.86.056312.

Das, D., & Carnahan, J. W. (2001). Moderate volatility analyte transport behavior with membrane desolvation reversed-phase liquid chromatography-helium microwave-induced plasma atomic emission spectroscopy. *Analytica Chimica Acta, 444*(2), 229–239. https://doi.org/10.1016/S0003-2670(01)01176-X.

de Andrade, M. F., Silva, I. J. S., Pimentel, M. F., Paim, A. P. S., Cervera, M. L., & de la Guardia, M. (2020). Ultrasonic nebulization inductively coupled plasma optical emission spectrometry method for wine analysis. *Spectrochimica Acta Part B: Atomic Spectroscopy, 170.*

Duyck, C., Miekeley, N., Porto da Silveira, C. L., & Szatmari, P. (2002). Trace element determination in crude oil and its fractions by inductively coupled plasma mass spectrometry using ultrasonic nebulization of toluene solutions. *Spectrochimica Acta, Part B: Atomic Spectroscopy, 57*(12), 1979–1990. https://doi.org/10.1016/S0584-8547(02)00171-4.

Greda, K., Jamroz, P., & Pohl, P. (2016). Ultrasonic nebulization atmospheric pressure glow discharge—Preliminary study. *Spectrochimica Acta Part B: Atomic Spectroscopy, 121,* 22–27. https://doi.org/10.1016/j.sab.2016.04.008.

Grindlay, G., Maestre, S., Mora, J., Hernandis, V., & Gras, L. (2005). A microwave assisted desolvation system based on the use of a TM 010 cavity for inductively coupled plasma based analytical techniques. *Journal of Analytical Atomic Spectrometry, 20*(5), 455–461. https://doi.org/10.1039/b418282k.

Grotti, M., Lagomarsino, C., & Frache, R. (2004). A new nebulization device with exchangeable aerosol generation mode as a useful tool to investigate sample introduction processes in inductively coupled plasma atomic emission spectrometry. *Spectrochimica Acta Part B: Atomic Spectroscopy, 59*(7), 1001–1006. https://doi.org/10.1016/j.sab.2004.05.003.

Hayashi, H., Furuzawa, S., Tanaka, T., & Hiraide, M. (2004). Low-pressure helium inductively coupled plasma mass spectrometry: Sample aerosol introduction through a capillary interface from a vibrating mesh nebulizer. *Journal of Analytical Atomic Spectrometry, 19*(6), 773–774. https://doi.org/10.1039/b310935f.

Heron, S. R., Wilson, R., Shaffer, S. A., Goodlett, D. R., & Cooper, J. M. (2010). Surface acoustic wave nebulization of peptides as a microfluidic interface for mass spectrometry. *Analytical Chemistry, 82*(10), 3985–3989. https://doi.org/10.1021/ac100372c.

Ho, J., Tan, M. K., Go, D. B., Yeo, L. Y., Friend, J. R., & Chang, H. C. (2011). Paper-based microfluidic surface acoustic wave sample delivery and ionization source for rapid and sensitive ambient mass spectrometry. *Analytical Chemistry, 83*(9), 3260–3266. https://doi.org/10.1021/ac200380q.

Hoenig, M., Doc;ekalová, H., & Baeten, H. (1998). Study of matrix interferences in trace element analysis of environmental samples by inductively coupled plasma atomic emission spectrometry with ultrasonic nebulization. *Journal of Analytical Atomic Spectrometry, 13*(3), 195–199. https://doi.org/10.1039/A708142A.

Jankowski, K., Karmasz, D., Ramsza, A., & Reszke, E. (1997). Characteristics of nebulizers for microwave induced plasma atomic emission spectrometry. II. Ultrasonic nebulizers. *Spectrochimica Acta Part B: Atomic Spectroscopy, 52*(12), 1813–1823. https://doi.org/10.1016/S0584-8547(97)00069-4.

Kahen, K., Jorabchi, K., & Montaser, A. (2006). Desolvation-induced non-linearity in the analysis of bromine using an ultrasonic nebulizer with membrane desolvation and inductively coupled plasma mass spectrometry. *Journal of Analytical Atomic Spectrometry, 21*(6), 588–591. https://doi.org/10.1039/b511811e.

Kooij, S., Astefanei, A., Corthals, G. L., & Bonn, D. (2019). Size distributions of droplets produced by ultrasonic nebulizers. *Scientific Reports, 9*(1). https://doi.org/10.1038/s41598-019-42599-8.

Kudo, T., Sekiguchi, K., Sankoda, K., Namiki, N., & Nii, S. (2017). Effect of ultrasonic frequency on size distributions of nanosized mist generated by ultrasonic atomization. *Ultrasonics Sonochemistry, 37,* 16–22. https://doi.org/10.1016/j.ultsonch.2016.12.019.

Kuo, Y. M., Chan, W. H., Lin, C. W., Huang, S. H., & Chen, C. C. (2019). Characterization of vibrating mesh aerosol generators. *Aerosol and Air Quality Research, 19*(8), 1678–1687. https://doi.org/10.4209/aaqr.2018.11.0436.

Liu, C., Zhu, Y., Zhou, Z., Yang, J., Qi, F., & Pan, Y. (2015). Ultrasonic nebulization extraction/low pressure photoionization mass spectrometry for direct analysis of chemicals in matrices. *Analytica Chimica Acta, 891*, 203–210. https://doi.org/10.1016/j.aca.2015.08.010.

Lu, Q., & Barnes, R. M. (1996). Evaluation of an ultrasonic nebulizer interface for capillary electrophoresis and inductively coupled plasma mass spectrometry. *Microchemical Journal, 54*(2), 129–143. https://doi.org/10.1006/mchj.1996.0086.

Masson, P., Vives, A., Orignac, D., & Prunet, T. (2000). Influence of aerosol desolvation from the ultrasonic nebulizer on the matrix effect in axial view inductively coupled plasma atomic emission spectrometry. *Journal of Analytical Atomic Spectrometry, 15*(5), 543–547. https://doi.org/10.1039/a909651e.

O'Brien, R. E., Ridley, K. J., Canagaratna, M. R., Jayne, J. T., Croteau, P. L., Worsnop, D. R., et al. (2019). Ultrasonic nenulization for the elemental analysis of microgram-level samples with offline aerosol mass spectrometry. *Atmospheric Measurement Techniques, 12*, 1659–1671. https://doi.org/10.5194/amt-12-1659-2019.

Pintabona, L., Astefanei, A., Corthals, G. L., & van Asten, A. C. (2019). Utilizing surface acoustic wave nebulization (SAWN) for the rapid and sensitive ambient ionization mass spectrometric analysis of organic explosives. *Journal of the American Society for Mass Spectrometry, 30*(12), 2655–2669. https://doi.org/10.1007/s13361-019-02335-y.

Rajan, R., & Pandit, A. B. (2001). Correlations to predict droplet size in ultrasonic atomisation. *Ultrasonics, 39*(4), 235–255. https://doi.org/10.1016/S0041-624X(01)00054-3.

Ribeiro, A. S., Gallindo Borges, D. L., Vieira, M. A., & Curtius, A. J. (2007). Coupling of ultrasonic nebulization to flame furnace atomic absorption spectrometry-new possibilities for trace element determination. *Microchemical Journal, 85*(2), 341–346. https://doi.org/10.1016/j.microc.2006.08.002.

Rodes, C., Smith, T., Crouse, R., & Ramachandran, G. (1990). Measurements of the size distribution of aerosols produced by ultrasonic humidification. *Aerosol Science and Technology, 13*(2), 220–229. https://doi.org/10.1080/02786829008959440.

Shiea, J., Chang, D. Y., Lin, C. H., & Jiang, S. J. (2001). Generating multiply charged protein ions by ultrasonic nebulization/multiple channel-electrospray ionization mass spectrometry. *Analytical Chemistry, 73*(20), 4983–4987. https://doi.org/10.1021/ac0101609.

Ślachciński, M. (2014). Recent achievements in sample introduction systems for use in chemical vapor generation plasma optical emission and mass spectrometry: From macro-to microanalytics. *Applied Spectroscopy Reviews, 49*(4), 271–321. https://doi.org/10.1080/05704928.2013.823547.

Song, L., You, Y., Perdomo, N. R., & Evans-Nguyen, T. (2020). Inexpensive ultrasonic nebulization coupled with direct current corona discharge ionization mass spectrometry for liquid samples and its fundamental investigations. *Analytical Chemistry, 92*(16), 11072–11079. https://doi.org/10.1021/acs.analchem.0c00524.

Tabibi, A., & Jafari, M. T. (2020). Ultrasonic piezoelectric nebulization of propoxur for the determination by corona discharge ionization ion mobility spectrometry. *Analytical Sciences, 36*(2), 227–232. https://doi.org/10.2116/analsci.19P262.

Tarr, M. A., Zhu, G., & Browner, R. F. (1993). Microflow ultrasonic nebulizer for inductively coupled plasma atomic emission spectrometry. *Analytical Chemistry, 65*(13), 1689–1695. https://doi.org/10.1021/ac00061a010.

Tveen-Jensen, K., Gesellchen, F., Wilson, R., Spickett, C. M., Cooper, J. M., & Pitt, A. R. (2015). Interfacing low-energy SAW nebulization with liquid chromatography-mass spectrometry for the analysis of biological samples. *Scientific Reports, 5*. https://doi.org/10.1038/srep09736.

Tyburska, A., Jankowski, K., Ramsza, A., Reszke, E., Strzelec, M., & Andrzejczuk, A. (2010). Feasibility study of the determination of selenium, antimony and arsenic in drinking and mineralwater by ICP-OES using a dual-flow ultrasonic nebulizer and direct hydride generation. *Journal of Analytical Atomic Spectrometry*, *25*(2), 210–214. https://doi.org/10.1039/B916729C.

Wang, Z., Cao, Y., Lu, Y., Zhang, F., Su, Y., & Guo, Y. (2020). Ultrasonic extraction and nebulization in real-time coupled with carbon fiber ionization mass spectrometry for rapid screening of the synthetic drugs adulterated into herbal products. *Analytica Chimica Acta*, *1136*, 62–71. https://doi.org/10.1016/j.aca.2020.08.045.

Yang, W., Zhang, H., Yu, A., & Jin, Q. (2000). Microwave plasma torch analytical atomic spectrometry. *Microchemical Journal*, *66*(1–3), 147–170. https://doi.org/10.1016/S0026-265X(00)00057-6.

Zhong, S., Zheng, R., Lu, Y., Cheng, K., & Xiu, J. (2015). Ultrasonic nebulizer assisted LIBS: A promising metal elements detection method for aqueous sample analysis. *Plasma Science and Technology*, *17*(11), 979–984. https://doi.org/10.1088/1009-0630/17/11/17.

SECTION 3

Applications

CHAPTER

10

Applications in single-particle inductively coupled plasma-mass spectrometry

Deyana Georgieva

Department of Analytical Chemistry and Computer Chemistry, University of Plovdiv "Paisii Hilendarski", Plovdiv, Bulgaria

10.1 Introduction

The growing impact of nanosized materials on human life and industry provoked the formation of a whole new section in analytical science, namely "analytical nanoscience and nanotechnology" (AN&N) (Laura Soriano, Zougagh, Valcárcel, & Ríos, 2018; López-Lorente & Valcárcel, 2014) where the nanoparticles and nanostructured materials are utilized as an analytical tool or they are the objects of analysis. In the second case, information on different measurands (i.e., chemical composition, structure, morphology, shape, size and size distribution, concentration, etc.) is considered as the goal of the analysis. To meet the needs for a variety of information related to nanoscale materials in recent years, many analytical methods have been developed (Mourdikoudis, Pallares, & Thanh, 2018) with the increasing use of plasma and laser-based techniques during the last decade (Galbács, Kéri, Kohut, Veres, & Geretovszky, 2021). Among the latter, inductively coupled plasma-mass spectrometry working in single-particle mode (spICP-MS) is pointed to as a powerful and versatile nanoparticle characterization method. The undoubted advantages of this technique such as low detection limits, high detection speed, and wide linear dynamic range allow the application of the commercial instruments already installed in the laboratories for the characterization of nanoscale materials. The intensive fundamental investigations in the last two decades on the methodological and metrological issues assist

Analytical Nebulizers
https://doi.org/10.1016/B978-0-323-91181-8.00013-8

in expanding the application of the method. The current status of the method development has been discussed in detail in several recently published reviews and book chapters on spICP-MS (Bolea et al., 2021; Flores et al., 2021; Laborda et al., 2021; Laborda, Gimenez-Ingalaturre, Bolea, & Castillo, 2020; Meermann & Nischwitz, 2018; Vidmar, Milačič, Ščančar, Goenaga-Infante, & Vidmar, 2021). However, the specific characteristics of nanoparticles as an analyte remain challenging and strongly influence results obtained from spICP-MS. Since the method is based on continuous nebulization of a dilute suspension of nanoparticles into a conventional ICP-MS instrument the main requirement for the sample introduction system is to ensure consistent entry of a single nanoparticle carried by aerosol droplets into the plasma discharge. Thus, the nebulization process and quality of obtained aerosol play a very important role in improving the spICP-MS analytical figures of merit. In this chapter, an overview of the spICP-MS method fundamentals as well as the influence of the nebulization process on obtained results is provided.

10.2 Fundamentals of spICP-MS

A series of seminal papers published by Degueldre, Favarger, and Bitea (2004) and Degueldre and Favarger (2003, 2004) are considered the starting point for developing and establishing the basic methodology of the spICP-MS technique as we know it now. If nanoparticles contain elements measurable by the ICP-MS, and diluted nanocolloid suspension is nebulized into plasma, a series of events over a continuous baseline is recorded in the time-resolved mode of measurement. The transformation of the nanoparticle carried by the aerosol droplet into the plasma and the resulted analytical signals are presented in Fig. 10.1. Depending on the detection frequency (dwell time t_d) the signals arising from NPs could be registered as pulses or as peaks. In the first case, t_d (typically in ms) exceeds the duration of the ion cloud produced from single NP in plasma (Fig. 10.1A), which results in the registration of a single integrated signal. Working with much shorter dwell times (below 100 μs) allows detection of several points per NP event (Fig. 10.1B).

The intensity (or peak area) of the registered signal is related to the mass of the analyte in the introduced NP, and if additional information such as shape, composition, analyte mass fraction, and density of bulk material are available, the size of the nanoparticles can be calculated. In most cases, nanomaterials are assumed to be spherical in shape and an equivalent diameter is used for the assessment of size of the nanomaterials. The number of registered NP events (Fig. 10.1C) is proportional to the particle number concentration in the sample. Additional information about the

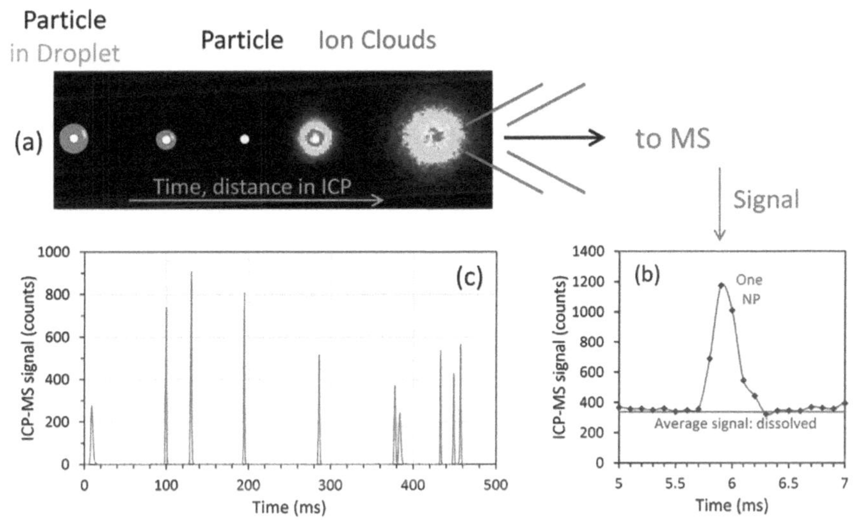

FIG. 10.1 Transformation of the nanoparticle in ICP and observed signals in time-resolved mode of analysis. (A) Processes to convert a particle suspended in the liquid into ions in the ICP. (B) A signal from one nanoparticle and the same element in solution (dissolved). (C) Signals from 10 nanoparticles. *From Olesik, J. (2020). Single particle ICP-MS: From engineered nanoparticles to natural nanoparticles.* Spectroscopy, *35(7), 20–22. https://www.spectroscopyonline.com/view/single-particle-icp-ms-engineered-nanoparticles-natural-nanoparticles.*

equilibria between the dissolved and particulate forms of analyte could be obtained from the registered baseline signal (Laborda, Jiménez-Lamana, Bolea, & Castillo, 2011).

In their critical review about the analytical application of spICP-MS Bolea et al. (2021) summarized different measurands and respective forms of the analyte (dissolved or particulate) used to obtain quantitative information about (i) characterization of NPs in terms of element content, size, and size distribution and (ii) concentration presented as particle number concentration or mass concentration of particulate/dissolved form of the element. The common workflow to obtain desired information includes several steps as presented in Fig. 10.2.

The main steps in this workflow to obtain information about the characterization of the NPs are described in more detail in the following sections.

10.2.1 Separation of the signal domains

Firstly, the raw signals recorded in time-resolved mode (Fig. 10.2A) should be assigned to the analyte form which generates them, i.e., signals obtained from nanoparticles and those belonging to the baseline signal formed from the background and/or ionic form of the analyte. Correct

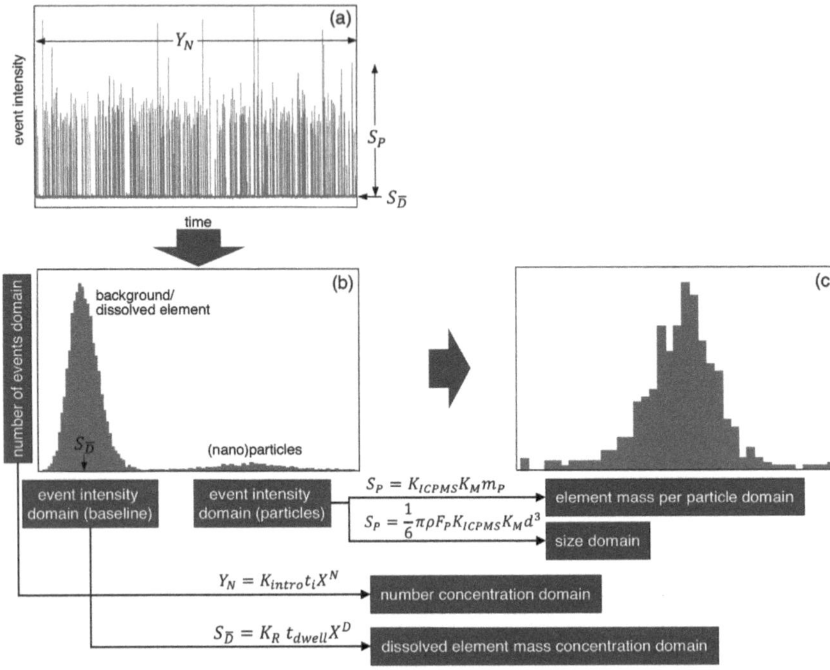

FIG. 10.2 Domains and transformations involved in SP-ICP-MS. (A) Time scan of a particle suspension containing dissolved forms of the element contained in the particles. (B) Event intensity frequency histogram of data from (A). (C) Size (or element mass per particle) distribution of nanoparticles calculated from the second intensity distribution in (B). *From Laborda, F., Gimenez-Ingalaturre, A. C., Bolea, E., & Castillo, J. R. (2020).* Spectrochimica Acta Part B: Atomic Spectroscopy, 169, *105883. https://doi.org/10.1016/j.sab.2020.105883.*

distinguishing of the intensity of nanoparticles from the baseline signal is one of the most critical steps in the spICP-MS analysis, and it is related to the limit of detection in terms of nanoparticle size and also the correct determination of the particle number/mass concentration. This implies the necessity of robust criteria for the determination of NP signals. However, in spICP-MS literature, there is no generally accepted approach for determining the signal threshold value (or so-called critical value), which separates both signal domains as illustrated in Fig. 10.2B.

In most cases, when particle events are recorded as pulses, a threshold criterion related to the baseline noise is used and its value depends both on the standard deviation of the baseline and on the accepted level of false positives. Several algorithms have been proposed in scientific literature to get this value based on baseline signal distribution. One of them suggested by Pace et al. (2011) is an iterative algorithm based on 3σ criteria for the removal of outliers from a normally distributed data set, assuming that the background signals have a (near) normal distribution. In order to remove the false positive data assigned to domain of NPs, some investigators use the same approach but recommended higher values of the coefficient—between 5 and 8 (Tuoriniemi, Cornelis, & Hassellöv, 2012). On the other hand, for the low-count baseline signals (detected by most commonly used quadrupole ICP-MS instruments equipped with electron multipliers), Laborda, Gimenez-Ingalaturre, Bolea, and Castillo (2019) and Laborda, Jiménez-Lamana, Bolea, and Castillo (2013) revealed that in this case, counting statistics dominate and the background more closely matches the Poisson distributions, and the usage of its parameters to determine the threshold value allows correct distinction between signal domains. Recently Mozhayeva and Engelhard (2019) also suggest, based on the Poisson statistics, a criterion to determine the threshold and to identify the beginning and end of NP signal events for data acquired with microsecond time resolution. The authors demonstrate its applicability and robustness for the correct determination of Ag NPs even in the case of high background levels due to the presence of ionic silver. For instruments equipped with TOF mass filters and a fast analog-to-digital conversion (ADC) detection system, some authors (Gundlach-Graham, Hendriks, Mehrabi, & Günther, 2018; Hendriks, Gundlach-Graham, & Günther, 2019) using Monte-Carlo simulations showed that detected low-count baseline signals are neither Gaussian nor Poisson distributed and application of the abovementioned iterative algorithms for the determination of threshold leads to an incorrect determination of its value. According to their investigations, compound Poisson background distributions are observed in this case. The authors demonstrate the robustness of the compound-Poisson-derived net critical values, used as a separator of the signal domains, obtained by the Monte-Carlo simulation, especially in cases when detected NPs have sizes close to the detection limit. Considering the above, it is clear that for the correct determination of the threshold value based on the background signal, it is imperative to know its distribution. Since the latter will depend on many factors such as the signal level, the instrumental parameters of the measurement (i.e., dwell time), and the type of detector system the appropriate approach for threshold evaluation should be chosen for every specific analysis. To avoid this Bi et al. (2014) proposed a completely different strategy based on the K-means clustering algorithm. The authors emphasized two advantages of this algorithm: (i) for the statistical estimation of the threshold value, both the background signal and the nanoparticle signal are taken into account, which increases accuracy in its determination compared with the unidirectional $n \times \sigma$ method discussed above and (ii) algorithm eliminates the ambiguity of the multiplier for σ. Nevertheless, as stated by the authors, the problems with this methodology, in cases when signal domains are overlapped, which often can be the case in real samples if the

NP sizes are close to the spICP-MS size detection limit or there is a high content of dissolved analyte, explain limited applicability of this approach.

10.2.2 Determination of the particle number concentration

In spICP-MS, after the discrimination of the signal domains, the number of particulate events registered for the given acquisition time could be related to particle number concentration in the sample as illustrated in Fig. 10.2 by the following equation:

$$Y_N = K_{intro} \times t_i \times X_N$$

where Y_N is number of particle events, t_i is acquisition time, X_N is the number concentration of the particle, and $K_{intro}(=\eta \times Q_{sam})$ is a factor related to the sample introduction, where η is the analyte transport efficiency and Q_{sam} is the sample introduction flow rate.

This relationship is valid if two requirements are met, namely: (i) there are provided conditions for the sequential entry of single particles into the plasma, and (ii) the effectiveness of the sample introduction system is evaluated correctly. The first condition is related to the general assumption that each recorded event represents a single particle and can be ensured by choosing an appropriate dilution factor, while the second requires an experimental assessment of the performance of the sample introduction system and transport losses.

The first prerequisite is directly related to the possibility of more than one nanoparticle being detected during a given dwell time. This probability can be predicted by the Poisson statistics and it is applied by many investigators (Ho, Lui, Lee, & Chan, 2013; Laborda et al., 2013; Olesik & Gray, 2012; Pace et al., 2012) but only for the current measurement conditions. Recently Kaynarova, Georgieva, and Stefanova (2022) proposed an Excel-based model that includes a module for dynamic calculation of dilution factor and probability for particle coincidence for dwell times in millisecond range. The model incorporates preliminary data for NP dispersion (such as the mass fraction of analyte in NPs, density of the bulk material, the total mass concentration of the suspension, and expected mean diameter of NPs) with instrumental parameters of the current measurement (sample flow rate, transport efficiency, acquisition time, and dwell time) included in an active spreadsheet where the equations connecting the input data are introduced. This allows to easily obtain information about appropriate conditions in which the analysis will be performed with an acceptable probability of particle matching. An advantage of the presented model is the possibility to use it for inverse calculation of probability for particle coincidence for experimentally obtained at given instrumental conditions number of NP peaks. In this case, calculated by the model, % probabilities for multiple events can be used for the correction of obtained particle number concentration in the tested sample.

It is worth mentioning that working with microsecond dwell times, achievable for time-of-light, double-focusing, as well as for the new generations of quadrupole ICP-MS instruments, reduces the above-discussed possibility for particle matching (Abad-Álvaro et al., 2016; Strenge & Engelhard, 2016) together with the opportunity to monitor more than one isotope per NP, improvement of size detection limits due to very low baseline signals. However, this is also associated with some difficulties related to the very low (often at a level of units of counts) signals, which requires a special attention for the correct estimation of peaks broadening (Kaňa, Loula, Koplík, Vosmanská, & Mestek, 2019). The latter is also associated with the deterioration of the unique spICP-MS possibility for the estimation of the distribution of the analyte between the particulate and dissolved form.

Another parameter related but not limited only to the particle number concentration determination is transport efficiency (TE). It is often considered as the most critical parameter in spICP-MS, as inaccuracy in its establishment leads to biases and uncertainties in the number concentration data. Several different methodologies for evaluating transport efficiency have been proposed in the scientific periodical, and each of them has its advantages and limitations. The necessity of the use of the nanoparticles certified reference materials for TE determination divided the methods into two main groups. In the first group two approaches can be combined, as suggested by Pace et al. (2011), which are related to the use of NP reference material with known number concentration (particle-frequency method) or known size (particle-size method). These methods are well recognized and implemented in the specialized software for spICP-MS analysis of commercial ICP-MS instruments. The second group united, so-called methods free of NP reference materials which are based on waste collection (Pace et al., 2011; Tuoriniemi et al., 2012; Tuoriniemi, Cornelis, & Hassellöv, 2014) and dynamic mass flow approach (Cuello-Nuñez et al., 2020). As a result of their comparative study Pace et al. (2011) reported that both RM-based methods give comparable results for TE assessment, till the waste collection method gives a biased result. However, a later work by Liu, Murphy, Winchester, and Hackley (2017) showed that determining the transport efficiency by the particle frequency method poses several problems. In most cases, commercially available reference materials certify the mean size and mass concentration of an element represented in the NP instead of the particle number concentration. Calculation of particle number concentration from this data may be incorrect if the element is also present in the liquid phase as an ionic form (Loula et al., 2019; Stefanova, Kaynarova, & Georgieva, 2019). Some authors (Liu et al.,

2017) have also reported that losses of NPs due to surface adsorption onto the sample vessel or sample introduction system may also lead to inconsistencies in the assessment of transport efficiency. A recent comprehensive investigation (Geiss et al., 2022) summarizes the essential factors and sources of bias in TE which must be taken into account when using RM-based methods for its determination. Besides the already mentioned problems related to the particle-frequency method, the authors also pointed out that when using the particle-size method, possible sources for biased results could be assumptions for identical behavior of the analyte in ionic solution and particles during the processes of nebulization, atomization, and ionization. A number of investigations (Ho et al., 2013; Ho, Lee, & Chan, 2015; Lee & Chan, 2015) show that this is not always the case for nanoparticles of larger size and/or composed of materials with high boiling points. In this case, the short residence time in plasma does not provide effective conversion of the analyte in NP into single charged ions. The other problem of the particle-size method is related to the lack of a strictly monodisperse RM of certified size (Meermann & Nischwitz, 2018), although Geiss et al. (2022) pointed out that using nanoparticle standard with narrow and Gaussian distribution and correct fitting model of experimental data leads to a negligible variation in the determination of TE. Contrariwise, Tuoriniemi et al. (2014) insistently recommend the waste collection (WC) method, based on the abovementioned issues with RM-based methods for TE evaluation. According to the authors, the observed biases in TE by the WC method reported elsewhere (Pace et al., 2011) are due to analyte partitioning effects during the nebulization process. The authors claimed that after applied correction of this effect, the WC method can be used as RM-free alternative which extends the application of spICP-MS.

A comprehensive study by Loula et al. (2019) investigated the influence of the instrumental and sample parameters on the transport efficiency and sensitivity of Ag NP detection. Expectedly, the parameters that significantly affect the transport efficiency are the flow rates of the nebulizer gas and the sample. Reducing the sample flow rate approximately 10 times leads to the TE enhancement up to almost 30%. However, despite the high transport efficiency, the overall effect on the total number of detected particles is negative at this low sample uptake rate. This necessitates the search for a reasonable compromise between sufficient sample throughput and acceptable transport efficiency.

All discussed approaches for determining transport efficiency use either water-suspended nanoparticle reference materials or aqueous solutions of the analyte. In most cases of spICP-MS analysis, this is not a problem due to the need to work with huge dilution factors (10^5–10^6). For some samples, however, sample preparation involves treatment with surfactants or enzymes (Flores et al., 2021) in order to extract and stabilize the nanoparticles in the analyzed suspension. This would lead to a difference in the transport efficiency of the nanoparticles in the sample and requires measures to correct TE for the matrix effect.

10.2.3 Particle size determination by spICP-MS

In contrast to electron microscopic techniques (TEM and SEM) for the analysis of nanosized materials, based on the visual detection of nanoparticles and determination of their morphology (size and shape), spICP-MS does not provide a direct measurement of nanoparticle dimensions. The intensity of registered transient NP signals is related to the content of the measured element included in the composition of the nanoparticle. This necessitates choice of an appropriate calibration strategy to link the signal intensity with the nanoparticle's mass and/or size. Two common approaches exist based on either (i) calibration by NPs standards or (ii) employing dissolved standard solutions of the monitored element and transport efficiency determination.

The first approach is considered more intuitive because it gives a direct correlation between the measured signal and the particle size. As indisputable advantages of this calibration should be pointed out: (i) the identity of the form in the samples and calibration standards and (ii) the relative independence of the calibration function from the transport efficiency of the used sample introduction system. However, its application is limited due to requirements of nanoparticle standards for the same element composition, geometry as well as density of the targeted nanoparticle. Since the signals generated by NPs depend on the vaporization and ionization efficiency in the plasma, which is related to the particle size, thus to ensure the correct experimental description of the signal vs mass or signal vs diameter function, the calibration standards must include representatives of NPs with small, medium, and large diameters, covering the entire studied interval. Unfortunately, there is still a scarcity of various RM that meet the stated prerequisites. Tables with the currently available nanoparticle RM are summarized in the publications of Laborda and Meermann (Laborda et al., 2021; Meermann & Nischwitz, 2018). Therefore, at present, applying this calibration strategy would drastically limit the applicability of the spICP-MS.

To overcome the above-pointed issues, a more feasible approach is proposed (Pace et al., 2011) based on the calibration with the dissolved ionic standards. In this case, the concentration of the analyte in the standard solution is related to the mass of the analyte delivered to the detector in a specific dwell time through the following equation:

$$m = \eta \times Q_{sam} \times t_d \times C$$

where m is the mass in fg, Q_{sam} is the sample flow rate in mL ms^{-1}, t_d is the dwell time in ms, η is the transport efficiency, and C is the ionic standard concentrations in fg mL^{-1}.

It is clear from the presented equation that the required constancy in the mass of analyte delivered per dwell time is dependent on sample introduction system performance and correct determination of TE discussed in the previous section.

To obtain signal vs size calibration function, the mass of analyte delivered per dwell time is converted into diameter (in most cases assuming a spherical shape) by the cubic relation between the two quantities. For the reliable application of this calibration approach, it is also mandatory to provide evidence of comparability of the behavior of the analyte, in terms of ionization and mass transport efficiency, if it is introduced through an ionic solution in calibrants or is present in particle form in samples. The incomplete particle vaporization and nonlinear detector response in the pulse counting mode are identified as the two major causes of nonlinear calibration curves at high particle mass (Lee & Chan, 2015) and limitation of the calibration range. The observed effect with incomplete vaporization will also depend on the NP material composition and its thermal properties and could lead to a systematic negative error in the estimation of their size (Ho et al., 2015, 2013; Lee & Chan, 2015).

10.3 Application and performance of different sample introduction systems for spICP-MS analysis

10.3.1 Pneumatic nebulizers/spray chamber systems

As mentioned earlier in the introduction part, one of the reasons for the wide application of the spICP-MS is the possibility to use the commercial instruments already installed in the laboratories. Since they are equipped with a standard sample introduction system, in most of the published studies dedicated to spICP-MS, a different type of pneumatic nebulizer (PN) is used for the purposes of analysis. An overview of their application for the characterization of different types of NPs is presented in Table 10.1.

It is obvious from the data presented in Table 10.1 that the so-called "conventional" type of sample introduction system is characterized with low analyte transport efficiency typically below 10%. In their recent investigation Olesik, Hartshorne, Casey, Linard, and Dettman (2021) compared the performance of four types of pneumatic nebulizers, namely MiraMist, Meinhard TR-30, Meinhard HEN-170, and Cetac MCN in double pass Scott spray chamber concerning transport efficiency and stability. The results of the investigation show that droplet-droplet collision/coalescence is a key process that controls analyte transport efficiency, while

TABLE 10.1 Examples for application of the pneumatic nebulizer/spray chamber systems for nanoparticles characterization by spICP-MS.

Type of nebulizer/ spray chamber	NP's analyzed	Mesurand	Transport efficiency	Ref
MicroMist/Scott double pass	Ag	NPs size particle number concentration	Not reported value	Pace et al. (2012)
Concentric PFA PTFE nebulizer/cyclonic	Ag	Particle number concentration	2.6%	Abad-Álvaro et al. (2016)
V-groove/Scott double pass	Au	NP size	2.1%	Lee and Chan (2015)
Glass nebulizer/ conical spray	Au CeO_2	NP size	1.7%	Bi et al. (2014)
PTFE concentric nebulizer/glass cyclonic	Ag	NPs size Particle number concentration	7.8%	Kaňa et al. (2019)
MicroMist/Scott double pass	Ag	NP size	9.8%	Kaynarova et al. (2022)
MicroMist/cyclonic	Pt	NP size Particle number concentration	10.6% ± 0.1%	Lin, Miyashita, Inagaki, Liu, and Hsu (2019)
Glass concentric slurry nebulizer/cyclonic	Ag Au Al	NP size Particle number concentration	3.15% ± 0.11%	Laborda et al. (2019)
Glass concentric Slurry nebulizer/Cyclonic	Ag Au	NP size Particle number concentration	2.99%	Laborda et al. (2013)
Microflow PFA concentric nebulizer/ glass impact bead spray chamber cooled	Ag	NP size	3.73% ± 0.01% size method 2.91% ± 0.11% frequency method	Liu et al. (2017)
MicroMist/cyclonic	Ag Au	NP size Particle number concentration	4.7%	Tuoriniemi et al. (2014)

Continued

TABLE 10.1 Examples for application of the pneumatic nebulizer/spray chamber systems for nanoparticles characterization by spICP-MS—cont'd

Type of nebulizer/ spray chamber	NP's analyzed	Mesurand	Transport efficiency	Ref
Total consumption sample introduction system	Ag, Pt	NP size Particle number concentration	82%–87% 93.3%	Miyashita et al. (2017)
Total consumption sample introduction system	Pt	NP size Particle number concentration	99.4±1.2%	Lin et al. (2019)

the impact of the aerosol solvent evaporation on the spray chamber is observed at sample flow rates less than 50–100 μL min^{-1}. The general trend observed for all the studied nebulizers is that decreasing the sample flow rate up to 10 μL min^{-1} increases the transport efficiency to 50%–60%. Although no significant difference is detected in TE between "conventional" and "micronebulizers" (compared at the same sample and nebulizer flow rates), the latest provide more stable aerosol production and therefore more stable ICP-MS signals that are critical for spICP-MS analysis. As a result of the conducted research, the authors concluded that despite the possible increase in the transport efficiency of pneumatic nebulizers, the level of aerosol loading is restricted by cooling or extinguishing the plasma. The plasma is the limiting factor on sample transport rates in pneumatic nebulizer/spray chamber introduction systems, if a desolvation system is not applied. A similar result about the enhancement of transport efficiency on low sample flow rates is reported when performance of four type spray chambers combined with MicroMist nebulizer are compared (Todolí, Maestre, Mora, Canals, & Hernandis, 2000).

In an attempt to achieve high values on TE, Miyashita et al. (2017) employed a total consumption sample introduction system, consisting of a large-bore high-performance concentric nebulizer and a custom-made small-volume on-axis cylinder chamber, for the applications of spICP-MS. By this system, working on the sample flow rate of 8.6 μL min^{-1}, the authors declared an enhancement of the TE up to 93% for Pt NPs and near 85% for Ag NPs. In a later investigation (Lin et al., 2019) the same group conducted a study comparing the analytical performance of their system with a conventional MicroMist nebulizer/cyclonic spray chamber and the same high-performance concentric nebulizer (HPCN) with a heated cyclonic spray chamber combined with a desolvation system. Both systems based on HPCN show similar results (above 99%) for the TE of Pt NPs, which is almost 10 times higher than the "conventional" sample introduction systems. The relative independence of the observed TE from the matrix for the combination HPCN/desolvation system makes it a suitable choice for the spICP-MS application.

10.3.2 Microdroplet generation systems

Monodisperse microdroplet generation (MDG) systems are well-known tools for fundamental studies of processes in flames (Hieftje & Malmstadt, 1969) and inductively coupled plasma combined with either optical emission or mass spectrometric detection (Olesik, 1997). Several unique possibilities offered by MDG, such as the controllable frequency of generation of microdroplets with tuneable size and repeatable volume carried with high and countable transport efficiency to the plasma, make them an attractive sample introduction choice for the characterization of NPs. Garcia, Murtazin, Groh, Horvatic, and Niemax (2010) first reported the application of the MDG for the characterization of Au and SiO_2 nano- and microparticles by ICP-OES. The authors demonstrated the possibility of introducing a specific mass of the analyte by microdroplets from dissolved standards for calibration in particle size characterization. Later Gschwind et al. (Gschwind et al., 2011; Gschwind, Hagendorfer, Frick, & Günther, 2013; Koch et al., 2013) proposed several modifications of MDG systems, among them using helium as the carrier gas, to improve their performance for spICP-MS applications. By the modified setup, authors achieved TE above 95% for droplets of size <43 μm at relatively high droplet generation frequency (Gschwind et al., 2013). As a result of their comparative study (Gschwind, de Lourdes Aja Montes, & Günther, 2015) on the determination of particle number concentration of Ag and Au NPs by using PN and MDG introductions, the authors reported that both systems have advantages and disadvantages. The authors illustrated the importance of the correct determination of TE when PN is used for a sample introduction, since this may lead to underestimation of PNC by over 40%. Compared to PN, MDG showed a higher overall recovery and TE. The use of MDG requires less dilution of the samples, and because of the less interaction with different materials (no nebulizer or spray chamber) losses of NPs due to sorption are minimized. However, the obtained results are less reproducible, as well as a longer measurement time needed, which is not favorable due to the more expensive analysis and intrinsic for ICP-MS sensitivity drift.

It should be emphasized that the MDG is still highly specialized equipment and is prone to greater risk of contamination and clogging compared to a PN. The necessity of a specially designed interface, for drying the generated pL volume droplets and to transport the resulting metal-containing particles to the ICP, also adds complexity to their application for routine use.

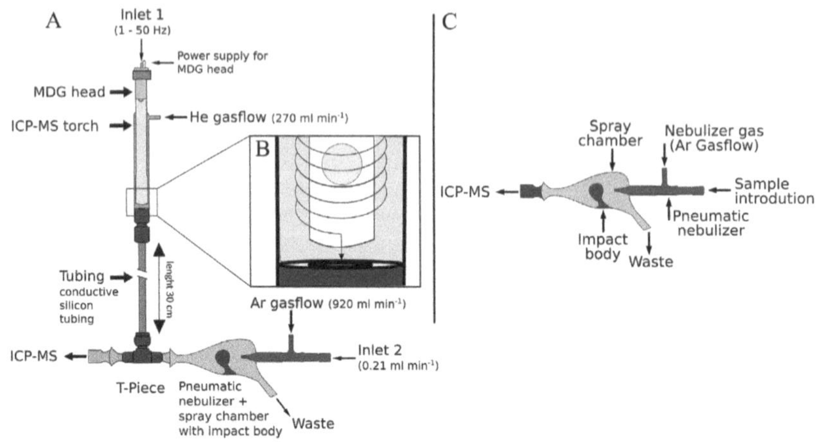

FIG. 10.3 Design of the PN-MDG interface setup (not drawn to scale). (A) Overview of the introduction setup including gas and liquid flow rates. (B) Inset: zoomed view of the He gas helical "curtain" flow preventing collision of droplets with interface surfaces. (C) Standard setup for spICP-MS measurements: PN inserted into an impact bead spray chamber connected directly to an ICP-MS. *From Rosenkranz, D., Kriegel, F. L., Mavrakis, E., Pergantis, S. A., Reichardt, P., Tentschert, J., et al. (2020).* Analytica Chimica Acta, 1099, *16–25. https://doi.org/10.1016/j.aca.2019.11.043.*

10.3.3 Combined PN/MDG systems

In an attempt to employ the benefits of both sample introduction systems, recently, several investigations have utilized a combined PN/MDG sample introduction system (Fig. 10.3) for the characterization of the NPs by quadrupole (Rosenkranz et al., 2020), time-of-flight (Gundlach-Graham & Mehrabi, 2020; Hendriks et al., 2019; Hendriks, Gundlach-Graham, & Günther, 2018), or sector field (Ramkorun-Schmidt, Pergantis, Esteban-Fernández, Jakubowski, & Günther, 2015) instruments.

In all cases, the dissolved analyte standard is introduced by the MDG system and the NP suspension by pneumatic nebulization. The opportunity to perform online calibration by ionic standards, for NP size determination, was reported for TOF instruments (Gundlach-Graham & Mehrabi, 2020; Hendriks et al., 2018; Hendriks, Ramkorun-Schmidt, et al., 2019). The fast detection capabilities enable the exploitation of the multielement capabilities of the method in a single nanoparticle measurement mode, allowing the addition of a tracer to ionic calibration standard to distinguish analyte signals from standards and NP ones. In this setup, microdroplets and NP suspension are introduced concurrently into the ICP, which ensures calibrant microdroplets and analyte NPs experience the same plasma conditions and obtained signals are subjected to the same matrix-related enhancement or suppression. Thus, online microdroplet calibration can be used to account for plasma- and MS-related matrix effects (Hendriks, Ramkorun-Schmidt, et al., 2019).

As an additional benefit, the flexibility of operation of this sample introduction system allows using microdroplets as NP proxies for fundamental studies of the effects of instrument parameters and data analysis procedures on the results produced by sp-ICP-MS (Gundlach-Graham & Mehrabi, 2020).

10.4 Future perspectives

Although Bolea et al. (2021), in their recent review, considered the possibility of spICP-MS reaching their "maturity" level and becoming a routine technique for nanoparticle characterization, further research is still needed to improve the analytical characteristics of the method. The reduced number of commercially available NP reference materials that can be utilized as calibration standards and quality control complicates the validation of methods and the traceability of obtained results. Currently, accessible reference materials are mainly on pristine nanoparticles of specific composition, which requires often to be applied an indirect calibration, based on the determination of the nebulization efficiency and the use of dissolved standards, to obtain quantitative information about the size and size distribution of studied nanoparticles. In this case, the performance of the sample introduction system, in view of the achievable aerosol transport efficiency, robustness, and matrix sensitivity, is of great importance to receive unbiased results. The desired quality of the sample introduction system for spICP-MS is the possibility for the production of small-size near monodisperse aerosol droplets, which would allow a reasonable increase in transport efficiency without compromising sample throughput. Operating at better sensitivity and stability will also reduce the obtained size detection limits. The high matrix tolerance is another important parameter of the sample introduction system since the sensitivity and the transport efficiency are highly influenced by the matrix. Therefore, accurate characterization of the real samples always requires considering this effect and using suitable strategies to deal with it. The development of the instrumental technique that allows a significant increase in the frequency of detection (especially for quadrupole ICP-MS) and enables the monitoring of more than one isotope in spICP-MS mode provides an excellent opportunity to investigate the applicability of approaches such as internal standardization or isotope dilution for matrix effect correction. The application of sample introduction systems by which it is possible to perform this correction in online mode also will expand and facilitates spICP-MS application. The above comments clearly emphasize the importance of sample introduction to obtain reliable results by spICP-MS. Therefore, the efforts for further implementation of the sample introduction systems with improved characteristics can significantly evolve the potential of spICP-MS future applications.

References

Abad-Álvaro, I., Peña-Vázquez, E., Bolea, E., Bermejo-Barrera, P., Castillo, J. R., & Laborda, F. (2016). Evaluation of number concentration quantification by single-particle inductively coupled plasma mass spectrometry: Microsecond vs. millisecond dwell times. *Analytical and Bioanalytical Chemistry, 408*(19), 5089–5097. https://doi.org/10.1007/s00216-016-9515-y.

Bi, X., Lee, S., Ranville, J. F., Sattigeri, P., Spanias, A., Herckes, P., et al. (2014). Quantitative resolution of nanoparticle sizes using single particle inductively coupled plasma mass spectrometry with the K-means clustering algorithm. *Journal of Analytical Atomic Spectrometry, 29*(9), 1630–1639. https://doi.org/10.1039/C4JA00109E.

Bolea, E., Jimenez, M. S., Perez-Arantegui, J., Vidal, J. C., Bakir, M., Ben-Jeddou, K., et al. (2021). Analytical applications of single particle inductively coupled plasma mass spectrometry: A comprehensive and critical review. *Analytical Methods, 13*(25), 2742–2795. https://doi.org/10.1039/D1AY00761K.

Cuello-Nuñez, S., Abad-Álvaro, I., Bartczak, D., del Castillo Busto, M. E., Ramsay, D. A., Pellegrino, F., et al. (2020). The accurate determination of number concentration of inorganic nanoparticles using spICP-MS with the dynamic mass flow approach. *Journal of Analytical Atomic Spectrometry, 35*(9), 1832–1839. https://doi.org/10.1039/C9JA00415G.

Degueldre, C., & Favarger, P. Y. (2003). Colloid analysis by single particle inductively coupled plasma-mass spectroscopy: A feasibility study. *Colloids and Surfaces A: Physicochemical and Engineering Aspects, 217*(1–3), 137–142. Elsevier https://doi.org/10.1016/S0927-7757(02)00568-X.

Degueldre, C., & Favarger, P. Y. (2004). Thorium colloid analysis by single particle inductively coupled plasma-mass spectrometry. *Talanta, 62*(5), 1051–1054. https://doi.org/10.1016/j.talanta.2003.10.016.

Degueldre, C., Favarger, P. Y., & Bitea, C. (2004). Zirconia colloid analysis by single particle inductively coupled plasma-mass spectrometry. *Analytica Chimica Acta, 518*(1–2), 137–142. https://doi.org/10.1016/j.aca.2004.04.015.

Flores, K., Turley, R. S., Valdes, C., Ye, Y., Cantu, J., Hernandez-Viezcas, J. A., et al. (2021). Environmental applications and recent innovations in single particle inductively coupled plasma mass spectrometry (SP-ICP-MS). *Applied Spectroscopy Reviews, 56*(1), 1–26. https://doi.org/10.1080/05704928.2019.1694937.

Galbács, G., Kéri, A., Kohut, A., Veres, M., & Geretovszky, Z. (2021). Nanoparticles in analytical laser and plasma spectroscopy-a review of recent developments in methodology and applications. *Journal of Analytical Atomic Spectrometry, 36*(9), 1826–1872. https://doi.org/10.1039/d1ja00149c.

Garcia, C. C., Murtazin, A., Groh, S., Horvatic, V., & Niemax, K. (2010). Characterization of single Au and SiO_2 nano- and microparticles by ICP-OES using monodisperse droplets of standard solutions for calibration. *Journal of Analytical Atomic Spectrometry, 25*(5), 645–653. https://doi.org/10.1039/B921041E.

Geiss, O., Bianchi, I., Bucher, G., Verleysen, E., Brassinne, F., Mast, J., et al. (2022). Determination of the transport efficiency in spICP-MS analysis using conventional sample introduction systems: An interlaboratory comparison study. *Nanomaterials, 12*(4). https://doi.org/10.3390/nano12040725.

Gschwind, S., de Lourdes Aja Montes, M., & Günther, D. (2015). Comparison of sp-ICP-MS and MDG-ICP-MS for the determination of particle number concentration. *Analytical and Bioanalytical Chemistry, 407*(14), 4035–4044. https://doi.org/10.1007/s00216-015-8620-7.

Gschwind, S., Flamigni, L., Koch, J., Borovinskaya, O., Groh, S., Niemax, K., et al. (2011). Capabilities of inductively coupled plasma mass spectrometry for the detection of nanoparticles carried by monodisperse microdroplets. *Journal of Analytical Atomic Spectrometry, 26*(6), 1166–1174. https://doi.org/10.1039/C0JA00249F.

Gschwind, S., Hagendorfer, H., Frick, D. A., & Günther, D. (2013). Mass quantification of nanoparticles by single droplet calibration using inductively coupled plasma mass spectrometry. *Analytical Chemistry, 85*(12), 5875–5883. https://doi.org/10.1021/ac400608c.

Gundlach-Graham, A., Hendriks, L., Mehrabi, K., & Günther, D. (2018). Monte Carlo simulation of low-count signals in time-of-flight mass spectrometry and its application to single-particle detection. *Analytical Chemistry, 90*(20), 11847–11855. https://doi.org/10.1021/acs.analchem.8b01551.

Gundlach-Graham, A., & Mehrabi, K. (2020). Monodisperse microdroplets: A tool that advances single-particle ICP-MS measurements. *Journal of Analytical Atomic Spectrometry, 35*(9), 1727–1739. https://doi.org/10.1039/D0JA00213E.

Hendriks, L., Gundlach-Graham, A., & Günther, D. (2018). Analysis of inorganic nanoparticles by single-particle inductively coupled plasma time-of-flight mass spectrometry. *Chimia, 72*(4), 221. https://doi.org/10.2533/chimia.2018.221.

Hendriks, L., Gundlach-Graham, A., & Günther, D. (2019). Performance of sp-ICP-TOFMS with signal distributions fitted to a compound Poisson model. *Journal of Analytical Atomic Spectrometry, 34*(9), 1900–1909. https://doi.org/10.1039/C9JA00186G.

Hendriks, L., Ramkorun-Schmidt, B., Gundlach-Graham, A., Koch, J., Grass, R. N., Jakubowski, N., et al. (2019). Single-particle ICP-MS with online microdroplet calibration: Toward matrix independent nanoparticle sizing. *Journal of Analytical Atomic Spectrometry, 34*(4), 716–728. https://doi.org/10.1039/C8JA00397A.

Hieftje, G. M., & Malmstadt, H. V. (1969). A new approach to flame spectrometric analysis utilizing isolated droplets of sample solution. *Analytical Chemistry, 41*(13), 1735–1744. https://doi.org/10.1021/ac60282a022.

Ho, K. S., Lee, W. W., & Chan, W. T. (2015). Effects of ionization potential of an element and boiling point of the corresponding oxide on the sensitivity of ICP-MS. *Journal of Analytical Atomic Spectrometry, 30*(10), 2066–2073. Royal Society of Chemistry https://doi.org/10.1039/c5ja00137d.

Ho, K. S., Lui, K. O., Lee, K. H., & Chan, W. T. (2013). Considerations of particle vaporization and analyte diffusion in single-particle inductively coupled plasma-mass spectrometry. *Spectrochimica Acta - Part B Atomic Spectroscopy, 89*, 30–39. https://doi.org/10.1016/j.sab.2013.08.012.

Kaňa, A., Loula, M., Koplík, R., Vosmanská, M., & Mestek, O. (2019). Peak bordering for ultrafast single particle analysis using ICP-MS. *Talanta, 197*, 189–198. https://doi.org/10.1016/j.talanta.2019.01.030.

Kaynarova, L. I., Georgieva, D. L., & Stefanova, V. M. (2022). An approach to estimate the contribution of signal noise to the diameter uncertainty of individual silver nanoparticles and resolution of spICP-MS analysis. *Journal of Analytical Atomic Spectrometry, 37*, 1484–1500. https://doi.org/10.1039/D2JA00039C.

Koch, J., Flamigni, L., Gschwind, S., Allner, S., Longerich, H., & Günther, D. (2013). Accelerated evaporation of microdroplets at ambient conditions for the on-line analysis of nanoparticles by inductively-coupled plasma mass spectrometry. *Journal of Analytical Atomic Spectrometry, 28*(11), 1707–1717. https://doi.org/10.1039/C3JA50052G.

Laborda, F., Gimenez-Ingalaturre, A. C., Bolea, E., & Castillo, J. R. (2019). Single particle inductively coupled plasma mass spectrometry as screening tool for detection of particles. *Spectrochimica Acta Part B: Atomic Spectroscopy, 159*, 105654. https://doi.org/10.1016/j.sab.2019.105654.

Laborda, F., Gimenez-Ingalaturre, A. C., Bolea, E., & Castillo, J. R. (2020). About detectability and limits of detection in single particle inductively coupled plasma mass spectrometry. *Spectrochimica Acta Part B: Atomic Spectroscopy, 169*, 105883. https://doi.org/10.1016/j.sab.2020.105883.

Laborda, F., Gimenez-Ingalaturre, A. C., Bolea, E., Milačič, R., Ščančar, J., Goenaga-Infante, H., et al. (2021). *Analysis and characterisation of metal-based nanomaterials. Vol. 93* (pp. 35–67). Elsevier. https://doi.org/10.1016/bs.coac.2021.02.012.

Laborda, F., Jiménez-Lamana, J., Bolea, E., & Castillo, J. R. (2011). Selective identification, characterization and determination of dissolved silver(i) and silver nanoparticles based on single particle detection by inductively coupled plasma mass spectrometry. *Journal of Analytical Atomic Spectrometry, 26*(7), 1362–1371. https://doi.org/10.1039/C0JA00098A.

Laborda, F., Jiménez-Lamana, J., Bolea, E., & Castillo, J. R. (2013). Critical considerations for the determination of nanoparticle number concentrations, size and number size distributions by single particle ICP-MS. *Journal of Analytical Atomic Spectrometry, 28*(8), 1220–1232. https://doi.org/10.1039/C3JA50100K.

Laura Soriano, M., Zougagh, M., Valcárcel, M., & Ríos, Á. (2018). Analytical nanoscience and nanotechnology: Where we are and where we are heading. *Talanta, 177*, 104–121. https://doi.org/10.1016/j.talanta.2017.09.012.

Lee, W.-W., & Chan, W.-T. (2015). Calibration of single-particle inductively coupled plasma-mass spectrometry (SP-ICP-MS). *Journal of Analytical Atomic Spectrometry, 30*(6), 1245–1254. https://doi.org/10.1039/C4JA00408F.

Lin, F., Miyashita, S., Inagaki, K., Liu, Y.-H., & Hsu, I.-H. (2019). Evaluation of three different sample introduction systems for single-particle inductively coupled plasma mass spectrometry (spICP-MS) applications. *Journal of Analytical Atomic Spectrometry, 34*(2), 401–406. https://doi.org/10.1039/C8JA00295A.

Liu, J., Murphy, K. E., Winchester, M. R., & Hackley, V. A. (2017). Overcoming challenges in single particle inductively coupled plasma mass spectrometry measurement of silver nanoparticles. *Analytical and Bioanalytical Chemistry, 409*(25), 6027–6039. https://doi.org/10.1007/s00216-017-0530-4.

López-Lorente, Á. I., & Valcárcel, M. (2014). Analytical nanoscience and nanotechnology. In *Vol. 66. Comprehensive analytical chemistry* (pp. 3–35). Elsevier B.V. https://doi.org/10.1016/B978-0-444-63285-2.00001-8.

Loula, M., Kaňa, A., Koplík, R., Hanuš, J., Vosmanská, M., & Mestek, O. (2019). Analysis of silver nanoparticles using single-particle inductively coupled plasma–mass spectrometry (ICP-MS): Parameters affecting the quality of results. *Analytical Letters, 52*(2), 288–307. https://doi.org/10.1080/00032719.2018.1459657.

Meermann, B., & Nischwitz, V. (2018). ICP-MS for the analysis at the nanoscale—A tutorial review. *Journal of Analytical Atomic Spectrometry, 33*(9), 1432–1468. https://doi.org/10.1039/C8JA00037A.

Miyashita, S., Mitsuhashi, H., Fujii, S., Takatsu, A., Inagaki, K., & Fujimoto, T. (2017). High transport efficiency of nanoparticles through a total-consumption sample introduction system and its beneficial application for particle size evaluation in single-particle ICP-MS. *Analytical and Bioanalytical Chemistry, 409*(6), 1531–1545. https://doi.org/10.1007/s00216-016-0089-5.

Mourdikoudis, S., Pallares, R. M., & Thanh, N. T. K. (2018). Characterization techniques for nanoparticles: Comparison and complementarity upon studying nanoparticle properties. *Nanoscale, 10*(27), 12871–12934. https://doi.org/10.1039/c8nr02278j.

Mozhayeva, D., & Engelhard, C. (2019). A quantitative nanoparticle extraction method for microsecond time resolved single-particle ICP-MS data in the presence of a high background. *Journal of Analytical Atomic Spectrometry, 34*(8), 1571–1580. https://doi.org/10.1039/C9JA00042A.

Olesik, J. W. (1997). Investigating the fate of individual sample droplets in inductively coupled plasmas. *Applied Spectroscopy, 51*(5), 158A–175A. https://doi.org/10.1366/0003702971940792.

Olesik, J. W., & Gray, P. J. (2012). Considerations for measurement of individual nanoparticles or microparticles by ICP-MS: Determination of the number of particles and the analyte mass in each particle. *Journal of Analytical Atomic Spectrometry, 27*(7), 1143–1155. https://doi.org/10.1039/C2JA30073G.

Olesik, J. W., Hartshorne, J. A., Casey, N., Linard, E., & Dettman, J. R. (2021). Further insight into analyte transport processes and water vapor, aerosol loading in ICP-OES and ICP-MS. *Paul Farnsworth: Setting High Standards in Science and Life, 176*, 106038. https://doi.org/10.1016/j.sab.2020.106038.

Pace, H. E., Rogers, N. J., Jarolimek, C., Coleman, V. A., Gray, E. P., Higgins, C. P., et al. (2012). Single particle inductively coupled plasma-mass spectrometry: A performance evaluation and method comparison in the determination of nanoparticle size. *Environmental Science & Technology, 46*(22), 12272–12280. https://doi.org/10.1021/es301787d.

Pace, H. E., Rogers, N. J., Jarolimek, C., Coleman, V. A., Higgins, C. P., & Ranville, J. F. (2011). Determining transport efficiency for the purpose of counting and sizing nanoparticles via single particle inductively coupled plasma mass spectrometry. *Analytical Chemistry, 83*(24), 9361–9369. https://doi.org/10.1021/ac201952t.

Ramkorun-Schmidt, B., Pergantis, S. A., Esteban-Fernández, D., Jakubowski, N., & Günther, D. (2015). Investigation of a combined microdroplet generator and pneumatic nebulization system for quantitative determination of metal-containing nanoparticles using ICPMS. *Analytical Chemistry, 87*(17), 8687–8694. https://doi.org/10.1021/acs.analchem.5b01604.

Rosenkranz, D., Kriegel, F. L., Mavrakis, E., Pergantis, S. A., Reichardt, P., Tentschert, J., et al. (2020). Improved validation for single particle ICP-MS analysis using a pneumatic nebulizer/microdroplet generator sample introduction system for multi-mode nanoparticle determination. *Analytica Chimica Acta, 1099*, 16–25. https://doi.org/10.1016/j.aca.2019.11.043.

Stefanova, V. M., Kaynarova, L. I., & Georgieva, D. L. (2019). General characteristics of silver nanoparticles analysis by single particle inductively coupled plasma mass spectrometry. *Bulgarian Chemical Communications, 51*, 77–84. https://bcc.iche.bas.bg/BCC_Volumes/Volume_51_Special_D_2019/BCC-51-D-2019-77-84-Stefanova-17.pdf.

Strenge, I., & Engelhard, C. (2016). Capabilities of fast data acquisition with microsecond time resolution in inductively coupled plasma mass spectrometry and identification of signal artifacts from millisecond dwell times during detection of single gold nanoparticles. *Journal of Analytical Atomic Spectrometry, 31*(1), 135–144. https://doi.org/10.1039/C5JA00177C.

Todolí, J.-L., Maestre, S., Mora, J., Canals, A., & Hernandis, V. (2000). Comparison of several spray chambersoperating at very low liquid flow ratesin inductively coupled plasma atomic emission spectrometry. *Fresenius' Journal of Analytical Chemistry, 368*(8), 773–779. https://doi.org/10.1007/s002160000583.

Tuoriniemi, J., Cornelis, G., & Hassellöv, M. (2012). Size discrimination and detection capabilities of single-particle ICPMS for environmental analysis of silver nanoparticles. *Analytical Chemistry, 84*(9), 3965–3972. https://doi.org/10.1021/ac203005r.

Tuoriniemi, J., Cornelis, G., & Hassellöv, M. (2014). Improving the accuracy of single particle ICPMS for measurement of size distributions and number concentrations of nanoparticles by determining analyte partitioning during nebulisation. *Journal of Analytical Atomic Spectrometry, 29*(4), 743–752. https://doi.org/10.1039/C3JA50367D.

Vidmar, J., Milačič, R., Ščančar, J., Goenaga-Infante, H., & Vidmar, J. (2021). *Analysis and characterisation of metal-based nanomaterials. Vol. 93* (pp. 345–380). Elsevier. https://doi.org/10.1016/bs.coac.2021.02.008.

CHAPTER

11

Application of online sample introduction systems for single-cell analysis with ICP-MS

Roberto Álvarez-Fernández García, Mario Corte-Rodríguez, Paula García-Cancela, Jörg Bettmer, and María Montes-Bayón

Department of Physical and Analytical Chemistry, Faculty of Chemistry and Health Research Institute of Asturias (ISPA), University of Oviedo, Oviedo, Spain

11.1 Introduction

Heterogeneity between individual cells is obvious in tissues and organisms, but it can also be observed in populations of isogenic cells that have been cultured under identical conditions (Altschuler & Wu, 2010). In fact, the cell-to-cell variation on the genes, transcripts, and proteins can be the cause of several diseases, including cancer, neurological and developmental disorders (Shapiro, Biezuner, & Linnarsson, 2013). Similarly, metal control is tightly regulated by cells (homeostasis) due to their role in processes such as cell respiration, growth, enzymatic reactions, cell proliferation, and immune function. Alteration on this homeostasis can be associated with the initiation of important pathological conditions like diabetes or cancer (Ganz & Nemeth, 2011; Jansen, Karges, & Rink, 2009; Lichten & Cousins, 2009; Turski & Thiele, 2009). Thus, the measurement of cell-to-cell transient changes in metal ion concentrations may lead to the generation of "signatures" representing how cells control levels of metal ions and how they influence cellular function. On the other hand, the widespread use of metallodrugs (e.g., cisplatin and related compounds, arsenic, or gold-based preparations) to treat numerous diseases

Analytical Nebulizers
https://doi.org/10.1016/B978-0-323-91181-8.00011-4

implies the incorporation of these metallic structures into cells. The intercellular variation in the uptake of these structures can be essential to evaluate the initiation of, for instance, drug resistance mechanisms. To explore all these scenarios, the development of analytical/biotechnological strategies that enable the survey of heterogeneity between cells and within populations, previously assumed to be homogeneous, is an active area of investigation.

Elemental analysis in small objects (e.g., nanoparticles) has been successfully accomplished by inductively coupled plasma-mass spectrometry (ICP-MS) since the work of Degueldre et al., which established the considerations to conduct such type of experiment called single-particle-ICP-MS (sp-ICP-MS) (Degueldre & Favarger, 2003). The subsequent extension to single-cell analysis has been relatively recent, although the potential of the technique for such measurements was provided by Li, Armstrong, and Houk (2005). This development coincided with the renaissance of time-of-flight (TOF) mass analyzers in connection with ICP ionization offering fast scanning capabilities for quasisimultaneous multielement/isotope detection maintaining high sensitivity. In addition, an important development of high efficiency and low consumption sample introduction systems (extensively covered in the first chapters of this book) provided the ideal platform for the development of numerous applications of single-cell-ICP-MS (sc-ICP-MS). Pending future improvements in miniaturization, integration and detection sensitivity, the analysis of single cells by ICP-MS is nowadays an almost mature topic with increasing applications in all kinds of disciplines (e.g., clinical, microbiological, or environmental) that will be revised in the present chapter.

11.2 Fundamentals

11.2.1 Considerations for sc-ICP-MS experiments

Conceptually, single-cell and single-particle analyses via ICP-MS are based on the same principle: the formation of a transient cloud of ions (in the range of several hundred microseconds) corresponding to the individual object (particle or cell). The formed ions are then further separated into the mass analyzer as function of their mass-to-charge ratio and then detected with generally short integration times ($<10\,\mu s$). However, the complexity affecting both kinds of measurement is diverse. The more remarkable differential aspects are given below:

1. Nanoparticles analyzed by sp-ICP-Ms are generally composed by single metals (Ag, Au), metal oxides (Fe_2O_3, TiO_2, ZrO_2, CuO, ZnO), and in the more complex case, mixtures of two or three elements (CdSe,

CdS, Fe/Si); meanwhile, cells are composed of many different elements coexisting within this microentity.
2. In general, elemental distribution in the studied nanoparticles is relatively homogeneous, while the cellular composition is highly heterogeneous (the cell wall is very different in elemental composition to the nucleus or the cytosol).
3. Nanoparticles are robust in terms of transport and nebulization, while cells are extremely sensitive to such processes and to the surrounding environment.

Lastly, the size of nanoparticles suitable for sp-ICP-MS, although strongly depending on the composition of the particle, ranges from a few to several hundred nm. However, cells are generally larger (from 1 to 30–50 μm, depending on the origin such as human or other mammalian cell lines, bacteria, yeast, or unicellular algae) and more prone to deposit on the container. In addition, parameters like cell morphology or stability, determined by the composition and structure of their cellular envelope, have an influence on their analysis strongly affecting their intact transport into the plasma.

Among the cell models available, bacterial cells are particularly suitable for sc-ICP-MS analysis. They are prokaryotic unicellular microorganisms that comprise a very heterogeneous domain of different morphologies and exhibit smaller sizes ranging between 0.2 and 2 μm. For example, *Escherichia coli* has an approximate diameter of 0.5 μm and a cell volume of 0.6–0.7 μm^3 (Kubitschek, 1990). The bacterial cytoplasm is surrounded by an envelope consisting of cytoplasmatic membrane, cell wall and, in some cases, a second outer membrane. The main component of this cell wall is a polysaccharide of thickness 3–6 nm in the case of Gram-negative and going up to 10–40 nm in Gram-positive bacteria (Turner, Vollmer, & Foster, 2014; Vollmer & Seligman, 2010). Such a structure confers bacterial cells robustness to be subjected to pneumatic nebulization when introduced into the ICP-MS. In addition, their small sizes permit their transportation with high efficiencies maintaining cell integrity.

Human and animal cell cultures are widely used in biomedical research, since they preserve their physiological, biochemical, and genetic properties and allow cellular research without the systemic variations and ethical aspects affecting studies that use complete organisms. However, each cell type has its own features that can be very different depending on their origin. Firstly, the cell size can be a differential feature. For instance, among human cells, the sizes range from about 7 μm for red blood cells up to 150 μm in some ovules. Nevertheless, all of them have common structural features. Their cellular envelope is constituted by a cellular membrane that surrounds the cytoplasm containing the cell organelles.

Since most human and animal cells grow adherent to a solid surface, the first challenge for their handling requires the dissembling from the solid surface preserving their structure and avoiding the formation of aggregates. This is normally done either mechanically or by the addition of low concentrations of trypsin, an enzyme that hydrolyzes cell adhesion proteins to the surface. Both types of treatment can also cause cell damage and need to be carefully conducted (Huang et al., 2010). Once cells are suspended, they have to be immediately analyzed to prevent cell degradation. However, among eukaryotic cells, not all show the same sensitivity toward manipulation. For instance, yeast cells, being eukaryotic unicellular microorganisms that belong to the fungi kingdom, have been successfully analyzed using sc-ICP-MS due to the presence of a second layer, apart from the lipid bilayer membrane, which is the cell wall that makes them mechanically resistant. Also, plant cells can be larger than animal cells ranging from 10 to 100 μm. Beyond size, the main structural differences between plant and animal cells lie in a few additional structures found in plant cells. These structures include chloroplasts, the cell wall, and vacuoles that increase their robustness.

Finally, algae are a very heterogeneous phylogenetic group of organisms. They can be unicellular or multicellular and include green algae, brown algae, red algae, and various groups of unicellular organisms such as dinoflagellates or diatoms, reflecting the huge diversity of the group. Unicellular algae are microorganisms of a variable size (from a few micrometers to millimeters) that usually have a cell wall (except for euglena). The structure and composition of the cell walls of algae vary considerably but are mainly composed of cellulose fibers and, to a less extent, other polysaccharides (Domozych et al., 2012). These walls remain intact even after the death of the alga making them suitable for sc-ICP-MS analysis.

Although some types of cells can be analyzed in their native form in a suspension-based approach, as previously stated, the cell robustness is a crucial feature to determine whether a specific cell type is susceptible to direct analysis in its native form or needs a chemical fixation to avoid, for example, the unwanted cell lysis due to changes in the osmotic pressure prior to the measurement. In any case, the chemical fixation methods for cells in suspension are simpler and less prone to modify cell status than those applied to fix cells on solid surfaces for direct solid analysis or imaging. Cell fixation is very useful when there is a time lapse between cell collection and analysis to avoid cell degradation or when the sample preparation can compromise cell structure. This strategy helps to preserve the architecture and composition of cells since it provides mechanical rigidity and prevents cell decomposition and autolysis.

11.2.2 Mass spectrometric features

The detection of single events (nanoparticles or biological cells) requires highly sophisticated mass analyzers as the ion clouds produced by the sought small objects (cells, particles) have a duration in the range of the few hundred μs. Using inductively coupled plasma as a highly efficient ion source, different mass analyzers are nowadays available, such as quadrupoles, sector fields, and time-of-flight systems. All of them are capable of detecting single cells with some advantages and/or restrictions as summarized in Fig. 11.1. The most decisive features of their suitability are categorized in detection sensitivity (limits of detection), detection speed, and multiisotope detection capabilities.

11.2.3 Transport efficiency in sc-ICP-MS

Transport efficiency is a key parameter in sp-ICP-MS as well as in sc-ICP-MS. The determination of this parameter is another differential aspect between the analysis of individual particles and individual cells with ICP-MS. In the previous chapter of this book, the authors report on how to calculate transport efficiencies for nanoparticles. However, in the absence of certified solutions with a well-established cell number concentration, as in the case of nanoparticles, transport efficiency calculations basically rely on two types of measurements: (i) the monitoring of polystyrene bead standards containing known concentrations of metal isotopes and (ii) the measurement of the cell suspension using cell counting systems (e.g., flow cytometry, cell counters, etc.) prior to sc-ICP-MS analysis of constitutive elements. In the first case, the important differences regarding density between cells and polystyrene beads impede the direct comparison of transport efficiencies. On the other hand, flow cytometry provides a more accurate cell counting method than other methods such as hemocytometry by measuring the forwarded and scattered light of the

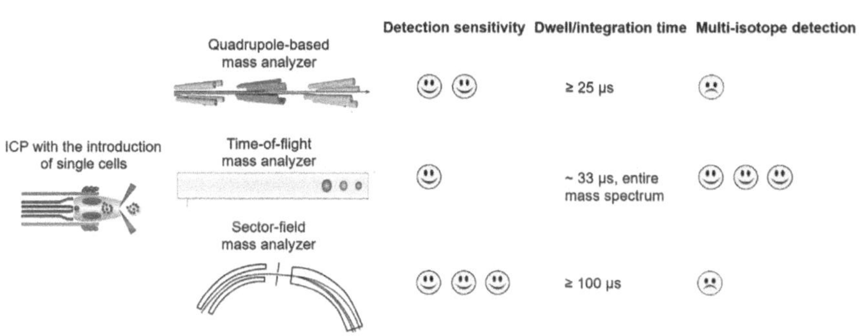

FIG. 11.1 Overview on mass analyzers commonly used for single-cell ICP-MS analysis.

individual cells when passing through a laser beam and is an adequate way to know the cell number density.

However, the comparison of such a number with the frequency of the events by ICP-MS must be carefully conducted, since only elements that do not exhibit a differential uptake (among cell populations) could be, in principle, used for data normalization. In this regard, essential elements that constitute the backbone of cells (e.g., P, S, Fe, Cu, Zn) would be the traces to follow for the evaluation of the frequency of events and thus the transport efficiency. The main limitation is associated with the fact that all these elements suffer from important polyatomic interferences, often showing high backgrounds that can hamper their determination. These limitations have been overcome using triple quadrupole mass analyzers (ICP-TQ-MS), particularly for the measurement of P and S on their respective oxides (Álvarez-Fernández García et al., 2020).

11.2.4 Quantitative analysis

Data evaluation in single-cell analysis follows the same principles as those already established for single-particle analysis, which have already been discussed by Degueldre et al. (Degueldre & Favarger, 2003; Degueldre, Favarger, & Bitea, 2004; Degueldre, Favarger, Rossé, & Wold, 2006; Degueldre, Favarger, & Wold, 2006) in the previous chapter. In this regard, all the equations governing the sp-ICP-MS analysis can be adapted to the analysis of single cells by ICP-MS. The data processing requires a previous discrimination between the data corresponding to cellular events and the background signal, proportional to the concentration of the dissolved forms of the element. This implies the determination of an intensity threshold value above which the data is considered as cell events. The threshold can be set empirically, but different algorithms have also been developed, based on iterative procedures that use as a criterion to set the threshold value as the sum of the average background plus n-fold the standard deviation (σ). The events complying with the criteria are filtered and the procedure is repeated until no new data points meet the requirement. Finally, the data points n-σ above the mean are the countable data (cell) events. Commonly, the 3σ criterion has been applied to set the threshold values (Laborda, Jiménez-Lamana, Bolea, & Castillo, 2011; Pace et al., 2012). However, other authors have also used 5σ (Tuoriniemi, Cornelis, & Hassellöv, 2012) or other methods that involve signal deconvolution algorithms (Cornelis & Hassellöv, 2014).

Once the cell events are filtered as described, signal calibration strategies are needed to transform the intensity of each event into the mass of the monitored element per cell. For this aim, it is common to apply an external calibration using liquid standards that are measured under the same experimental conditions as cell suspensions following the principles of single-particle analysis. After obtaining the calibration curve, the mass of element per cell can then be calculated by relating the height or area of each cell event to that obtained by the external calibration, and following the equation:

$$m_c = (\eta \cdot F \cdot t \cdot I)/b \tag{11.1}$$

where m_c is the mass of the element in the cell, F is the sample flow rate, t is the dwell time, I is the net event intensity (or area), b is the slope of the calibration curve using external liquid standards and η is the transport efficiency of the inorganic standard solutions (normally different than 100%). It is important to remark that the transport efficiency of cells cannot be used to estimate η, since cells can be transported to the plasma in a very different manner (seen in the previous section), due to their different physical features (size, shape, robustness, etc.). Therefore, some authors estimate η based on the measurement of nanoparticles by sp-ICP-MS, since due to their much smaller size (usually 30 or 60 nm nanoparticles vs ~10 μm cells), it can be assumed that nanoparticles and liquid standards behave more equally in their transport to the plasma.

11.3 Applications

The adequately diluted cell suspensions need to be introduced into the ICP-MS and analyzed in a time-resolved manner. For this task, pneumatic nebulizers have been mostly used, being robust, stable, easy-handling, and low-cost alternatives. In fact, different combinations of pneumatic systems, in terms of different designs of nebulizers and spray chambers, have been applied for the analysis of individual cells in a plethora of applications. Some of them will be covered in the following sections.

11.3.1 Single-cell analysis of constitutive elements

The endogenous trace metals (e.g., Fe, Cu, Zn, Mn, and Co) and nonmetals (e.g., P and S) are essential constituents in cells, which are found in metalloproteins, metalloenzymes, nucleic acids, and adenosine triphosphate (ATP) among others. They play critical roles in numerous biological processes, such as oxidative-reductive balance, gene expression, cell cycle regulation, and energy metabolism. The contents of these essential elements in biological tissues or cells are associated with various pathophysiological conditions. For instance, the cellular requirement for Fe is directly correlated with the cell type, the rate of cell growth, and the stage of cell differentiation. Similarly, several tissue abnormalities and disease states in humans have been reported to be associated with either reduced or elevated levels of Cu.

Regarding the analysis of essential elements by sc-ICP-MS, it is noteworthy that the first studies using Meinhard-type nebulizers in combination with Scott-type spray chambers are already from the year 1994, when the group of Nomizu et al. described the analysis of Ca in single mammalian cells of 10–20 μm diameter (Nomizu et al., 1994). The authors reported a transport efficiency of approximately 0.1%. With a similar setup (PFA nebulizer and cyclonic spray chamber), the group of Feng et al. evaluated the contents and distribution patterns of the essential mineral elements Fe, Cu, Zn, Mn, P, and S in human cancer cells, including human cervical carcinoma (HeLa), human lung carcinoma epithelial cell line (A549), and human bronchial epithelial cells (16HBE). The main objective was to address the cellular heterogeneity to try to understand the behavior of mineral elements in tumor cells and further distinguish the tumor cells from the normal ones. Transport efficiencies were, in this case, similar to those reported previously (about 0.1%–0.2%) (Wang et al., 2015). The same group employed a d-CMN nebulizer to also include Co in the analysis of the same cell lines. In this case, the analysis showed that several subpopulations were coexisting within cancer cell lines, especially in HeLa cells, based on the differences in the elemental contents of Cu, Fe, Zn, and P, while normal 16HBE cells were homogeneous regarding the elemental content of the monitored elements except for Zn (Wang et al., 2017).

Many studies have been published on the application of sc-ICP-MS to monitor essential elements in single yeast cells, mainly due to the high stability of these cells as mentioned in the previous section. In this regard, the high-performance concentric nebulizer (HPCN) designed by Groombridge et al. was successfully applied to the analysis of Mg, P, Ca, Mn, Fe, Cu, and Zn in single *Saccharomyces cerevisiae cells* (Groombridge et al., 2013). This group reported transport efficiencies up to 75%, probably ascribed to the specific setup in combination with the analyzed cell model. Another approach using a concentric nebulizer for the analysis of single *Saccharomyces cerevisiae* cells was applied in 2019 by the group of Liu et al. to the determination of their content on K, Mg, Zn, Mn, and Cu. For the quantification, dissolved metal standard solutions were used for calibration, and the authors found good agreement between the elemental contents obtained by single-cell analysis and those obtained by conventional ICP-MS after cell digestion (Liu, Xue, Chen, & Li, 2019).

The robustness of the cell structure was also exploited by other authors to conduct sc-ICP-MS experiments to determine essential elements in samples of microbial origin like the cyanobacterium *Synechocystis* (Miyashita et al., 2014), and the bacteria *Streptomyces coelicolor* (González-Quiñónez et al., 2019), in the latter case with the final aim to obtain information about the specific mechanisms buffering cytosolic Cu and the biochemical pathways modulated by the element.

11.3.2 The incorporation of metal species in single cells

Although some investigations have been focused on the basic uptake studies of metal or metalloid species into cells, many applications of sc-ICP-MS refer to the incorporation of metallodrugs and/or nanoparticles into cellular structures that will be addressed later within this chapter. The first elemental analysis of single cells by ICP-MS was performed in 2005 for the determination of U incorporation into bacteria by means of a microconcentric nebulizer. This work detected single-cell events from the U-containing bacteria and quantified the U content by calibration with inorganic standards (Li et al., 2005). Recent studies also reported the quantitative analysis of As in human epithelial lung adenocarcinoma cells (A549) after incubation with arsenite. In this procedure, the loss of cytosolic As was minimized by avoiding the application of cell fixation methods that can cause cell membrane permeabilization (Meyer, López-Serrano, Mitze, Jakubowski, & Schwerdtle, 2018).

As in the case of the analysis of constitutive elements, some authors have focused their research on the study of Se incorporation in individual yeast cells (Álvarez-Fernández García et al., 2020). The authors applied a battery of analytical techniques for the detection of potentially present Se-containing nanoparticles in *Saccharomyces cerevisiae*. As the first step, the intracellular incorporation of Se was checked using sc-ICP-MS. This technique provided a fast tool to distinguish between incorporated and extracellular Se, and the combination of specific lysis protocols and sc-ICP-MS allowed the detection of biogenically formed Se-containing nanoparticles (Figs. 11.2 and 11.3). Such experiments were further confirmed by the inclusion of a suitable HPLC-ICP-MS method.

11.3.2.1 Metallodrugs in single cells

The use of cisplatin and other Pt-related drugs as effective chemotherapy for treating various malignancies is well established. Although its antineoplastic effects were initially ascribed to its ability to generate nuclear DNA lesions, increasing evidences also associate its mode of action with the alteration of both nuclear and cytoplasmic signaling pathways (Cohen & Lippard, 2001). However, the positive chemotherapeutic effect of cisplatin in most cancers (e.g., ovarian, prostate, etc.) is hampered by the inherent and acquired drug resistance, a multifactorial and still not well characterized process (Galluzzi et al., 2014). In fact, several mechanisms have been suggested to participate in conferring Pt-resistant properties to a tumor cell such as genetic alterations in genes involved in drug uptake (Ishida, Lee, Thiele, & Herskowitz, 2002). Therefore, the study of the cellular uptake of metallodrugs at the single-cell level for the elucidation of drug action, drug safety, and developed resistance is an important topic of research.

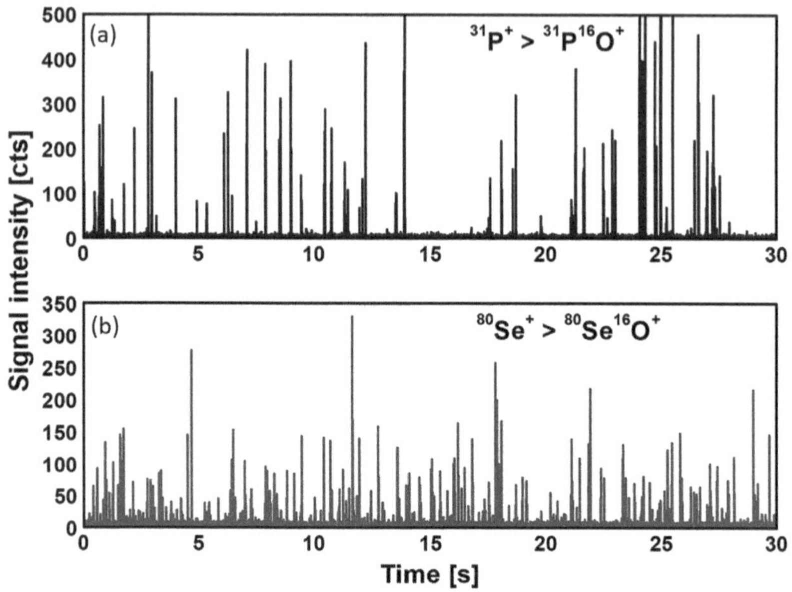

FIG. 11.2 Typical single-cell ICP-MS experiments on selenized yeast. (A) and (B) show the analysis of intact yeast cells monitoring phosphorus (as a cell marker) and selenium, respectively.

In this regard, an EnyaMist nebulizer was used for the determination of the intracellular Pt concentration in two different ovarian cancer cell lines, one of them sensitive (A2780) and one resistant (A2780cis) to cisplatin, after being exposed to this drug. The Pt mass per cell was dependent on the cisplatin dose and a great difference was found between the sensitive and resistant cells, making this technique potentially useful to distinguish different cell lines by means of their sensitivity to the drug. Additionally, the higher heterogeneity of the cisplatin-sensitive cells may point to resistance mechanisms related to drug influx inhibition (Corte Rodríguez, Álvarez-Fernández García, Blanco, Bettmer, & Montes-Bayón, 2017).

One of the limitations in the use of cisplatin is its low penetration into cells. In addition, some cells develop resistance, a multifactorial event that significantly decreases the intracellular cisplatin concentration. To circumvent these limitations, recent studies are focused on the use of nanocarriers that permit, among others, to achieve a higher drug uptake. Recently, a nanodelivery system including metallic nanoparticles has been used to immobilize a cisplatin (IV) prodrug. This prodrug was an octahedral platinum (IV) complex containing two additional chloride ligands in axial positions that facilitate the attachment of the prodrug to the nanoparticles. Upon entrance into the cell cytosol, further reduction of the Pt (IV)

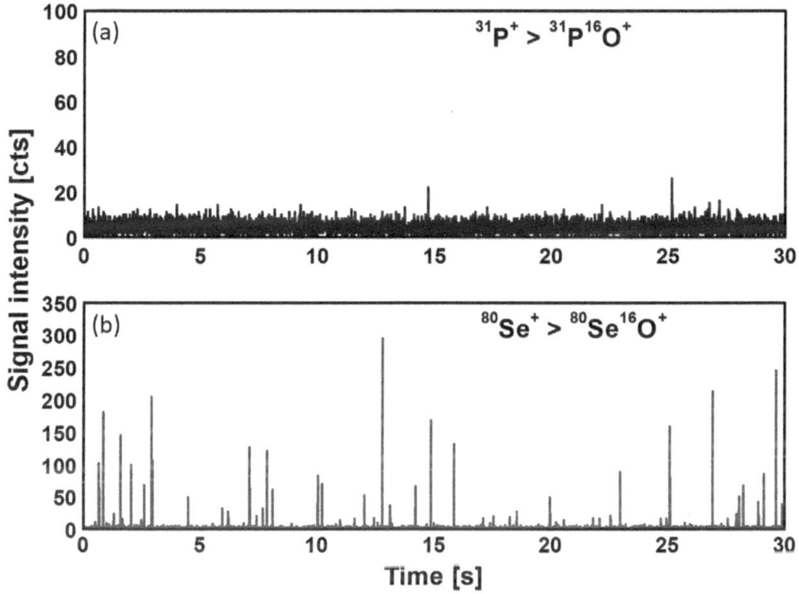

FIG. 11.3 Typical single-cell ICP-MS experiments on selenized yeast. (A) Phosphorus and (B) selenium show the corresponding results after cell lysis. Whereas the disappearance of phosphorus signals demonstrated the efficiency of the cell lysis procedure, the presence of selenium signals revealed the presence of selenium-containing nanoparticles. *(A and B) Reproduced from Álvarez-Fernández García, R., Corte-Rodríguez, M., Macke, M., Leblanc, K. L., Mester, Z., et al. (2020). Addressing the presence of biogenic selenium nanoparticles in yeast cells: Analytical strategies based on ICP-TQ-MS.* Analyst, *145(4), 1457–1465. https://doi.org/10.1039/c9an01565e, with permission from the Royal Society of Chemistry.*

center to cisplatin (II) by small biomolecules like glutathione or ascorbate occurs and the functional cisplatin (II) can start acting toward the biological target. The incorporation of these nanostructures to cisplatin-sensitive and -resistant cell lines was addressed using sc-ICP-MS employing a Glass Expansion commercial sample introduction system for single-cell analysis with promising results (Turiel-Fernández, Gutiérrez-Romero, Corte-Rodriguez, Bettmer, & Montes-Bayón, 2021).

Similar experiments were conducted by Zheng et al., who also evaluated a new nanomaterial as an antitumor agent. The endohedral hydroxylated metallofullerenol nanoparticle, $Gd@C_{82}(OH)_{22}$, was used to treat HeLa and 16HBE cells for 2 and 24 h, which were then analyzed by sc-ICP-MS using a MicroMist microconcentric nebulizer. The data showed that the response of equal cells of the same line was different in comparison to cisplatin under the same exposure time or dose. This study demonstrated that the antitumor mechanisms of cisplatin and the nanomaterial $Gd@C_{82}(OH)_{22}$ were different, causing different cytotoxicity and a tumor inhibition at a lower dose for the Gd@$C_{82}(OH)_{22}$ (Zheng et al., 2015).

Only the uptake of Pt drugs has been investigated by sc-ICP-MS. Some of the first studies using the Bi-based drug CBS (colloidal Bi subcitrate) were published already in 2011 by Tsang et al. demonstrating the incorporation of Bi in *Helicobacter pylori* after exposure to the metallodrug by sc-ICP-MS using a microconcentric nebulizer, which allowed to evaluate the time- and dose-related bismuth uptake. This study showed that Bi was linearly increased for concentrations of CBS lower than 20 $\mu g\,mL^{-1}$, reaching a maximum at higher concentrations (Tsang, Ho, Sun, & Chan, 2011).

At the end of the 1990s, clinical studies in China showed that As_2O_3 was an effective and relatively safe drug in the treatment of acute promyelocytic leukemia (APL). Zhou et al. studied the cytotoxicity of As_2O_3 in cell cultures of leukemia (NB4 and HL60) that were further analyzed by sc-ICP-MS employing a microconcentric nebulizer (Zhou, Li, & Sun, 2017). The analysis showed differential intracellular contents of As after treating the cells with increasing concentrations of the drug. A higher accumulation of As in NB4 cells was also observed, explaining a higher sensitivity of this cell line. Another As-based drug, ZIO-101, was employed by the same group to again treat NB4 and HL60 cells and analyze them, using in this case a glass concentric nebulizer (Zhou, Wang, Tse, Li, & Sun, 2018). The cells were synchronized and treated with either As_2O_3 or ZIO-101 at different phases of their cell cycle (G1, S, and G2/M), showing a higher uptake of the As drugs at the G2/M and G1 phases, whereas the S phase showed the lowest incorporation. These results proved that the activity of these drugs was cell cycle-dependent.

Curcumin is another universal anticancer agent whose activity is improved by including curcumin in a Co complex. By monitoring Co, Sun et al. could study the cellular trafficking and metabolism of the complex in HepG2 and MCF-7 cancer cells by sc-ICP-MS (Sun et al., 2019). The amount of the incorporated curcumin-Co complex was increased with the increased drug concentration until a concentration threshold causing the uptake to slow down. This work also studied the excretion of the complex through the Co remaining in the culture medium, finding a similar trend as that observed for the uptake.

11.3.2.2 Nanoparticles in single cells

The analysis of nanoparticles incorporated by cells is often carried out, as in the previously explained case of constitutive elements, in bulk. This means digesting a (usually high) number of cells, which are then analyzed by ICP-MS, and the results referred to the total biomass as an average of the cell population. But, as discussed previously in this chapter, the information about heterogeneity and cell-to-cell variations is lost with this method. Therefore, appropriate single-cell analysis methodologies that permit the quantification of nanoparticles incorporated by individual cells can play an important role in the understanding of dose-response relationships in nanoparticle toxicity studies.

Two main aspects hamper the correct evaluation of nanoparticle uptake by individual cells using sc-ICP-MS: (i) when cells are collected, even after several washing steps, part of the noninternalized nanoparticles might not be removed with the risk of contributing to the background noise of the measurement; and (ii) some of these particles might aggregate yielding intensities similar to those provided by the internalization of nanoparticles by the cells. The only way to solve these two problems would be the simultaneous measurement of a constitutive element (e.g., P, S, etc.) and the metal/metalloid of the nanoparticle. The use of a fast analyzer (e.g., TOF) would permit to conduct such measurements and to discriminate between signals corresponding to aggregated nanoparticles and cells. However, even with quadrupole-based instruments, several examples have been found in which sc-ICP-MS is used to determine nanoparticle uptake.

Due to their wide use in areas such as drug delivery, cell/tissue imaging, and quantitation of biomarkers, Au nanoparticles have been deeply studied by this sp- and sc-ICP-MS. Merrifield et al. studied the Au mass in the unicellular alga *Cryptomonas ovate* exposed to nanoparticulate and ionic Au (Merrifield, Stephan, & Lead, 2018). This work could quantify the fraction of cells containing more than one nanoparticle. The results showed a dose- and time-dependent accumulation for both forms of Au. However, a higher percentage of the cell population accumulated Au after exposure to Au nanoparticles as compared to ionic Au at equivalent concentrations, meaning that nanoparticles are either more readily taken or more slowly excreted.

Au nanoparticles in individual MCF-7 cells were also quantified using a MicroCross droplet generator. This work used a nanoparticle suspension as a calibration standard to calculate the number of nanoparticles internalized by each cell, which ranged from 130 to 584 particles per MCF-7 cell (Wei et al., 2018). This group also developed a single-cell sample introduction system for the analysis of Au in human leukemic cells (K562). This system employed a spiral-helix tubing array to focus the single cells before being sprayed by a concentric nebulizer into the spray chamber and plasma (Wei et al., 2019). The focus of this work was to study the uptake of Au nanoparticles, and a significant heterogeneity of the Au content in the individual cells was observed and ascribed to their different behavior regarding this uptake, probably due to the different positions in the cell division cycle or different cell physiological states at the time of the exposure.

A more standard sample introduction system with a PFA microflow nebulizer was used by Wang et al. to study the uptake of two different modified Au nanoparticles (Au NPs@citric acid and Au NPs@DNA) of different sizes (15, 30, and 60nm) in single HeLa cells. In this case, Zn was used as a cell marker that allowed the detection of all cells and, therefore, calculate the percentage of cells containing nanoparticles. The uptake of nanoparticles was higher for longer incubation times and concentrations. When comparing both types of nanoparticles, the Au NPs@DNA were more readily taken by the cells at low concentrations and short incubation times, while the cells needed higher concentrations and longer incubation times to finally achieve a greater uptake of Au NPs@citric acid (Wang et al., 2019). Therefore, the authors could conclude that the surface modification influenced the cellular uptake and internalization pathways for which they proposed a clathrin-mediated endocytosis for the 15 and 30nm Au NPs@DNA.

The other particles studied by the same authors were ZnO nanoparticles using a droplet chip coupled to a concentric nebulizer. The authors found four different incorporation levels of ZnO nanoparticles within the single cells: those which had incorporated no nanoparticles (71%), and those incorporating one, two, and three nanoparticles (29%), showing the heterogeneity of the HepG2 cells used for this study (Wang, Chen, He, & Hu, 2017).

Ag nanoparticles, widely used due to their antibacterial properties, have been also studied using sc-ICP-MS. López-Serrano et al. used a MicroMist concentric nebulizer to analyze suspensions of Ag nanoparticles by ICP-MS (López-Serrano Oliver et al., 2018). Using standard nanoparticles, it was possible to calibrate the system and perform a quantitative analysis of the nanoparticles and quantify the number of particles incorporated in individual THP-1 monocytes incubated with different concentrations of Ag nanoparticles for different times. A time- and concentration-dependent uptake was observed for these cells. However, macrophages showed a higher nanoparticle uptake that, in this case, was only concentration-dependent but independent from exposure time.

Similarly, Yu et al. evaluated the uptake of ionic and nanoparticulated (51nm) Ag in single HepG2 cells. They used a droplet-based microfluidic device, emloying low-cost, off-the-shelf modules to generate monodisperse droplets. The cells were embedded in the droplets, which were formed by contact of two flowing immiscible fluid phases. A pneumatic nebulizer was then used to introduce the droplets into the ICP-MS (Yu, Chen, He, Wang, & Hu, 2019). In this case, the cells incubated with nanoparticles showed a more heterogeneous incorporation of Ag than those incubated with ionic Ag, and the percentage of Ag incorporated by cells was, in both cases, almost 100%.

Finally, the biogenic synthesis of nanoparticles is a particular case of cells, which were not previously exposed to nanoparticles, containing them. An example has been given before reporting the formation of Se nanoparticles in selenized yeast (see Section 11.3.2). Furthermore, a recent work has been conducted on bacteria and focused on the synthesis of endogenous Cu nanoparticles. Actinomycetes are important industrial bacteria producing two-thirds of clinical antibiotics and forming Cu nanoparticles. Cu modulates the secondary metabolism in Streptomyces, but the relationship between endogenous Cu nanoparticles and antibiotic production has been recently evaluated. For this aim, the incorporation of Cu in individual spores of *Streptomyces coelicolor* exposed to different Cu concentrations (40, 80, and 160 μM) has been evaluated. Once the intracellular Cu incorporation was obtained, a mechanical lysis of the spores and a single-particle ICP-MS detection in combination with Transmission Electron Microscopy (TEM) were performed to prove the presence of the nanoparticles. Secondary metabolism is known to be activated under stressful conditions and cytosolic Cu seem to be one of the signals triggering antibiotic production. Nanoparticle formation prevents Cu toxicity and contributes to modulate secondary metabolism (submitted for publication).

11.3.3 Analysis of biomarkers by sc-ICP-MS

Beside elemental composition, the cell-to-cell variation on the proteins can be the cause of several diseases, including cancer, neurological and developmental disorders. In addition, some proteins can serve as specific biomarkers with a prognostic or predictive value of a certain pathological condition, or the performance of a given therapy. To identify those proteins, it is important to have the capability to measure, simultaneously and quantitatively, a large number of proteins or other biomarkers that may be present in a complex sample. This challenging task can be accomplished by a combination of strategies: on the one hand, the possibility of having several antibodies against specific antigenic proteins that can be labeled with different metal (isotopic) probes and, on the other hand, the capabilities to simultaneously measure all the probes within a single cell using, for example, the mass cytometry (CyTOF) technology. The advantage of this approach lies in the large number of available elements and isotopes (potentially $>$100) found in low abundance in biological systems, which will allow multiple tags to be used simultaneously.

In this sense, sc-ICP-MS has been used for a variety of studies often applied to the monitoring of individual biomarkers (based on the same principle but using quadrupole-based mass analyzers). Among them, Zhou et al., who evaluated the exposure of leukemia cells to As_2O_3, also studied apoptosis and cell differentiation using two lanthanide tags (DOTA-Nd and DOTA-Dy) to label antibodies against an apoptosis-related protein, Annexin-V, and a differentiation-related protein, CD11b, respectively (Zhou et al., 2017). By this strategy, it was possible to evaluate the apoptosis and differentiation status by sc-ICP-MS by measuring Nd and Dy, respectively, in order to correlate it with the As exposure.

Nd-labeled antibodies were used with sc-ICP-MS detection and a high-performance concentric nebulizer to quantify the transferrin receptor 1 (TfR1), which is a breast cancer biomarker. The detection was performed by monitoring P as a cell marker and quantifying the Nd-labeled antibody as a reporter of the TfR1 expression on the cell surface (Corte-Rodríguez, Blanco-González, Bettmer, & Montes-Bayón, 2019). After proper characterization of the stoichiometry of the antibody, it was possible to quantify the number of TfR1 expressed by each cell. This biomarker was quantified in two breast cancer cell lines (MCF-7 and MDA-MB-231). The most malignant phenotype (MDA-MB-231) showed significantly higher expression levels of this biomarker.

Liang et al. achieved the simultaneous recognition of five different bacterial strains (*Escherichia coli, Staphylococcus aureus, Listeria monocytogenes, Shigella dysenteriae,* and *Vibrio parahaemolyticus*) by using generic Eu-tagged antibodies for cell counting and five lanthanide-tagged antibodies, one for each cell type (Liang et al., 2019). They additionally applied this labeling strategy to analyze the response of *Staphylococcus aureus* isogenic bacteria after being treated with nanoparticulated Ag and vancomycin.

As previously stated, the possibility to conduct multiple isotope monitoring in a single cell relies on the possibility of having a high-sensitivity and fast-scanning mass spectrometric detector. The simultaneous analysis of biomarkers in single cells by using metal-labeled antibodies and mass spectrometric detection has been coined as mass cytometry (Bandura et al., 2009). In principle, mass cytometry originates from inductively coupled plasma TOF mass spectrometry (ICP-TOF-MS), which is broadly used for quantifying isotopic contents of elements having low ionization potentials in the periodic table. The commercial version of a mass cytometer, which was first introduced to the market by DVS Science Inc., now Fluidigm, is called CyTOF. Mass cytometry has been regarded as the 'next generation' of flow cytometry because it uses metal-isotope-tagged antibodies as reporting probes. Using this advance, CyTOF overcomes the inherent limits of spectral overlap observed with conventional fluorescence-based cytometry and principally enables simultaneous measurements of >100 parameters in single cells. However, since the working principles are similar to those described for the monoplex assays, we will not cover these aspects in the present chapter.

11.4 Future perspectives

The analysis of small objects like individual nanoparticles and individual biological cells has matured during the last years. The working principle has been set and it can be expected that further instrumental developments will allow in the near future even improved detection limits for most elements, faster detection, and faster data acquisition for the simultaneous determination of elements of interest in single cells. However, even the best instrumental and methodological development for single cell analysis will not have a broader impact if other disciplines such as toxicology, pharmacology, biochemistry, etc. are not aware of such a powerful tool to solve their specific problems.

Acknowledgments

The authors of this chapter gratefully acknowledge the support of several organizations and institutions to conduct research in the field of single-cell ICP-MS throughout the following projects: Ministerio de Economía y Competitividad (MINECO-17-CTQ2016-80069-C2-1-R and MCIU-19-RTI2018-094605-B-I00) and Principado de Asturias (FC-GRUPIN-IDI/2018/000242). MCR acknowledges a Sara Borrell postdoc research contract from Instituto de Salud Carlos III (CD19/00249). Furthermore, we kindly acknowledge the instrumental support from Thermo Fisher Scientific (Bremen, Germany) and Teledyne Cetac (Omaha, United States).

References

Altschuler, S. J., & Wu, L. F. (2010). Cellular heterogeneity: Do differences make a difference? *Cell, 141*(4), 559–563. https://doi.org/10.1016/j.cell.2010.04.033.

Álvarez-Fernández García, R., Corte-Rodríguez, M., Macke, M., Leblanc, K. L., Mester, Z., Montes-Bayón, M., et al. (2020). Addressing the presence of biogenic selenium nanoparticles in yeast cells: Analytical strategies based on ICP-TQ-MS. *Analyst, 145*(4), 1457–1465. https://doi.org/10.1039/c9an01565e.

Bandura, D. R., Baranov, V. I., Ornatsky, O. I., Antonov, A., Kinach, R., Lou, X., et al. (2009). Mass cytometry: Technique for real time single cell multitarget immunoassay based on inductively coupled plasma time-of-flight mass spectrometry. *Analytical Chemistry, 81*(16), 6813–6822. https://doi.org/10.1021/ac901049w.

Cohen, S. M., & Lippard, S. J. (2001). Cisplatin: From DNA damage to cancer chemotherapy. *Progress in Nucleic Acid Research and Molecular Biology, 67*, 93–130. https://doi.org/10.1016/S0079-6603(01)67026-0.

Cornelis, G., & Hassellöv, M. (2014). A signal deconvolution method to discriminate smaller nanoparticles in single particle ICP-MS. *Journal of Analytical Atomic Spectrometry, 29*(1), 134–144. https://doi.org/10.1039/C3JA50160D.

Corte Rodríguez, M., Álvarez-Fernández García, R., Blanco, E., Bettmer, J., & Montes-Bayón, M. (2017). Quantitative evaluation of cisplatin uptake in sensitive and resistant individual cells by single-cell ICP-MS (SC-ICP-MS). *Analytical Chemistry, 89*(21), 11491–11497. https://doi.org/10.1021/acs.analchem.7b02746.

Corte-Rodríguez, M., Blanco-González, E., Bettmer, J., & Montes-Bayón, M. (2019). Quantitative analysis of transferrin receptor 1 (TfR1) in individual breast cancer cells by means of labeled antibodies and elemental (ICP-MS) detection. *Analytical Chemistry, 91*(24), 15532–15538. https://doi.org/10.1021/acs.analchem.9b03438.

Degueldre, C., & Favarger, P. Y. (2003). Colloid analysis by single particle inductively coupled plasma-mass spectroscopy: A feasibility study. *Colloids and Surfaces A: Physicochemical and Engineering Aspects, 217*(1–3), 137–142. Elsevier https://doi.org/10.1016/S0927-7757(02)00568-X.

Degueldre, C., Favarger, P. Y., & Bitea, C. (2004). Zirconia colloid analysis by single particle inductively coupled plasma-mass spectrometry. *Analytica Chimica Acta, 518*(1–2), 137–142. https://doi.org/10.1016/j.aca.2004.04.015.

Degueldre, C., Favarger, P. Y., Rossé, R., & Wold, S. (2006). Uranium colloid analysis by single particle inductively coupled plasma-mass spectrometry. *Talanta, 68*(3), 623–628. https://doi.org/10.1016/j.talanta.2005.05.006.

Degueldre, C., Favarger, P. Y., & Wold, S. (2006). Gold colloid analysis by inductively coupled plasma-mass spectrometry in a single particle mode. *Analytica Chimica Acta, 555*(2), 263–268. https://doi.org/10.1016/j.aca.2005.09.021.

Domozych, D. S., Ciancia, M., Fangel, J. U., Mikkelsen, M. D., Ulvskov, P., & Willats, W. G. T. (2012). The cell walls of green algae: A journey through evolution and diversity. *Frontiers in Plant Science, 3*. https://doi.org/10.3389/fpls.2012.00082.

Galluzzi, L., Vitale, I., Michels, J., Brenner, C., Szabadkai, G., Harel-Bellan, A., et al. (2014). Systems biology of cisplatin resistance: Past, present and future. *Cell Death & Disease, 5*. https://doi.org/10.1038/cddis.2013.428.

Ganz, T., & Nemeth, E. (2011). Hepcidin and disorders of iron metabolism. *Annual Review of Medicine, 62*, 347–360. https://doi.org/10.1146/annurev-med-050109-142444.

González-Quiñónez, N., Corte-Rodríguez, M., Álvarez-Fernández-García, R., Rioseras, B., López-García, M. T., Fernández-García, G., et al. (2019). Cytosolic copper is a major modulator of germination, development and secondary metabolism in *Streptomyces coelicolor*. *Scientific Reports, 9*(1). https://doi.org/10.1038/s41598-019-40876-0.

Groombridge, A. S., Miyashita, S. I., Fujii, S. I., Nagasawa, K., Okahashi, T., Ohata, M., et al. (2013). High sensitive elemental analysis of single yeast cells (*Saccharomyces cerevisiae*) by time-resolved inductively-coupled plasma mass spectrometry using a high efficiency cell introduction system. *Analytical Sciences, 29*(6), 597–603. https://doi.org/10.2116/analsci.29.597.

Huang, H. L., Hsing, H. W., Lai, T. C., Chen, Y. W., Lee, T. R., Chan, H. T., et al. (2010). Trypsin-induced proteome alteration during cell subculture in mammalian cells. *Journal of Biomedical Science, 17*(1). https://doi.org/10.1186/1423-0127-17-36.

Ishida, S., Lee, J., Thiele, D. J., & Herskowitz, I. (2002). Uptake of the anticancer drug cisplatin mediated by the copper transporter Ctr1 in yeast and mammals. *Proceedings of the National Academy of Sciences of the United States of America, 99*(22), 14298–14302. https://doi.org/10.1073/pnas.162491399.

Jansen, J., Karges, W., & Rink, L. (2009). Zinc and diabetes—Clinical links and molecular mechanisms. *Journal of Nutritional Biochemistry, 20*(6), 399–417. https://doi.org/10.1016/j.jnutbio.2009.01.009.

Kubitschek, H. E. (1990). Cell volume increase in *Escherichia coli* after shifts to richer media. *Journal of Bacteriology, 172*(1), 94–101. https://doi.org/10.1128/jb.172.1.94-101.1990.

Laborda, F., Jiménez-Lamana, J., Bolea, E., & Castillo, J. R. (2011). Selective identification, characterization and determination of dissolved silver(i) and silver nanoparticles based on single particle detection by inductively coupled plasma mass spectrometry. *Journal of Analytical Atomic Spectrometry, 26*(7), 1362–1371. https://doi.org/10.1039/c0ja00098a.

Li, F., Armstrong, D. W., & Houk, R. S. (2005). Behavior of bacteria in the inductively coupled plasma: Atomization and production of atomic ions for mass spectrometry. *Analytical Chemistry, 77*(5), 1407–1413. https://doi.org/10.1021/ac049188l.

Liang, Y., Liu, Q., Zhou, Y., Chen, S., Yang, L., Zhu, M., et al. (2019). Counting and recognizing single bacterial cells by a lanthanide-encoding inductively coupled plasma mass spectrometric approach. *Analytical Chemistry, 91*(13), 8341–8349. https://doi.org/10.1021/acs.analchem.9b01130.

Lichten, L. A., & Cousins, R. J. (2009). Mammalian zinc transporters: Nutritional and physiologic regulation. *Annual Review of Nutrition, 29*, 153–176. https://doi.org/10.1146/annurev-nutr-033009-083312.

Liu, Z., Xue, A., Chen, H., & Li, S. (2019). Quantitative determination of trace metals in single yeast cells by time-resolved ICP-MS using dissolved standards for calibration. *Applied Microbiology and Biotechnology, 103*(3), 1475–1483. https://doi.org/10.1007/s00253-018-09587-w.

López-Serrano Oliver, A., Baumgart, S., Bremser, W., Flemig, S., Wittke, D., Grützkau, A., et al. (2018). Quantification of silver nanoparticles taken up by single cells using inductively coupled plasma mass spectrometry in the single cell measurement mode. *Journal of Analytical Atomic Spectrometry, 33*(7), 1256–1263. https://doi.org/10.1039/c7ja00395a.

Merrifield, R. C., Stephan, C., & Lead, J. R. (2018). Quantification of au nanoparticle biouptake and distribution to freshwater algae using single cell—ICP-MS. *Environmental Science and Technology, 52*(4), 2271–2277. https://doi.org/10.1021/acs.est.7b04968.

Meyer, S., López-Serrano, A., Mitze, H., Jakubowski, N., & Schwerdtle, T. (2018). Single-cell analysis by ICP-MS/MS as a fast tool for cellular bioavailability studies of arsenite. *Metallomics, 10*(1), 73–76. https://doi.org/10.1039/c7mt00285h.

Miyashita, S., Groombridge, A. S., Fujii, S., Minoda, A., Takatsu, A., Hioki, A., et al. (2014). Highly efficient single-cell analysis of microbial cells by time-resolved inductively coupled plasma mass spectrometry. *Journal of Analytical Atomic Spectrometry, 29*(9), 1598–1606. https://doi.org/10.1039/C4JA00040D.

Nomizu, T., Kaneco, S., Tanaka, T., Ito, D., Kawaguchi, H., & Vallee, B. T. (1994). Determination of calcium content in individual biological cells by inductively coupled plasma atomic emission spectrometry. *Analytical Chemistry, 66*(19), 3000–3004. https://doi.org/10.1021/ac00091a004.

Pace, H. E., Rogers, N. J., Jarolimek, C., Coleman, V. A., Higgins, C. P., & Ranville, J. F. (2012). Correction to determining transport efficiency for the purpose of counting and sizing nanoparticles via single particle inductively coupled plasma mass spectrometry. *Analytical Chemistry, 84*(10), 4633. https://doi.org/10.1021/ac300942m.

Shapiro, E., Biezuner, T., & Linnarsson, S. (2013). Single-cell sequencing-based technologies will revolutionize whole-organism science. *Nature Reviews Genetics, 14*(9), 618–630. https://doi.org/10.1038/nrg3542.

Sun, Q. X., Wei, X., Zhang, S. Q., Chen, M. L., Yang, T., & Wang, J. H. (2019). Single cell analysis for elucidating cellular uptake and transport of cobalt curcumin complex with detection by time-resolved ICPMS. *Analytica Chimica Acta, 1066*, 13–20. https://doi.org/10.1016/j.aca.2019.03.062.

Tsang, C. N., Ho, K. S., Sun, H., & Chan, W. T. (2011). Tracking bismuth antiulcer drug uptake in single *Helicobacter pylori* cells. *Journal of the American Chemical Society, 133*(19), 7355–7357. https://doi.org/10.1021/ja2013278.

Tuoriniemi, J., Cornelis, G., & Hassellöv, M. (2012). Size discrimination and detection capabilities of single-particle ICPMS for environmental analysis of silver nanoparticles. *Analytical Chemistry, 84*(9), 3965–3972. https://doi.org/10.1021/ac203005r.

Turiel-Fernández, D., Gutiérrez-Romero, L., Corte-Rodriguez, M., Bettmer, J., & Montes-Bayón, M. (2021). Ultrasmall iron oxide nanoparticles cisplatin (IV) prodrug nanoconjugate: ICP-MS based strategies to evaluate the formation and drug delivery capabilities in single cells. *Analytica Chimica Acta, 1159*. https://doi.org/10.1016/j.aca.2021.338356.

Turner, R. D., Vollmer, W., & Foster, S. J. (2014). Different walls for rods and balls: The diversity of peptidoglycan. *Molecular Microbiology, 91*(5), 862–874. https://doi.org/10.1111/mmi.12513.

Turski, M. L., & Thiele, D. J. (2009). New roles for copper metabolism in cell proliferation, signaling, and disease. *Journal of Biological Chemistry, 284*(2), 717–721. https://doi.org/10.1074/jbc.R800055200.

Vollmer, W., & Seligman, S. J. (2010). Architecture of peptidoglycan: More data and more models. *Trends in Microbiology, 18*(2), 59–66. https://doi.org/10.1016/j.tim.2009.12.004.

Wang, H., Chen, B., He, M., & Hu, B. (2017). A facile droplet-chip-time-resolved inductively coupled plasma mass spectrometry online system for determination of zinc in single cell. *Analytical Chemistry, 89*(9), 4931–4938. https://doi.org/10.1021/acs.analchem.7b00134.

Wang, H., Chen, B., He, M., Li, X., Chen, P., & Hu, B. (2019). Study on uptake of gold nanoparticles by single cells using droplet microfluidic chip-inductively coupled plasma mass spectrometry. *Talanta, 200*, 398–407. https://doi.org/10.1016/j.talanta.2019.03.075.

Wang, H., Wang, M., Wang, B., Zheng, L., Chen, H., Chai, Z., et al. (2017). Interrogating the variation of element masses and distribution patterns in single cells using ICP-MS with a high efficiency cell introduction system. *Analytical and Bioanalytical Chemistry, 409*(5), 1415–1423. https://doi.org/10.1007/s00216-016-0075-y.

Wang, H., Wang, B., Wang, M., Zheng, L., Chen, H., Chai, Z., et al. (2015). Time-resolved ICP-MS analysis of mineral element contents and distribution patterns in single cells. *Analyst, 140*(2), 523–531. https://doi.org/10.1039/c4an01610f.

Wei, X., Zhang, X., Guo, R., Chen, M. L., Yang, T., Xu, Z. R., et al. (2019). A spiral-helix (3D) tubing array that ensures ultrahigh-throughput single-cell sampling. *Analytical Chemistry, 91*(24), 15826–15832. https://doi.org/10.1021/acs.analchem.9b04122.

Wei, X., Zheng, D. H., Cai, Y., Jiang, R., Chen, M. L., Yang, T., et al. (2018). High-throughput/high-precision sampling of single cells into ICP-MS for elucidating cellular nanoparticles. *Analytical Chemistry, 90*(24), 14543–14550. https://doi.org/10.1021/acs.analchem.8b04471.

Yu, X., Chen, B., He, M., Wang, H., & Hu, B. (2019). 3D droplet-based microfluidic device easily assembled from commercially available modules online coupled with ICPMS for determination of silver in single cell. *Analytical Chemistry, 91*(4), 2869–2875. https://doi.org/10.1021/acs.analchem.8b04844.

Zheng, L. N., Wang, M., Zhao, L. C., Sun, B. Y., Wang, B., Chen, H. Q., et al. (2015). Quantitative analysis of Gd@$C_{82}(OH)_{22}$ and cisplatin uptake in single cells by inductively coupled plasma mass spectrometry. *Analytical and Bioanalytical Chemistry, 407*(9), 2383–2391. https://doi.org/10.1007/s00216-014-8422-3.

Zhou, Y., Li, H., & Sun, H. (2017). Cytotoxicity of arsenic trioxide in single leukemia cells by time-resolved ICP-MS together with lanthanide tags. *Chemical Communications, 53*(20), 2970–2973. https://doi.org/10.1039/c7cc00143f.

Zhou, Y., Wang, H., Tse, E., Li, H., & Sun, H. (2018). Cell cycle-dependent uptake and cytotoxicity of arsenic-based drugs in single leukemia cells. *Analytical Chemistry, 90*(17), 10465–10471. https://doi.org/10.1021/acs.analchem.8b02444.

CHAPTER

12

Nonconventional applications of nebulizers: Nanomaterials synthesis

Zahra Amouzegar[a], *Fariba Mollarasouli*[b], *Sepideh Asadi*[a], *Ali Fathi*[a], *Mazaher Ahmadi*[a,c], *Abbas Afkhami*[a,d], *and Tayyebeh Madrakian*[a,c]

[a]Department of Analytical Chemistry, Faculty of Chemistry, Bu-Ali Sina University, Hamedan, Iran, [b]Department of Chemistry, Yasouj University, Yasouj, Iran, [c]Autophagy Research Center, Shiraz University of Medical Sciences, Shiraz, Iran, [d]D-8 International University, Hamedan, Iran

12.1 Introduction

Nanomaterials are the critical elements in nanotechnology. They have been studied from many aspects through time. The size and the shape of nanomaterials cause considerable diversity in the physicochemical properties of nanomaterials. Manipulating the properties and creating the novel characterization for nanomaterials is possible by the selection of appropriate synthesis methods. This feature makes a special nanoparticle that can be used in various fields, including pollutants removal, (bio)sensor design, etc. (Kolahalam et al., 2019; Madrakian, Afkhami, Zadpour, & Ahmadi, 2015; Madrakian, Haghshenas, Ahmadi, & Afkhami, 2015; Madrakian, Haryani, Ahmadi, & Afkhami, 2015; Madrakian, Zadpour, Ahmadi, & Afkhami, 2015).

Currently, researchers use various methods and techniques to achieve the desired goal and synthesize the desired nanoparticles (Zhang, Zhou, Wang, Li, & Biswas, 2021). There are three main methods for nanomaterials synthesis; biological methods, chemical methods, and physical

https://doi.org/10.1016/B978-0-323-91181-8.00002-3

methods. The biological methods exploit the microorganisms (bacteria, fungi, or algae), biological templates, and different plant parts for converting the solution of metal salts to nanomaterials. The chemical and physical methods include a variety of "top-down" and "bottom-up" synthesis approaches. The "top-down" approach converts the larger materials to smaller particles, and the "bottom-up" approaches condense nanoparticles (liquid or gaseous phase) to form more extensive materials (Kolahalam et al., 2019).

One of the most powerful methods in the synthesis of nanomaterials is the aerosol-based method, which deals with a set of chemical reactions and physical processes. This chapter emphasizes the fundamental information (mechanisms, merits, and drawbacks) of aerosol-assisted nanomaterials synthesis methods. In the next step, some applications are introduced. Finally, we present a future perspective on this area of research.

12.2 Fundamentals

12.2.1 Liquid-phase aerosol-assisted nanomaterials synthesis

By nebulizing solvents of a particular chemical formula, different methods such as pneumatic nebulization, ultrasonic nebulization, electrosynthesis nebulization, and spray pyrolysis can be used to produce aerosol particles (Kim, Kulkarni, Ha, Shin, & Kim, 2017; Wang et al., 2008). The role of nebulization in determining particle size is critical. In the nebulization process, the original precursor (in the solution/colloid/slurry form) is fed into a special nebulizer to generate many microsized droplets. Several types of driving forces, such as electrostatic, gravity, centrifugal, and ultrasonic forces, can be used to aid the nebulization operation, depending on the droplet size requirements (Nie et al., 2018). In this technique, the liquid feed is converted to small droplets by a nozzle of liquid nebulization. Various nebulizers (including hydraulic friction, pneumatic, ultrasonic nozzle, rotary disk nebulizers, etc.) can be used and yield fine droplets of different sizes. The fundamental theory behind nanoparticle synthesis is that the solvent in the nebulized droplets evaporates, causing the droplets to freeze and create nanomaterials (Gautam, Kim, & Yong, 2021; Yu Vasilyev, Domrina, Kaufman, & Maiorova, 2019).

12.2.2 Pneumatic nebulization

The essential premise of pneumatic nebulization is high relative viscosities. Recently, it is suggested that this parameter be 1–500 cp, which would be between the liquid and the gas form. Such status creates a significant shear force across a liquid surface, causing the liquid to split up into minute droplets. The gas flow carries the droplets away, resulting in an aerosol stream (Nie et al., 2018; Wilkinson, Smith, Kay, & Harris, 2019). The spray stream is then directed onto an impactor plate, where larger droplets are deposited and returned to the liquid reservoir or are broken up, exiting from the nebulizers through the outflow stream. A design of the pneumatic nebulizer is shown in Fig. 12.1A. During the nebulization, a carrier gas is ejected from the top of the ink supply channel (Fig. 12.1A (1)). This creates a low-static-pressure zone, which attracts the ink to the carrier gas flow. The topmost layer of the liquid is sheared when it reaches the carrier gas stream, resulting in a series of polydisperse droplets (Fig. 12.1A (2)). Large, high-inertia droplets of this stream collide with the spray chamber's sidewalls and return to the reservoir (Fig. 12.1A (3)), while lower-inertia droplets remain as an aerosol (Fig. 12.1A (4)). Following the nebulization, the formed aerosol is transferred to the virtual impactor, which separates the droplets depending on their inertia using a region of stagnant flow (Fig. 12.1B) (Wilkinson et al., 2019). Small droplets that have no inertia radiate out of the mainstream to overcome this area. The geometric parameters of the nebulizer, as well as the features of the liquid and carrier gas, may influence the size of the aerosol droplets. To manage the size of the droplets, qualities of the precursor solution, such as surface tension (σ), viscosity (μ), and density (ρ), should be addressed in addition to the nebulizer's parameters (Nandiyanto & Okuyama, 2011; Nie et al., 2018; Wang et al., 2008).

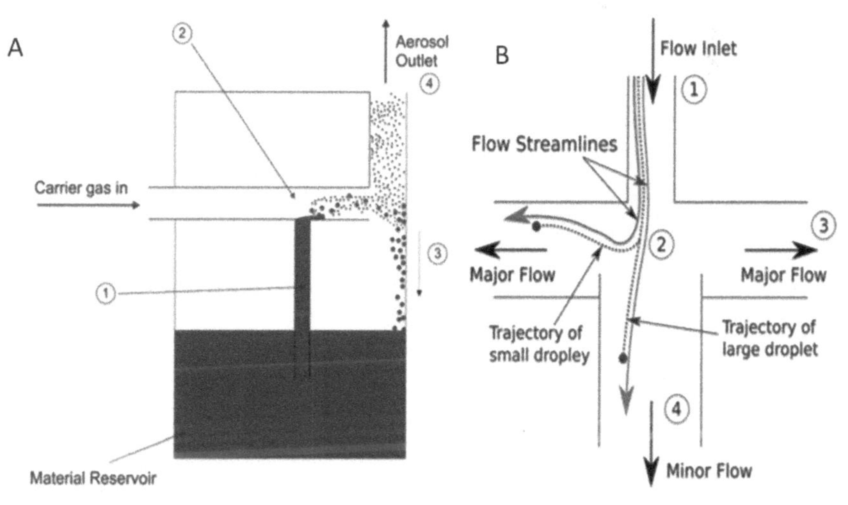

FIG. 12.1 Scheme of (A) pneumatic nebulization, and (B) a virtual impactor. *Reproduced from Wilkinson, N. J., Smith, M. A. A., Kay, R. W., & Harris, R. A. (2019). A review of aerosol jet printing—A nontraditional hybrid process for micro-manufacturing.* International Journal of Advanced Manufacturing Technology, 105, *4599–4619.*

Pneumatic nebulizers create primary droplets with different sizes ranging from 1 to 10 μm and concentrations ranging from 5 to 50 g^{-3}, depending on geometric parameters and operating circumstances. Pneumatic nebulization produces polydisperse particles (δg of 1.4 or more) with a moderate charge level that varies depending on particle dimension and solute concentration (Tsai, Lin, Deshpande, & Liu, 2005). Another critical stage in the removal of solvent from droplets and transformation into particles is the droplet-to-particle conversion. In theory, after droplets form, they reach the droplet-to-particle conversion chamber via the carrier gas flow, where the aerosol droplets evaporate and solute condensation takes place within the droplet. Precipitation begins from the outer shell and then decomposes when the precursor concentration approaches saturation at the droplet surface. Nucleation and growth take place later to produce a particle. The carrier gas is utilized to aid in the formation of the aerosol and to carry the droplets of the aerosol to the reaction zone. An inert gas (N_2, Ar) and compressed air are the most usual; depending on the chemical components, hot gas or steam might be used (Nie et al., 2018).

12.2.3 Ultrasonic nebulization

When a liquid in the form of a thin film is allowed to run across a vibrating surface (with a frequency greater than 20 kHz), it breaks up into smaller droplets. This phenomenon, known as ultrasonic nebulization, originated in late-19th-century acoustical physics, namely the work of Wood and Loomis (Wood & Loomis, 1927). Fig. 12.2 shows the process of ultrasonic nebulization. To explain the process of liquid fragmentation during ultrasonic nebulization, there are two primary ideas: the capillary wave theory and the cavitation theory. Taylor's instability criteria (Taylor, 1950) are used to support the capillary wave concept. The capillary wave

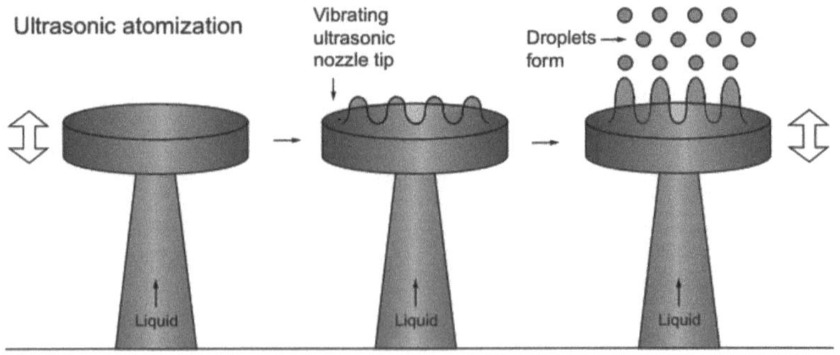

FIG. 12.2 Scheme of the ultrasonic nebulization process. *Reproduced from Gogate, P. R. (2015). The use of ultrasonic atomization for encapsulation and other processes in food and pharmaceutical manufacturing, in:* Power ultrasonics *(pp. 911–935), Elsevier.*

theory is supported by the high association between capillary wavelength and mean droplet size. Capillary waves were formed with crests and troughs on a vibrating surface. Nebulization occurs when unstable oscillations break the peaks of capillary waves at the surface away from the bulk liquid. In this example, drop creation occurs near the crest, and drop size is proportional to wavelength. Capillary wavelength normally decreases as the frequency of irradiation increases; hence, tiny droplets form at higher frequencies of the ultrasound (Gogate, 2015). The hypothesis of cavitation is commonly applied to systems with high energy intensity ($W m^{-2}$) and high frequency (16 kHz–2 MHz). Cavitation bubbles develop in the liquid film when it is sonicated, assuming the liquid layer on the vibrating surface has a minimal thickness. High-intensity hydraulic shock occurs during the implosive collapse of these holes, particularly those near the liquid's surface. These hydraulic shocks cause the liquid coating to disintegrate and the droplets to be ejected directly (Avvaru, Patil, Gogate, & Pandit, 2006).

One of the benefits of the ultrasonic nebulization technique over traditional nebulization techniques is how it regulates the density and particle size of fog separately. This is impossible to do using other regularly utilized nebulization procedures. The density of fog may be changed in ultrasonic nebulization by altering the airflow past the liquid surface. Ultrasonic nebulizers may produce very monodisperse particles (δg of less than 1.02) with much lower particle number concentrations when compared to pneumatic nebulizers. The frequent drawbacks of pneumatic nebulizers (such as the possibility of clogging, vast droplet dispersion, and a lack of control over the mean droplet size) can be avoided by using ultrasonic energy to produce droplets with a somewhat uniform size distribution. Spray congealing, which uses ultrasonic energy to produce spherical microparticles or nanoparticles with high size distribution and encapsulation efficiency, is another key technology based on a similar regulating mechanism. Ultrasonic nebulization is a durable and unique single-step method with scale-up potential (Avvaru et al., 2006; Biskos, Vons, Yurteri, & Schmidt-ott, 2008; Gogate, 2015; Lang, 1962).

12.2.4 Electrosynthesis nebulization

Electrohydrodynamic nebulization (EHDA) is a well-known method for producing monodispersed small droplets from a liquid under the influence of electrical forces. William Gilbert was the first to notice and publish the phenomenon of EHDA, commonly known as electrospray. However, it has been utilized to create and treat micro/nanostructured materials since the 1990s. Because of the simplicity and flexibility of the experimental setup, EHDA has been successfully employed to generate particulate materials with controlled compositions, structures, and sizes. Though EHDA has a wide range of applications, including raindrops in thunderclouds, combustion, agricultural spraying, and electrospray ionization mass spectroscopy, it has gained a lot of interest in the 1990s for creating and digesting micro-/nanoparticulate materials. Droplets with mean diameters ranging from nanometers to several micrometers may be formed by adjusting the liquid flow velocity and the electrostatic potential between the liquid and the counter electrode. A syringe pump, a syringe, a metal needle serving as a nozzle, a high voltage power supply, and a grounded substrate functioning as a collector comprise the basic EHDA setup. EHDA droplets have a high electric charge in the range of the Rayleigh charge limit. When the droplets evaporate, they surpass the Rayleigh charge limit, fracturing into smaller droplets. The droplets must be totally or partially neutralized to avoid Rayleigh disintegration and to manage particle size distribution. The usage of the chamber reduces solvent evaporation and promotes the creation of smaller particles with finer surface morphology. Briefly, EHDA is the process by which a liquid jet decomposes into small droplets under the influence of an external electric field. The competition of the electric stress and the surface tension stress on the liquid-gas interface, as well as the kinetic energy of the liquid leaving the nozzle, determine the various spraying modes observed (e.g., dripping, microdripping, simple-jet, single cone spraying (stable cone-jet), and multiple cones spraying) (Xie, Jiang, Davoodi, Srinivasan, & Wang, 2015). Extremely low flow rates are necessary to create very tiny particles. The lower labor rates restrict the industrial application of EHDA. By merely raising the flow rate, it is difficult to increase the throughput of material generated from a single nozzle. As a result, rather than upscaling, out-scaling is required to increase the number of injection ports. Increasing the number of capillary nozzles appears to be a straightforward and efficient technique to increase the number of droplets. Out-scaling, on the other hand, is hampered by fluctuations in electric field intensity and, as a result, droplet size from nozzle to nozzle (Biskos et al., 2008).

12.2.5 Spray pyrolysis

Spray pyrolysis is dependent on the production of an aerosol from a precursor solution, as well as regulation over the decomposition of the aerosol in a tubular flow reactor in general, and phase transformation. This procedure typically consists of three steps: nebulization, droplet-to-particle transformation, and product selection. When an aerosol-generation unit is integrated with a heat treatment, an efficient technique called aerosol spray pyrolysis may be generated. Aerosol spray can be used to produce high-purity particles in a cost-effective, continuous, and time-efficient manner. Because of its quick and simple method, aerosol-spray pyrolysis has gained a lot of attention to produce electrode materials for energizers. This approach may generate dry powder from a liquid or a slurry with ease. It also creates spherical and homogeneous particles that have a narrow particle size distribution and multicomponent materials are formed that can be used in applications such as battery production because they are free from aggregation. Aerosol-spray pyrolysis is a possible alternative to conventional synthetic processes (hydrothermal and precipitation approaches). It is a continuous, simple, and low-cost process with a short reaction time, suited for industrial scaling-up synthesis. However, it suffers from poor thin-film quality, heat breakdown, and vapor convection (Tahir et al., 2020). Dissolving the proper metal salts in the suitable solvent produces a precursor solution to prepare the particles for decomposition by heat decomposition. A nebulizer (for example, an ultrasonic nebulizer) converts the precursor solution into droplets which are then supplied to a heated reactor. Inside the droplet, solute deposition, drying, heat reaction/decomposition, solvent evaporation, or curing may occur for the formation of the final product (Wang et al., 2008; Widiyastuti, Wang, Lenggoro, Iskandar, & Okuyama, 2007). Depending on the temperature employed to condition the droplets, the techniques employed are spray pyrolysis, spray drying, or freeze-drying. Spray pyrolysis involves heating nebulized particles in a furnace, a flame, or a laser beam. The quick evaporation of the droplets' solvent and the commencement of interparticle reactions result in the best product, caused by high temperatures (Biskos et al., 2008; Lenggoro, Hata, Iskandar, Lunden, & Okuyama, 2000).

12.2.6 Gas-phase aerosol-assisted nanomaterial synthesis

As an aerosol-assisted approach in nanomaterial synthesis, the gas-phase route is based on the generation of the precursor vapor. In this route, the precursor vapor can be a flow of the input gas, evaporated liquid, or sublimated solid. As a result of the chemical reactions in the reactor, supersaturation occurs. It is an important driving force for the nucleation step of nanosized particles. The formed nuclei are the centers of condensation and coagulation (surface growth). Generally, the process conditions (initial concentration, temperature profile, and residence time) have a significant effect on the size and morphology of the final nanoparticles (Bonpua, Yagües, Aleshin, Dasappa, & Camacho, 2019; Kim, Song, Kim, Hwang, & Park, 2018; Zhang et al., 2021). Also, a wide range of nanomaterials [pure metal (Byeon & Roberts, 2012; Lafont, Simonin, Tabrizi, & Schmidt-Ott, 2009), oxides (Li, Ren, Biswas, & Tse, 2016; Rudin, Tsougeni,

Gogolides, & Pratsinis, 2012), alloy (Lafont et al., 2009; Tabrizi, Xu, Van Der Pers, & Schmidt-Ott, 2010), carbon-based materials (Li et al., 2016), silicone-based materials (Gr, Richard, Walther, & Peukert, 2012), semiconductors (Biskos et al., 2008), carbides (Herrmann, Standke, Höhn, Himpel, & Gestrich, 2014), etc. (Zhang et al., 2021)] and nanostructures [nanoparticle (Kim et al., 2018), nanoflakes (Chen et al., 2019), nanospheres (Krishnan, Verheijen, Gert, Palasantzas, & Kooi, 2013), nanotube (Ding et al., 2020), nanowire (Cvelbar, 2011; Heurlin, Magnusson, Lindgren, Ek, & Wallenberg, 2012), etc. (Zhang et al., 2021)] are synthesized by this method. The final products are applicable for photocatalysis, solar cells, energy storage, etc.

Today, different reactors have been created for gas-phase aerosol-assisted nanomaterial synthesis. Commonly, the developed approach is named based on reactor usage. Therefore, they are called furnace generation, glowing-wire aerosol generation, spark-discharge generation, plasma, and laser ablation (Biskos et al., 2008; Zhang et al., 2021). For material synthesis, there is not any monopolization of these reactors (Zhang et al., 2021) and it is possible to combine them with other nebulizers, probes, or aerosol sources (Attoui, 2018; Byeon & Roberts, 2012). To produce high-quality nanosized materials, gas-phase approaches need supernumerary considerations relating to desired nanomaterials and reactor type.

12.2.7 Furnace generation

The Furnace aerosol generator, as a widely used reactor, employs the external energy (for example, electricity) source to produce heat. Such an energy source provides two advantages: tunable temperature and controllable residence time. These two factors cause good control over the size and morphology of the product (Biskos et al., 2008; Djenadic & Winterer, 2017; Zhang et al., 2021). Also, this reactor utilizes an inert carrier gas to convey fed precursors (precursor feeding). Here the physical state of the feed can be vapor or liquid. In the vapor-fed type, the vapor precursor is directly transferred to the furnace room where the nucleation step occurs. Clean diluting gas with a high flow rate is used to rapidly quench particle growth. The formation of particles takes place based on a mechanism called chemical vapor synthesis (CVS, gas to particle mechanism) (Biswas, Wang, & Attoui, 2018; Swihart, 2003; Zhang et al., 2021). For the liquid-fed furnace reactor, the method of precursor entry to the furnace is different. At first, the precursors are nebulized into fine droplets. Then the formed droplets are transferred to the furnace room and the following pathway proceeds according to the temperature. At high temperatures, precursor droplets convert into the vapor phase and the nanoparticle formation takes place based on the gas to particle mechanism. If the temperature is not high enough to convert droplets into vapor, nanoparticles form via the liquid to particle pathway (Zhang et al., 2021). This mechanism is related to the spray pyrolysis method described earlier.

Following the agglomeration that occurred in the high concentration, the morphology of nanomaterials (with a size larger than a few tens of nanometers) tends to be irregular. This is a limitation of furnace aerosol generation, where nanomaterials with a uniform morphology are required. The next limitation of this generator is related to the melting point (T_{mp}) or the corrosion condition of material applied in the inner wall of the generator. These parameters should be appropriate to avoid the diffusion of vapors/droplets into the furnace space and to prevent the contamination of the synthesized product (Biskos et al., 2008; Hee, Cheol, Soo, Ho, & Soo, 2006).

12.2.8 Glowing-wire aerosol generation

Glowing-wire generators (GWG) use a glowing (heated) wire coated with the precursor to evaporate them. Then the precursor vapor is thermally quenched via passing a gas stream and the nucleation occurs. Therefore, nanoparticles are formed through the rapid condensation of vapor in the supersaturated state. The particles produced by GWG typically range between a few angstroms and a few nanometers, offering unexpected behaviors. The glowing-wire technique can precisely control the property and impurity of the product, which is an advantage for GWG (Biswas et al., 2018; Ghosh, Tripathi, Joshi, Mayya, & Khan, 2020). For example, the nanoparticle size can be controlled by regulating the wire dimensions, electrical current, and heating potency (Chen & Zhang, 2017). By using this method, access to very high-purity nanomaterial is possible. The production of clean particles has an undeniable impact in some applications. The synthesis rate of the nanomaterial is about one microgram per hour. Therefore, the applications of nanomaterials produced by GWG are limited to those that require a small scale such as a gas sensor (Biskos et al., 2008).

12.2.9 Spark-discharge generation

The generation of nanoparticles via the spark occurs between electrodes by discharging the repeated sparks. In this way, about a few milligrams of the nanoparticles are created in every trigger. Each spark is the plasma medium between electrodes with a life span of about a few microseconds and a temperature of about 20,000 Kelvin or more. Following the generation of each spark, hot spots are generated between the electrodes.

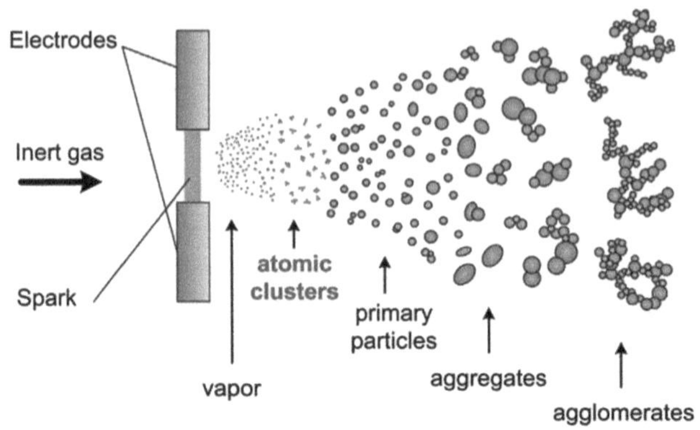

FIG. 12.3 The state of nanomaterial formation by spark discharge. *Reproduced from Pfeiffer, T. V., Feng, J., & Schmidt-Ott, A. (2014). New developments in spark production of nanoparticles.* Advanced Powder Technology, 25, *56–70.*

Therefore, a volume of vapor is produced in the surrounding space of electrodes. Due to the extinction of each spark, cooling the formed vapor occurs by flowing the surrounding gas and adiabatic expansion. Usually, a flow of inert gas is used to carry the formed nanoparticles or clusters away from the space between two electrodes. Depending on the concentration of the evaporated precursor, the final product can include atomic clusters, particles, aggregates, and agglomerates. The concentration of the primary aerosol, which depends on the spark energy and the gas flow rate in the space between the electrodes, determines the type of the product. Fig. 12.3 shows the state of nanomaterial formation by spark discharge (Pfeiffer, Feng, & Schmidt-Ott, 2014).

In a classical circuit, the energy generated per spark is irregular and it is considered a drawback. Another limitation of this circuit is related to its operation at high frequencies ($> 500\,Hz$). However, it is feasible to overcome these drawbacks by using a switched pulse generator (Pfeiffer et al., 2014). The nanomaterials produced by spark discharge and laser ablation are very similar. However, the lack of a laser (as an expensive component) is an advantage of this method. The potential of scaling-up, high-purity products, and environmentally friendly processes are the other advantages of this method (Biskos et al., 2008; Pfeiffer et al., 2014).

12.2.10 Plasma

Plasma is a quasi-neutral medium inclusive of photons, electrons, ions, and atoms (grounded and excited states) (Pankaj & Keener, 2017). Plasma-based processes can convert solid, liquid, or gaseous precursors to nanomaterials. Generally, there are different types of plasma for the synthesis of nanomaterial: hot (thermal) plasma, cold (nonthermal) plasma (Zhang et al., 2021), microplasmas (Mariotti & Sankaran, 2010), microwave plasmas (Dager, Baliyan, Kurosu, Maekawa, & Tachibana, 2020), remote plasma (Doan, Weeks, Agarwal, & van de Sanden, 2014), and low-pressure plasma (Chavée et al., 2021). Here we only give a brief description of the synthesis of nanomaterial using thermal and nonthermal plasma.

Thermal plasma is a flow of electrons and ionized gases in their thermodynamic equilibrium state. In this state, the temperature of the plasma region is about 10,000–20,000 K (Kim et al., 2018; Kim & Kim, 2019). The fabrication of a nanomaterials base on the thermal plasma is divided into bottom-up and top-down approaches (Fig. 12.4).

In the bottom-up approach, the injected precursor is immediately vaporized. Then the formed vapors encounter a speedy quench (about 106 $K s^{-1}$), where the plasma jet is expanded. The vapor supersaturation leads to homogeneous nucleation and nanoparticle formation. In the following, the growth of nanoparticles occurs through the surface reaction (cocondensation and coagulation). To adjust the nanoparticle size, the additional flow of quenching gas (Ar or He) is introduced downstream. Finally, the separation step by a cyclone separator, in situ filtrations, and their passivation (safe handling in the glove box if needed) is applied

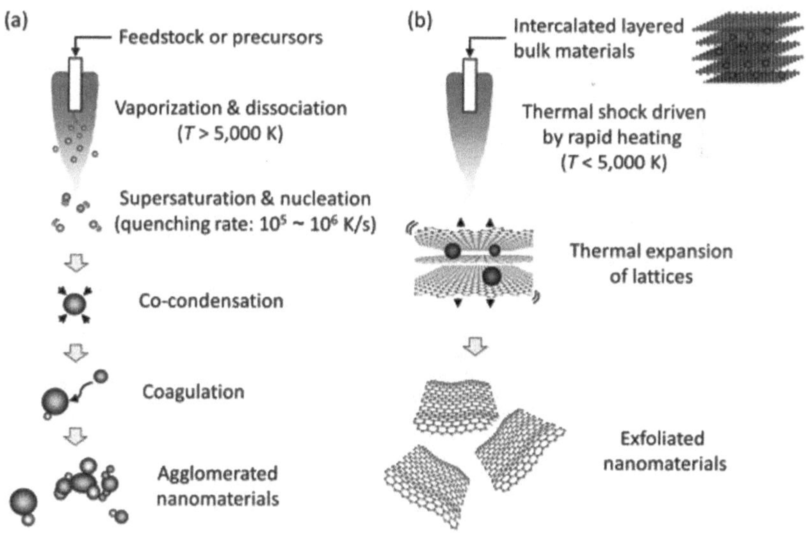

FIG. 12.4 A scheme of nanomaterial fabrication by a thermal plasma method. (A) Bottom-up and (B) top-down. *Reproduced from Kim, K. S., & Kim, T. H. (2019). Nanofabrication by thermal plasma jets: From nanoparticles to low dimensional nanomaterials.* Journal of Applied Physics, 125, *1–26.*

to collect the desired materials. Also, the rapid heating induced by thermal plasma is able to synthesize nanomaterials via a top-down approach. This approach is appropriate for layered material where the weak van der Waals force is the retaining agent of the layers. Due to the heat of thermal plasma, the intercalants are interpolated between the layers and the exfoliation of materials occurs. This method results in the formation of nanoplates and smaller layers (Kim & Kim, 2019).

When different species of plasma medium are thermodynamically in a nonequilibrium state, it is called nonthermal plasma. In the nonthermal plasma, the temperature of free electrons is about several electron volts and the temperature of other plasma species is about the room temperature (maximum of 1000 K). Unlike in the past, the generation of nonthermal plasma can be done at atmospheric pressure and it does not necessarily require low-pressure conditions. In particular, the high energy of the electrons leads to the dissociation of covalently bounded precursors (Pankaj & Keener, 2017). Following the passage of precursor gas mixture through the reactor quartz tube, the nanomaterials arise by nucleation, condensation, and coagulation (Zhang et al., 2021).

12.2.11 Laser ablation

A laser beam is a powerful tool for nanomaterials synthesis. In this method, the focused laser beam on a spot of the solid surface results in rising temperature and forming the evaporated species. The collisions that occurred between them and the surrounding species lead to form the plasma and the ablation region. The characteristics of laser-induced plasma are determined by the parameters of the applied laser, the pressure, the target material, and the surrounding medium. Finally, the formed vapors are taken out from the ablation region by a carrier gas flow and cooled down to form the clusters. By further condensation and coagulation, the clusters convert to the nanoparticles (Kim, Osone, Kim, Higashi, & Seto, 2016). The effect of the surrounding medium is shown in Fig. 12.5.

This method requires high input energy and irradiates a small area of target materials. Typically, the rate of mass production is about 100 ng per pulse of the laser. Controlling the ablated product is accessible by changing the energy and the frequency of the laser pulse. Also, other parameters (solvent composition, laser wavelength, pulse duration, repetition rate, or energy fluence) affect this synthesis method and the product created (Biskos et al., 2008; Kim et al., 2016; Mostafa & Mwafy, 2020; Piotto, Litti, & Meneghetti, 2020).

In advanced methods of laser ablation, postprocessing (gas-phase annealing and particle classifying) can develop shape- and size-controlled nanoparticles. However, postprocessing reduces the production efficiency. To scale up the laser ablation method, it is necessary to produce

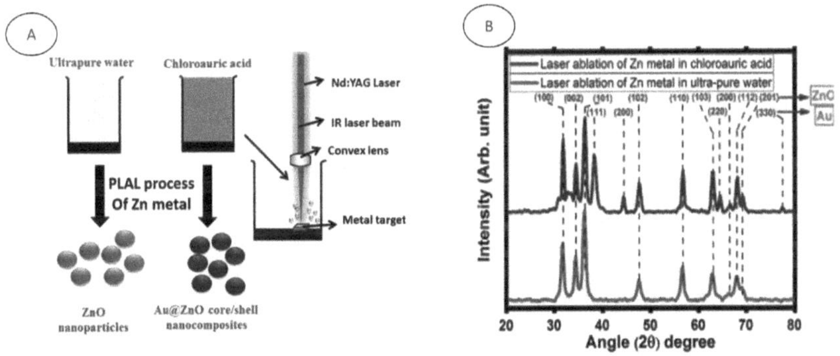

FIG. 12.5 (A) A scheme of the synthesis of ZnO nanoparticles and Au@ZnO nanocomposites by laser ablation in ultrapure water and chloroauric acid. (B) XRD of samples produced by laser ablation in different liquid media. *Reproduced from Mostafa, A. M., & Mwafy, E. A. (2020). Synthesis of ZnO and Au@ZnO core/shell nanocatalysts by pulsed laser ablation in different liquid media.* Journal of Materials Research and Technology, *9, 3241–3248.*

the desired nanoparticles without the need for the postprocessing step (Kim et al., 2016).

12.3 Applications

12.3.1 Metal nanoparticles

The use of aerosol-based methods leads to the production of nanomaterials with controlled physical and chemical properties. For example, in 2003, Backman et al. proposed a single-step method for the production of metal catalysts based on the aerosol process which included simple configurations consisting of a bubble generator and a tubular flow reactor. Systems' performance was evaluated by preparing Ag nanoparticles on TiO_2 support and the obtained powder was characterized by different methods. The results showed that the obtained powders have a high specific surface area (Backman, Tapper, & Jokiniemi, 2004).

Another application is the synthesis of Ag nanoparticles during the pyrolysis process using an ultrasonic nebulizer. The experimental data and their results explain the effect of feed solution concentration and applied ultrasonic power on the nanoparticle size. By changing the concentration of the solution and the power of the nebulizer, the average particle size in the range of about 20–300 nm is obtained (Pingali, Rockstraw, & Deng, 2005). The ultrasonic spray pyrolysis method was also used to prepare spherical cobalt powder particles using a hydrogen atmosphere. The Co particle size was controlled by a change in the initial solution concentration and nebulizer frequency. The results showed that a decrease in the concentration of the initial cobalt nitrate solution caused a decrease in the obtained particle size (Gürmen, Stopić, & Friedrich, 2006).

Au, Ag, and Au-Ag alloy nanoparticles were synthesized using microplasma under atmospheric pressure in polyvinyl alcohol hydrogel. Polyvinyl alcohol prevents agglomerating nanoparticles (Nolan et al., 2018). In 2012, Lin et al. used organic metal compounds to synthesize metal nanoparticles. The use of these compounds is especially suitable in plasma-based processes, as the energy of the electrons can be enough to break the organic bonds. In fact, vapors dissociation of metallic organic compounds by plasma leads to the synthesis of the corresponding carbon-free metal nanoparticles (Lin, Kumar, & Sankaran, 2012). Using these metallic organic precursors and plasma techniques, multimetallic nanoparticles were synthesized. In fact, multimetallic nanoparticles are a special group of nanoparticles that contain two or more metals. The presence of different metals together causes unique properties compared to separate compounds. The results showed that this is an economical method for the preparation of high-purity material. In this method, there is no need to use surfactants to control nucleation and particle growth (Lin & Sankaran, 2011; Tabrizi et al., 2010).

A group of transition metals (Co, Cr, Mn, Mo, Ni, Cu, Fe, etc.) was introduced in TiO_2 by an aerosol-based single-step process, using flame spray pyrolysis. The products were used to study the optical decomposition of acetonitrile. Their effect on catalytic activity was investigated. Based on the obtained results, Cr-doped TiO_2 shows high photocatalytic activity compared to other metals (Inturi, Boningari, Suidan, & Smirniotis, 2014).

12.3.2 Metal oxide nanoparticles

Like metals, many metal oxides have different catalytic applications (Marino, Heijden, Creyghton, & van der Heijden, 2007). Among metal oxides, magnetic nanoparticles have a wide range of applications, and because of their magnetic properties, it is possible to track and follow them by a magnetic field. Accordingly, magnetic nanoparticles have applications such as magnetic resonance imaging, drug delivery and cancer treatment, data storage, etc. (Ismail, Sulaiman, Abdulrahman, & Marzoog, 2015; Strobel & Pratsinis, 2009).

In 2015, Ismail et al. synthesized iron oxide nanoparticles by pulsed laser ablation in DMF and SDS solutions. Antibacterial properties of nanoparticles synthesized on two types of bacteria (Gram-positive and Gram-negative) were investigated. The results show that the antibacterial properties of synthesized iron nanoparticles are dependent on the preparation conditions. The obtained SEM images show the spherical shape of iron oxide nanoparticles in the approximate range of 50–60 nm in SDS and DMF solutions. This size range is highly dependent on the laser fluence. The higher the laser fluence, the more the energy is transferred to the target, and the material melts and the size gets bigger. As the concentration of the starting material increases, the color of the iron nanoparticles changes from light yellow to dark brown. Increasing the laser fluence leads to a redshift in the absorption spectrum of iron oxide nanoparticles. This shift can be related to defects in the network of the nanoparticles. In general, by controlling the reaction conditions, different sizes of iron oxide nanoparticles can be obtained (Ismail et al., 2015).

Another example of oxide nanoparticles formed by the pulsed laser ablation method is zinc oxide nanoparticles. The possibility of forming these oxide nanoparticles at room temperature and under different liquid phases was investigated. Nanoparticles synthesized in deionized water and isopropanol as the solvent have a spherical shape with a size of about 14–20 nm. Further, studies in acetone show that the particles obtained in this solvent are of two different types in both shape and size. The first group is spherical in shape and the second group is like platelet, which indicates the effect of the solvent on the size and shape of the synthesized nanoparticles. The different size distributions of nanoparticles can be attributed to differences in the absorption coefficient of solvents in laser light or differences in the chemical reaction that takes place in different solvents (Thareja & Shukla, 2007).

The synthesis of iron oxide nanoparticles by flame spray pyrolysis (FSP), as a fast method, has also been reported. In fact, by changing the fuel to air ratio and the valence state of the iron precursor, they achieved different products such as Fe_3O_4, Fe_2O_3, and FeO (Strobel & Pratsinis, 2009). The FSP method is also used for the synthesis of crystalline nanoparticles, for example, the synthesis of crystalline tin dioxide nanoparticles with a size of about 17 nm was successfully reported and used to sensing gases in atmospheric conditions (Sahm et al., 2004).

The use of ultrasonic spray pyrolysis (USP) for the synthesis of oxide nanoparticles leads to the formation of nanostructures with different shapes (Ebin, Arıg, Özkal, & Gürmen, 2012; Yudin et al., 2016). For example, NiO nanostructures were synthesized by the USP process. The results show that temperature plays an important role in the synthesis process. The temperature of the reactor and the concentration of the initial salt solution control the structure of the obtained product (Yudin et al., 2016). ZnO nanoparticles were synthesized by zinc nitrate under the influence of the air atmosphere at different temperatures. As the reaction temperature increased, the structure of the ZnO particles obtained from the sphere changed to a hexagon. ZnO spherical nanoparticles were obtained at lower temperatures of about 400°C with an approximate size of 25 nm. As the temperature rises, the accumulated nanoparticles also increase. In fact, the dominant and influential parameter in the USP method is temperature (Ebin et al., 2012; Yudin et al., 2016).

Preparation of spherical and uniform ZnO particles by the spray pyrolysis (SP) method with an ultrasonic nebulizer was also reported. The role of the solvent, flow rate, and solution concentration on the system was investigated. The results show that the use of alcohol as a solvent instead of water causes the formation of particles without a hollow shell layer. Also, finer particles are formed at higher concentrations and lower flow rates (Liu, Sakurai, Mizutani, & Kato, 1986).

Synthesis of manganese oxides for molecular sieve application was also performed by ultrasonic nebulization. This is considered a simple method in which it does not rely on heat and leads to the formation of particles with a uniform morphology and porous structure. Therefore, it is a suitable substance to be used as a catalyst. The result showed that the self-aggregation of the particles does not occur when the flow rate of the initial solution is high (Iyer et al., 2017).

Today, the use of gas sensors plays an important role in the early diagnosis of various diseases, especially cancer. The use of metal oxide-based gas sensors is limited due to the dependence of the sensor response to humidity. In 2018, Vasiliev et al. succeeded in the dry synthesis of SnO_2 nanoparticles in the gas phase by the spark-discharge process, which was less sensitive to water. Potentially, they have different uses in the future, for example, in the early detection of the fire (Vasiliev et al., 2018).

12.3.3 Carbonaceous nanomaterials

Carbon nanotubes (CNs) are a group of carbon materials that have received much attention due to their unique properties and numerous applications (Jo, Jeong, Bae, & Jeon, 2008; Lara-Romero et al., 2011). There are several methods for CNTs synthesis, and the pyrolysis method has been widely used due to its cost-effectiveness and the possibility of controlling the size. Various factors affect the synthesis, growth, and morphology of CNTs, which can be referred to as the concentration of the precursor, injection speed, time, carrier gas flow rate, substrate surface, temperature, etc. In fact, by controlling each of these factors, the growth and even the morphology of the created product can be affected. In this method, different substrates (Annu, Bhattacharya, Shukla, Singh, & Rhee, 2017) such as SiO_2, silicon, copper, etc. (Rosi, Noor, & Yulkifli, 2008; Zhu, Lin, & Guo, 2007) can be used. Choosing a good substrate creates an effective interaction between the substrate and the catalyst (Annu et al., 2017). A suitable catalyst for the synthesis of CNTs is ferrocene, which dissolves in xylene (Barreiro et al., 2007; Cele & Coville, 2009; Rosi et al., 2008). For CNTs synthesis, temperature plays a key role in the spray pyrolysis method. At high temperatures, hydrocarbons break down, allowing the catalyst to diffuse to the surface and synthesize better (Annu et al., 2017).

In 2015, Huang et al. synthesized carbon nanomaterials with different morphologies using the spray pyrolysis method, and they also looked at factors such as carbon precursor concentration and efficiency. The results show that carbon nano coils (CNCs) are formed at lower temperatures (600°C) and CNTs are formed at higher temperatures (700–800°C). In this case, a furnace was used for chemical vapor deposition vertically with three heating zones, and a nano palladium catalyst was also used. Interestingly, when the temperature was selected differently in three zones, it produced carbon nanomaterials with a specific shape and morphology. At a temperature of 600°C, CNTs with a spiral structure and a diameter of about 100–150nm were obtained, whereas CNTs at a temperature of 700°C had a hollow structure and a diameter of about 20nm. As the temperature increased, the CNTs became smaller in diameter and thinner in structure (Huang et al., 2015). Another factor affecting the diameter of nanotubes (NTs) is gas flow. For example, in 2007, Barreiro et al. synthesized single-walled carbon nanotubes (SWCNTs) at atmospheric pressures by aerosol-assisted chemical vapor deposition. As the total gas flow increased, the average diameter of the particles formed decreased. The use of sulfur along with ferrocene solution accelerates the growth and increases the efficiency of SWCNTs (Barreiro et al., 2007).

CNTs synthesis was also reported using the laser ablation process in flame. Iron ablation in a CO-enriched flame led to the formation of SWCNTs, while the ablation of Ni in the acetylene-enriched flame led to the formation of carbon nanofibers (Vander Wal, Berger, & Ticich, 2003).

Fullerenes are one of the carbon allotropes that have also been synthesized by the aerosol-based method (Joutsensaari et al., 1996, 2000). Fullerene nanocrystals were prepared by spray drying. As the temperature increases, the size of fullerenes decreases, and high temperatures lead to the formation of spherical fullerenes about 20nm in size (Joutsensaari et al., 1996).

A simple synthesis method for the preparation of graphene films was performed by the nebulizer (spray pyrolysis method) in 2016. In this method, dextrose and hydrazine hydrate were used to reduce graphene oxide to graphene. The transfer of the graphene oxide absorption peak from 230 to 290nm indicates the reduction of graphene oxide and the successful synthesis of graphene (Illakkiya, Rajalakshmi, & Oommen, 2016). In addition to spray pyrolysis-based methods, in 2019, Hsiung Tseng et al. used an electric spark-discharge-based method to reduce graphene oxide. In practice, a silver wire electrode was placed in graphene oxide (GO) solution, and as a result of discharge in the gap, silver nano atoms are suspended in the solution. Due to the ionization, Ag^+ ions are produced and reduce GO to rGO. Finally, the surface of the silver nanoparticles was covered with rGO (Tseng, Ku, Tien, & Stobinski, 2019).

In 2018, Shah et al. proposed a new and promising approach to synthesize graphene using mango peel. Mango peel is suggested as an available and inexpensive carbon source for the synthesis of graphene on a Cu substrate in the presence of plasma. Plasma exposure time is an important factor. Short-time exposure to plasma results in the formation of multilayer graphene. The formation of monolayer graphene is related to a longer period of time (Shah, Janneth, Carreon, Lopez-Miranda, & Carreon, 2018).

12.3.4 Nanoporous materials

Nanoporous materials with a specific size and shape have several applications due to their porosity, size distribution, and physical and optical properties, for example, lithium-ion batteries, sensors, catalysts, chromatography, supercapacitors, and even drug delivery (Pan & Wang, 2001; Zeng & Weber, 2014; Zhang, Du, Shiab, & Gao, 2012).

The chemical sol-gel method is commonly used to prepare porous particles. There is a lack of scalability and limitations in fast and continuous synthesis in this method. Therefore, spray-based methods are considered (Zeng & Weber, 2014). In 2011, Patil et al. synthesized nanoporous hollow SnO_2 nanostructures using the ultrasonic nebulization technique. These porous nanomaterials were used to make the sensor which had a high and fast response to hydrogen and showed good selectivity. The rapid response of these sensors can be attributed to the fast penetration of gases through the porous nanomaterials (Patil, Shinde, Bari, & Deo, 2011).

Porous silica particles were prepared by the SiO_2 precursor and spray drying process. Polyvinyl pyrrolidone (PVP) was used as the porous template. After removing it by the calcination process, porous particles were obtained. These porous particles can be used as catalysts for the Fischer-Tropsch synthesis after making the necessary modifications (Zeng & Weber, 2014). Nanoporous vanadium oxide thin films were synthesized using spray pyrolysis on a glass substrate. Aqueous ammonium metavanadate solution was used as the initial solution (Ingole & Lokhande, 2016). This method was also used to make porous, water-soluble, and thermally stable aluminum oxide nanoparticles (Kim, Liu, & Zachariah, 2002).

In 2018, Kyung Kim et al. synthesized porous TiO_2/SiO_2 nanocomposites using an aerosol-based method. The photocatalytic activity of these nanoparticles was investigated by measuring the optical decomposition of methylene blue and NO_x. The results show that this nanoporous composite has more photocatalytic activity than TiO_2 nanoparticles and TiO_2 nanoporous particles. The synthesized porous nanoparticles have a spherical shape and a pore diameter of about 1–100 nm. The synthesis of this photocatalyst is simple and more economical (Kim et al., 2011).

12.3.5 Nanocomposites and functional nanomaterials

Nanocomposites are materials with many applications and a combination of unusual properties and unique designs (Camargo, Satyanarayana, & Wypych, 2009). Nanomaterials have created a breakthrough in science due to the unique properties of these materials. Many studies and experiments are performed in this field (Mathew et al., 2020). Therefore, the use of new methods with appropriate efficiency and higher speed for the preparation of nanocomposites and functional nanomaterials is of particular importance. The preparation of these nanocomposites by traditional methods is very time-consuming and has a high waste. In the method used by two pneumatic nebulizers and aerosol jet printing, these problems have been eliminated (Craton, Albrecht, Chahal, & Papapolymerou, 2020).

Controlling the morphology of SiO_2 nanocomposites is of vital importance. A very practical way to do this is to use an ultrasonic spray device. An ultrasonic spray device is used to produce nanoparticles with smaller sizes and spherical morphology (Azhary et al., 2019).

The synthesis of very fine magnetic powders has received a lot of attention in recent years. One of these materials is magnesium ferrite nanocomposite which is used as a catalyst for ammonia synthesis and organic synthesis. Various methods are used to synthesize this nanocomposite, one of which is the use of an ultrasonic nebulizer followed by the reduction process (Bahgat, 2006).

Olivine structure lithium transition-metal phosphates, $LiMPO_4$ (M = Fe, Mn, Co, Ni), have received much attention in the next generations of lithium-ion batteries as cathode materials. They render high capacity as well as chemical and thermal stability. Many efforts have been made to increase the electrochemical properties of lithium manganese phosphate with carbon coating. The performance of the nanocomposite is not sufficient for use in commercial cells. A method that has been used for this synthesis and is more efficient than the previous methods is the use of the pyrolysis spray technique followed by dry ball milling. The efficiency of this method is due to the smaller particles and materials obtained (Doan & Taniguchi, 2011). Other nanocomposites that have been synthesized and prepared by the pyrolysis spray method include silver vanadate and ZnO-CuO (Saravanakkumar et al., 2018; Zheng et al., 2017).

One of the applications of the spark-discharge generation method is to produce an Mg-Ti nanocomposite. This nanocomposite is used to store hydrogen (Anastasopol et al., 2013). The spark-discharge technique is also used to prepare semiconductor metal nanocomposites, one of which is AuGe nanocomposites (Kala, Theissmann, & Kruis, 2013).

Nanocomposites, which are a combination of nanoparticles and polymers, are highly regarded for their optical, electrical, mechanical, and magnetic properties. One of these nanocomposites is PVK/AgNPs

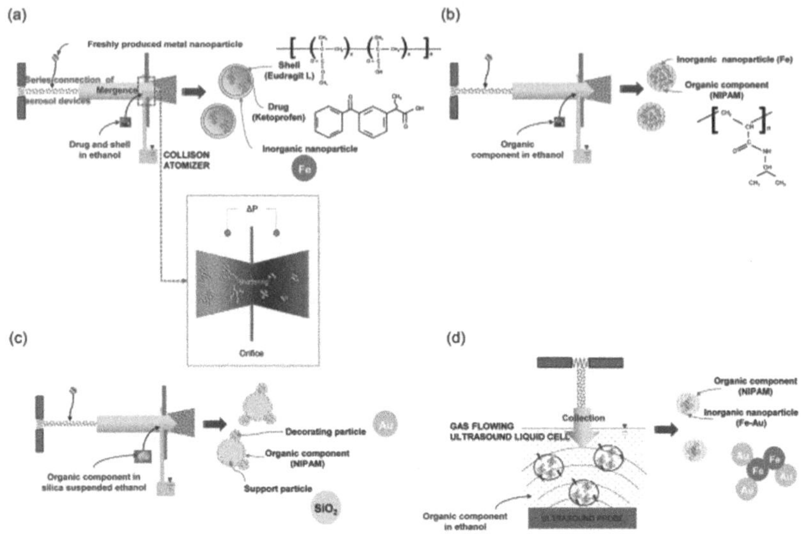

FIG. 12.6 Fabrication of (A) nanomagnetosols, (B, D) hybrid nanogels, and (C) nanoinorganics. (A–C) show serial connections of a spark generator and a collision nebulizer (on the fly manner). (D) shows a spark generator coupled to an ultrasound probe (aerosol into liquid configuration). *Reproduced from Byeon, J. H., & Roberts, J. T. (2012). Aerosol-based fabrication of biocompatible organic-inorganic nanocomposites.* ACS Applied Materials & Interfaces, 4, *2693–2698.*

[PVK: poly(n-vinylcarbazole)]. AgNPs are synthesized by the laser ablation method and precipitated in PVK solution (El-kader, Hakeem, Elashmawi, & Menazea, 2015).

In 2012, Byeon and Roberts fabricated several biocompatible organic-inorganic nanocomposites based on aerosol techniques (Fig. 12.6). These nanocomposites are nanomagnetosols (Fe nanoparticles, the drug ketoprofen, and Eudragit shells), hybrid nanogels (Fe nanoparticles and N-isopropylacrylamide, gel), and nanoinorganics (Au-immobilized Si). They used a serial aerosol reactor to produce aerosol metal nanoparticles embedded into organic matrices. The aerosol metal nanoparticles and organic matrices are merged in an "on the fly" manner. Also, they introduced "aerosol into liquid" configuration to produce a hybrid material (Fe-Au nanoparticles and NIPAM), in which metal aerosol was collected into a NIPAM/ethanol solution (Byeon & Roberts, 2012).

Recently, Tsai's research group synthesized several functional nanoparticles using the aerosol spray method (Chang, Lai, & Tsai, 2019; Liang, Lin, Chou, Wang, & Tsai, 2018; Lin, Chou, & Tsai, 2018). They developed a two-stage method to synthesize CuO_x-CeO_2 hybrid nanoparticles. The final product exhibits a tunable oxidation state and gives out the catalytic activity of methane combustion (Lin et al., 2018). Also, they synthesized a new Ni-CeO_x hybrid nanoparticle as a high-performance catalyst for dry reforming of methane with carbon dioxide (DRM) (Liang et al., 2018). The Ni-CeO_2-Al_2O_3 hybrid nanoparticle cluster is another nanopowder catalyst developed by them. They utilized the product to catalyze reductive amination of polypropylene glycol (Chang et al., 2019).

12.3.6 Nanostructured thin films and nanocoatings

Cuprous iodide (CuI) is a high-potential semiconductor that has many applications in electronics, organic catalysts, light-emitting diodes, and field displays. A CuI thin film is produced by various chemical and physical methods in nanometer dimensions. One of the methods used to generate CuI is a pneumatic nebulizer in the mister nebulizer (Amalina, Zainun, Rasheid, & Rusop, 2012; Lee, Leem, & Kim, 2008; Sirimanne, Soga, & Jimbo, 2003).

Tin oxide is an n-type semiconductor with high electrical conductivity. Many techniques such as evaporation, dispersion, etc. are used to produce tin oxide thin films, which have advantages and disadvantages. Due to its wide range of applications, cheapness, and reproducibility, the spray pyrolysis technique is used to synthesize tin oxide thin films (Rozati, 2006; Shadmania & Rozati, 2012).

CuO nanostructured films used as electrodes in lithium batteries are produced by a variety of methods. Due to its advantages (easy thickness control and no need for advanced equipment), the mainly used method is the pyrolysis spray (Morales, Sánchez, Martín, Ramos-Barrado, & Sánchez, 2004). Some other nanostructured thin films are semiconductor materials. These nanostructured thin films are also prepared by the pyrolysis spray method. The performance of such thin films can be optimized with temperature parameters and precursor concentration (Ranasinghe, Vequizo, & Yamakata, 2018).

Zirconium oxide is one of the known materials used as dielectric layers in semiconductor devices or as the basic material of oxygen ion conducting electrolytes in solid oxide fuel cells (SOFC). The use of plasma techniques to produce zirconia nanostructured thin films to control conditions such as high uniformity is to improve the density of films (Cho, Noh, & Lee, 2016; Ko, Kerman, & Ramanathan, 2012).

The preparation of nanostructures has received much attention due to their many applications. One of these nanostructures is nickel oxide. Nickel oxide nanocrystals are one of the main components of nanometer electronic devices. Nickel oxide thin film has attractive properties due to its excellent durability, electrochemical stability, low cost. One of the successful methods for the preparation of nickel oxide thin film is the pulsed laser deposition method (Sasi & Gopchandran, 2007). Another thin film produced by the pulsed laser deposition method is the Sb_2O_3 thin film (Xue & Fu, 2006).

12.4 Future perspectives

Regardless of the significant achievement, aerosol-assisted nanomaterials synthesis encounters some challenges. It is necessary to deeply understand the synthesis process, to pursue the formation mechanism of the nanomaterials step by step, and control the shape of manufactured powders. In addition, advanced characterization techniques are needed to enable the sampling and analysis of intermediates that are rapidly formed during a process at high temperatures.

For spray-based synthesis, the most challenging aspect is stoichiometric control, which is complicated by the varying solubility of the precursors. As a result of this, segregation can occur, restricted to the droplet dimension. Furthermore, the spray source's droplet production has a considerable impact on the product's characteristics, with bigger and less homogenous particles. Another difficulty is the generation of crashing or hollow particles, which is undesirable, especially in the creation of ready-to-press powders.

Scaling up the aerosol-assisted nanomaterials synthesis is another important challenge. Modeling a large-scale process must be studied in various views: the cost, performance, wastes, nanoparticle collection, environment, etc. However, there is a strong motivation to do so.

References

Amalina, M. N., Zainun, A. R., Rasheid, N. A., & Rusop, M. (2012). The electrical conductivity of copper (I) iodide (CuI) thin films prepared by mister atomizer. In *10th IEEE Int. Conf. Semicond. Electron.*

Anastasopol, A., Pfeiffer, T. V., Middelkoop, J., Lafont, U., Canales-Perez, R. J., Schmidt-Ott, A., et al. (2013). Reduced enthalpy of metal hydride formation for Mg–Ti nanocomposites produced by spark discharge generation. *Journal of the American Chemical Society, 135*, 7891–7900.

Annu, A., Bhattacharya, B., Shukla, P. K., Singh, K. P., & Rhee, H.-W. (2017). Carbon nanotube using spray pyrolysis: Recent scenario. *Journal of Alloys and Compounds, 691*, 970–982.

Attoui, M. (2018). Activation of sub 2 nm singly charged particles with butanol vapors in a boosted 3776 TSI CPC. *Journal of Aerosol Science, 126*, 47–57.

Avvaru, B., Patil, M. N., Gogate, P. R., & Pandit, A. B. (2006). Ultrasonic atomization: Effect of liquid phase properties. *Ultrasonics, 44*, 146–158. https://doi.org/10.1016/j.ultras.2005.09.003.

Azhary, S. Y., Purnama, D., Florena, F. F., Vanitha, M., Panatarani, M. C., & Joni, I. M. (2019). Synthesis and characterization of chitosan: SiO2 nanocomposite by ultrasonic spray drying. *IOP Conference Series: Materials Science and Engineering*. IOP Publishing.

Backman, U., Tapper, U., & Jokiniemi, J. K. (2004). An aerosol method to synthesize supported metal catalyst nanoparticles. *Synthetic Metals, 142*, 169–176.

Bahgat, M. (2006). Reduction behaviour of MgFe2O4 with the synthesis of Fe/MgO nanocomposite. *Minerals Processing and Extractive Metallurgy, 115*, 195–200.

Barreiro, A., Kramberger, C., Rümmeli, M., Grüneis, A., Grimm, D., Hampel, S., et al. (2007). Control of the single-wall carbon nanotube mean diameter in sulphur promoted aerosol-assisted chemical vapour deposition. *Carbon, 45*, 55–61.

Biskos, G., Vons, V., Yurteri, C. U., & Schmidt-ott, A. (2008). Generation and sizing of particles for aerosol-based nanotechnology. *Kona Powder and Particle Journal, 26*, 13–35.

Biswas, P., Wang, Y., & Attoui, M. (2018). Sub-2 nm particle measurement in high-temperature aerosol reactors : A review. *Current Opinion in Chemical Engineering, 21*, 60–66.

Bonpua, J., Yagües, Y., Aleshin, A., Dasappa, S., & Camacho, J. (2019). Flame temperature effect on sp 2 bonds on nascent carbon nanoparticles formed in premixed flames (T f, max > 2100 K): A Raman spectroscopy and particle mobility sizing study. *Proceedings of the Combustion Institute, 37*, 943–951.

Byeon, J. H., & Roberts, J. T. (2012). Aerosol-based fabrication of biocompatible organic–inorganic nanocomposites. *ACS Applied Materials & Interfaces, 4*, 2693–2698.

Camargo, P. H. C., Satyanarayana, K. G., & Wypych, F. (2009). Nanocomposites: Synthesis, structure, properties and new application opportunities. *Materials Research, 12*, 1–39.

Cele, L. M., & Coville, N. J. (2009). The negative effects of alcohols on carbon nanotube synthesis in a nebulised spray pyrolysis process. *Carbon, 47*, 1824–1832.

Chang, H.-Y., Lai, G.-H., & Tsai, D.-H. (2019). Aerosol route synthesis of Ni-CeO2-Al2O3 hybrid nanoparticle cluster for catalysis of reductive amination of polypropylene glycol. *Advanced Powder Technology, 30*, 2293–2298.

Chavée, L., Colomer, J., Nominé, A., Bruyère, S., Achour, A., Houssiau, L., et al. (2021). Pt x Cu y @ TiO 2 nanoparticles by low-pressure plasma synthesis. *Materials Letters, 291*, 129576–129579.

Chen, X., Wang, C., Song, M., Ma, J., Ye, T., & Xia, W. (2019). The morphological transformation of carbon materials from nanospheres to graphene nanoflakes by thermal plasma. *Carbon, 155*, 521–530.

Chen, H., & Zhang, Z. (2017). Condensation-type quasi monodisperse aerosol generator with WOx nanoparticles as nuclei produced by glowing wire method. *Powder Technology, 313*, 344–352.

Cho, G. Y., Noh, S., & Lee, Y. H. (2016). Properties of nanostructured undoped ZrO2 thin film electrolytes by plasma enhanced atomic layer deposition for thin film solid oxide fuel cells. *Journal of Vacuum Science & Technology A: Vacuum, Surfaces, and Films, 34*.

Craton, M. T., Albrecht, J. D., Chahal, P., & Papapolymerou, J. (2020). In situ nanocomposite fabrication for RF electronics applications with additive manufacturing. *IEEE Transactions on Microwave Theory and Techniques, 68*(5), 1646–1659.

Cvelbar, U. (2011). Towards large-scale plasma-assisted synthesis of nanowires U Cvelbar to cite this version : HAL id : HAL-00613275. *Journal of Physics D: Applied Physics, 44*, 174014.

Dager, A., Baliyan, A., Kurosu, S., Maekawa, T., & Tachibana, M. (2020). Ultrafast synthesis of carbon quantum dots from fenugreek seeds using microwave plasma enhanced decomposition : Application of C - QDs to grow fluorescent protein crystals. *Scientific Reports, 10*, 1–15.

Ding, E., Hussain, A., Ahmad, S., Zhang, Q., Liao, Y., Jiang, H., et al. (2020). High-performance transparent conducting films of long single- walled carbon nanotubes synthesized from toluene alone. *Nano Research, 13*, 112–120.

Djenadic, R., & Winterer, M. (2017). Control of nanoparticle agglomeration through variation of the time-temperature profile in chemical vapor synthesis. *Journal of Nanoparticle Research, 19*, 1–12. https://doi.org/10.1007/s11051-016-3714-5.

Doan, T. N., & Taniguchi, I. (2011). Cathode performance of LiMnPO4/C nanocomposites prepared by a combination of spray pyrolysis and wet ball-milling followed by heat treatment. *Journal of Power Sources, 196*, 1399–1408.

Doan, L., Weeks, S. L., Agarwal, S., & van de Sanden, M. C. M. (2014). Nucleation of silicon nanocrystals in a remote plasma without subsequent coagulation. *Journal of Applied Physics, 115,* 244301–244309. https://doi.org/10.1063/1.4885195.

Ebin, B., Arıg, E., Özkal, B., & Gürmen, S. (2012). Production and characterization of ZnO nanoparticles and porous particles by ultrasonic spray pyrolysis using a zinc nitrate precursor. *International Journal of Minerals, Metallurgy, and Materials, 19*(7), 651–656.

El-kader, F. H. A., Hakeem, N. A., Elashmawi, I. S., & Menazea, A. A. (2015). Synthesis and characterization of PVK/AgNPs nanocomposites prepared by laser ablation. *Spectrochimica Acta. Part A, Molecular and Biomolecular Spectroscopy, 138,* 331–339.

Gautam, M., Kim, J. O., & Yong, C. S. (2021). Fabrication of aerosol-based nanoparticles and their applications in biomedical fields. *Journal of Pharmaceutical Investigation,* 1–15. https://doi.org/10.1007/s40005-021-00523-1.

Ghosh, K., Tripathi, S. N., Joshi, M., Mayya, Y. S., & Khan, A. (2020). Particle formation from vapors emitted from glowing wires : Theory and experiments. *Aerosol Science and Technology, 54,* 243–261. https://doi.org/10.1080/02786826.2019.1688758.

Gogate, P. R. (2015). The use of ultrasonic atomization for encapsulation and other processes in food and pharmaceutical manufacturing. In *Power ultrasonics* (pp. 911–935). Elsevier. https://doi.org/10.1016/B978-1-78242-028-6.00030-2.

Gr, M., Richard, K., Walther, M., & Peukert, W. (2012). Process control strategies for the gas phase synthesis of silicon nanoparticles. *Chemical Engineering Science, 73,* 181–194. https://doi.org/10.1016/j.ces.2012.01.035.

Gürmen, S., Stopić, S., & Friedrich, B. (2006). Synthesis of nanosized spherical cobalt powder by ultrasonic spray pyrolysis. *Materials Research Bulletin, 41,* 1882–1890.

Hee, J., Cheol, H., Soo, H., Ho, J., & Soo, S. (2006). Metal nanoparticle generation using a small ceramic heater with a local heating area. *Journal of Aerosol Science, 37,* 1662–1670. https://doi.org/10.1016/j.jaerosci.2006.09.002.

Herrmann, M., Standke, G., Höhn, S., Himpel, G., & Gestrich, T. (2014). High-temperature corrosion of silicon carbide ceramics by coal ashes. *Ceramics International, 40,* 1471–1479.

Heurlin, M., Magnusson, M. H., Lindgren, D., Ek, M., & Wallenberg, L. R. (2012). Continuous gas-phase synthesis of nanowires with tunable properties. *Nature, 492,* 90–94.

Huang, W. K., Chung, K. J., Liu, Y. M., Ger, M. D., Pu, N. W., & Youh, M. J. (2015). Carbon nanomaterials synthesized using a spray pyrolysis method. *Vacuum, 118,* 94–99.

Illakkiya, J. T., Rajalakshmi, P. U., & Oommen, R. (2016). Nebulized spray pyrolysis: A new method for synthesis of graphene film and their characteristics. *Surface and Coatings Technology, 307,* 65–72.

Ingole, R., & Lokhande, B. (2016). Nanoporous vanadium oxide network prepared by spray pyrolysis. *Materials Letters, 168,* 95–98.

Inturi, S. N. R., Boningari, T., Suidan, M., & Smirniotis, P. G. (2014). Visible-light-induced photodegradation of gas phase acetonitrile using aerosol-made transition metal (V, Cr, Fe, Co, Mn, Mo, Ni, Cu, Y, Ce, and Zr) doped TiO2. *Applied Catalysis B: Environmental, 144,* 333–342.

Ismail, R. A., Sulaiman, G. M., Abdulrahman, S. A., & Marzoog, T. R. (2015). Antibacterial activity of magnetic iron oxide nanoparticles synthesized by laser ablation in liquid. *Materials Science and Engineering: C, 53,* 286–297.

Iyer, A., Kuo, C.-H., Saminda, D., Zhu, L., Rathnayake, D., He, J., et al. (2017). An ultrasonic atomization assisted synthesis of self-assembled manganese oxide octahedral molecular sieve nanostructures and their application in catalysis and water treatment. *Nanoscale, 9,* 5009–5018.

Jo, S., Jeong, H., Bae, S. R., & Jeon, S. (2008). Modified platinum electrode with phytic acid and single-walled carbon nanotube: Application to the selective determination of dopamine in the presence of ascorbic and uric acids. *Microchemical Journal, 88,* 1–6.

Joutsensaari, J., Ahonen, P. P., Kauppinen, E. I., Brown, D. P., Lehtinen, K. E. J., Jokiniemi, J. K., et al. (2000). Aerosol synthesis of fullerene nanocrystals in controlled flow reactor conditions. *Journal of Nanoparticle Research*, *2*, 53–74.

Joutsensaari, J., Ahonen, P., Tapper, U., Kauppinen, E. I., Laurila, J., & Kuokkal, V. T. (1996). Generation of nanophase fullerene particles via aerosol routes. *Synthetic Metals*, *77*, 85–88.

Kala, S., Theissmann, R., & Kruis, F. E. (2013). Generation of AuGe nanocomposites by co-sparking technique and their photoluminescence properties. *Journal of Nanoparticle Research*, *15*, 1–12.

Kim, S., Chang, H., Cho, K., Kil, D., Cho, S., Jang, H., et al. (2011). Enhanced photocatalytic property of nanoporous TiO2/SiO2 micro-particles prepared by aerosol assisted co-assembly of nanoparticles. *Materials Letters*, *65*, 3330–3332.

Kim, K. S., & Kim, T. H. (2019). Nanofabrication by thermal plasma jets : From nanoparticles to low-dimensional nanomaterials. *Journal of Applied Physics*, *125*, 1–26. https://doi.org/10.1063/1.5060977.

Kim, H.-U., Kulkarni, A., Ha, S., Shin, D., & Kim, T. (2017). Note: Electric field assisted megasonic atomization for size-controlled nanoparticles. *The Review of Scientific Instruments*, *88*, 076106. https://doi.org/10.1063/1.4984825.

Kim, S., Liu, B., & Zachariah, M. (2002). Synthesis of nanoporous metal oxide particles by a new inorganic matrix spray pyrolysis method. *Chemistry of Materials*, *14*, 2889–2899.

Kim, M., Osone, S., Kim, T., Higashi, H., & Seto, T. (2016). Synthesis of nanoparticles by laser ablation : A review. *Kona Powder and Particle Journal*, 1–11. https://doi.org/10.14356/kona.2017009.

Kim, B., Song, J., Kim, J., Hwang, J., & Park, D. (2018). Study on nanoparticle production process characteristics in transferred arc plasma system using heat and mass balance. *Advanced Powder Technology*, *29*, 3192–3198.

Ko, C., Kerman, K., & Ramanathan, S. (2012). Ultra-thin film solid oxide fuel cells utilizing un-doped nanostructured zirconia electrolytes. *Journal of Power Sources*, *213*, 343–349.

Kolahalam, L. A., Kasi Viswanath, I. V., Diwakar, B. S., Govindh, B., Reddy, V., & Murthy, Y. L. N. (2019). Review on nanomaterials: Synthesis and applications. *Materials Today: Proceedings*, *18*, 2182–2190. https://doi.org/10.1016/j.matpr.2019.07.371.

Krishnan, G., Verheijen, M. A., Gert, H., Palasantzas, G., & Kooi, B. J. (2013). Tuning structural motifs and alloying of bulk immiscible Mo–Cu bimetallic nanoparticles by gas-phase synthesis. *Nanoscale*, *5*(12), 5375–5383. https://doi.org/10.1039/c3nr00565h.

Lafont, E. M. K. U., Simonin, L., Tabrizi, N. S., & Schmidt-Ott, A. (2009). Synthesis of nanoparticles of Cu, Sb, Sn, SnSb and Cu2Sb by densification and atomization process. *Journal of Nanoscience and Nanotechnology*, *9*, 2546–2552.

Lang, R. J. (1962). Ultrasonic atomization of liquids. *The Journal of the Acoustical Society of America*, *34*, 6–8. https://doi.org/10.1121/1.1909020.

Lara-Romero, J., Calva-Yañez, J. C., López-Tinoco, J., Alonso-Nuñez, G., Jiménez-Sandoval, S., & Paraguay-Delgado, F. (2011). Temperature effect on the synthesis of multi-walled carbon nanotubes by spray pyrolysis of botanical carbon feedstocks: Turpentine, α-pinene and β-pinene. *Fullerenes, Nanotubes, and Carbon Nanostructures*, *19*, 483–496.

Lee, J.-H., Leem, D.-S., & Kim, J.-J. (2008). High performance top-emitting organic light-emitting diodes with copper iodide-doped hole injection layer. *Organic Electronics*, *9*, 805–808.

Lenggoro, I. W., Hata, T., Iskandar, F., Lunden, M. M., & Okuyama, K. (2000). An experimental and modeling investigation of particle production by spray pyrolysis using a laminar flow aerosol reactor. *Journal of Materials Research*, *15*, 733–743. https://doi.org/10.1557/JMR.2000.0106.

Li, S., Ren, Y., Biswas, P., & Tse, S. D. (2016). Flame aerosol synthesis of nanostructured materials and functional devices : Processing, modeling, and diagnostics. *Progress in Energy and Combustion Science*, *55*, 1–59.

Liang, T.-Y., Lin, C.-Y., Chou, F.-C., Wang, M., & Tsai, D.-H. (2018). Gas-phase synthesis of Ni–CeOx hybrid nanoparticles and their synergistic catalysis for simultaneous reforming of methane and carbon dioxide to syngas. *Journal of Physical Chemistry C, 122*, 11789–11798.

Lin, C.-Y., Chou, F.-C., & Tsai, D.-H. (2018). Mechanistic understanding of surface reduction of Cu–Ce–O hybrid nanoparticles for catalytic methane combustion. *Journal of the Taiwan Institute of Chemical Engineers, 92*, 80–90.

Lin, P. A., Kumar, A., & Sankaran, R. M. (2012). New insights into plasma-assisted dissociation of organometallic vapors for gas-phase synthesis of metal nanoparticles. *Plasma Processes and Polymers, 9*, 1184–1193.

Lin, P. A., & Sankaran, R. M. (2011). Plasma-assisted dissociation of organometallic vapors for continuous, gas-phase preparation of multimetallic nanoparticles. *Angewandte Chemie, International Edition, 50*(46), 10953–10956.

Liu, T. Q., Sakurai, O., Mizutani, N., & Kato, M. (1986). Preparation of spherical fine ZnO particles by the spray pyrolysis method using ultrasonic atomization techniques. *Journal of Materials Science, 21*(10), 3698–3702.

Madrakian, T., Afkhami, A., Zadpour, B., & Ahmadi, M. (2015). New synthetic mercaptoethylamino homopolymer-modified maghemite nanoparticles for effective removal of some heavy metal ions from aqueous solution. *Journal of Industrial and Engineering Chemistry, 21*, 1160–1166. https://doi.org/10.1016/j.jiec.2014.05.029.

Madrakian, T., Haghshenas, E., Ahmadi, M., & Afkhami, A. (2015). Construction a magneto carbon paste electrode using synthesized molecularly imprinted magnetic nanospheres for selective and sensitive determination of mefenamic acid in some real samples. *Biosensors & Bioelectronics, 68*, 712–718. https://doi.org/10.1016/j.bios.2015.02.001.

Madrakian, T., Haryani, R., Ahmadi, M., & Afkhami, A. (2015). Spectrofluorometric determination of venlafaxine in biological samples after selective extraction on the superparamagnetic surface molecularly imprinted nanoparticles. *Analytical Methods, 7*, 428–435. https://doi.org/10.1039/C4AY02144D.

Madrakian, T., Zadpour, B., Ahmadi, M., & Afkhami, A. (2015). Selective extraction and sensitive determination of mercury (II) ions by flame atomic absorption spectrometry after preconcentration on an ion-imprinted polymer-coated maghemite nanoparticles. *Journal of the Iranian Chemical Society, 12*, 1235–1243. https://doi.org/10.1007/s13738-015-0587-y.

Marino, E., Heijden, T., Creyghton, Y., & van der Heijden, A. (2007). Synthesis and coating of copper oxide nanoparticles using atmospheric pressure plasmas. *Surface and Coatings Technology, 201*, 9205–9208.

Mariotti, D., & Sankaran, R. M. (2010). Microplasmas for nanomaterials synthesis. *Journal of Physics D: Applied Physics, 43*, 323001–323021. https://doi.org/10.1088/0022-3727/43/32/323001.

Mathew, T., Aishwarya Sree, R., Aishwarya, S., Kounaina, K., Patil, A. G., Satapathy, P., et al. (2020). Graphene-based functional nanomaterials for biomedical and bioanalysis applications. *FlatChem, 23*, 100184.

Morales, J., Sánchez, L., Martín, F., Ramos-Barrado, J. R., & Sánchez, M. (2004). Nanostructured CuO thin film electrodes prepared by spray pyrolysis: A simple method for enhancing the electrochemical performance of CuO in lithium cells. *Electrochimica Acta, 49*, 4589–4597.

Mostafa, A. M., & Mwafy, E. A. (2020). Synthesis of ZnO and Au @ ZnO core/shell nanocatalysts by pulsed laser ablation in different liquid media. *Journal of Materials Research and Technology, 9*, 3241–3248.

Nandiyanto, A. B. D., & Okuyama, K. (2011). Progress in developing spray-drying methods for the production of controlled morphology particles: From the nanometer to submicrometer size ranges. *Advanced Powder Technology, 22*, 1–19. https://doi.org/10.1016/j.apt.2010.09.011.

Nie, P., Xu, G., Jiang, J., Dou, H., Wu, Y., Zhang, Y., et al. (2018). Aerosol-spray pyrolysis toward preparation of nanostructured materials for batteries and supercapacitors. *Small Methods, 2*, 1700272. https://doi.org/10.1002/smtd.201700272.

Nolan, H., Sun, D., Falzon, B. G., Chakrabarti, S., Padmanaba, D. B., Maguire, P., et al. (2018). Metal nanoparticle-hydrogel nanocomposites for biomedical applications—An atmospheric pressure plasma synthesis approach. *Plasma Processes and Polymers, 15*, 1800112.

Pan, Z. W., & Wang, Z. L. (2001). Nanobelts of semiconducting oxides. *Science, 291*, 1947–1949.

Pankaj, S. K., & Keener, K. M. (2017). Cold plasma : Background, applications and current trends. *Current Opinion in Food Science, 16*, 49–52.

Patil, L. A., Shinde, M. D., Bari, A. R., & Deo, V. V. (2011). Synthesis of SnO2 hollow microspheres from ultrasonic atomization and their role in hydrogen sensing. *Materials Science and Engineering B, 176*, 579–587.

Pfeiffer, T. V., Feng, J., & Schmidt-Ott, A. (2014). New developments in spark production of nanoparticles. *Advanced Powder Technology, 25*, 56–70.

Pingali, K. C., Rockstraw, D. A., & Deng, S. (2005). Silver nanoparticles from ultrasonic spray pyrolysis of aqueous silver nitrate. *Aerosol Science and Technology, 39*, 1010–1014.

Piotto, V., Litti, L., & Meneghetti, M. (2020). Synthesis and shape manipulation of anisotropic gold nanoparticles by laser ablation in solution. *Journal of Physical Chemistry C, 124*, 4820–4826.

Ranasinghe, C. S. K., Vequizo, J. J. M., & Yamakata, A. (2018). Fabrication of robust TiO2 thin films by atomized spray pyrolysis deposition for photoelectrochemical water oxidation. *Journal of Photochemistry and Photobiology A: Chemistry, 358*, 320–326.

Rosi, M., Noor, F. A., & Yulkifli, L. L. Z. (2008). Synthesis of carbon nanotubes from a mixture of ferrocene and benzene by spray pyrolisis. In *Proceeding 2nd Int. Conf. Math. Nat. Sci* (pp. 966–971).

Rozati, S. (2006). The effect of substrate temperature on the structure of tin oxide thin films obtained by spray pyrolysis method. *Materials Characterization, 56*, 150–153.

Rudin, T., Tsougeni, K., Gogolides, E., & Pratsinis, S. E. (2012). Flame aerosol deposition of TiO 2 nanoparticle films on polymers and polymeric microfluidic devices for on-chip phosphopeptide enrichment. *Microelectronic Engineering, 97*, 341–344.

Sahm, T., Mädler, L., Gurlo, A., Barsan, N., Pratsinis, S., & Weimar, U. (2004). Flame spray synthesis of tin dioxide nanoparticles for gas sensing. *Sensors and Actuators B: Chemical, 98*, 148–153.

Saravanakkumar, D., Sivaranjani, S., Kaviyarasu, K., Ayeshamariam, A., Ravikumar, B., Pandiarajan, S., et al. (2018). Synthesis and characterization of ZnO–CuO nanocomposites powder by modified perfume spray pyrolysis method and its antimicrobial investigation. *Journal of Semiconductors, 39*, 33001.

Sasi, B., & Gopchandran, K. (2007). Preparation and characterization of nanostructured NiO thin films by reactive-pulsed laser ablation technique. *Solar Energy Materials & Solar Cells, 91*, 1505–1509.

Shadmania, E., & Rozati, S. (2012). Characterization of Zn: SnO2 nanostructure thin film prepared by spray pyrolysis technique. In *4th Int. Conf. Nanostructures*.

Shah, J., Janneth, L.-M., Carreon, M. G., Lopez-Miranda, A., & Carreon, M. L. (2018). Plasma synthesis of graphene from mango peel. *ACS Omega, 3*, 455–463.

Sirimanne, P. M., Soga, T., & Jimbo, T. (2003). Identification of various luminescence centers in CuI films by cathodoluminescence technique. *Journal of Luminescence, 105*, 105–109.

Strobel, R., & Pratsinis, S. E. (2009). Direct synthesis of maghemite, magnetite and wustite nanoparticles by flame spray pyrolysis. *Advanced Powder Technology, 20*, 190–194.

Swihart, M. T. (2003). Vapor-phase synthesis of nanoparticles. *Current Opinion in Colloid & Interface Science, 8*, 127–133. https://doi.org/10.1016/S1359-0294.

Tabrizi, N. S., Xu, Q., Van Der Pers, N. M., & Schmidt-Ott, A. (2010). Generation of mixed metallic nanoparticles from immiscible metals by spark discharge. *Journal of Nanoparticle Research, 12*(1), 247–259.

Tahir, M. B., Rafique, M., Rafique, M. S., Nawaz, T., Rizwan, M., & Tanveer, M. (2020). Photocatalytic nanomaterials for degradation of organic pollutants and heavy metals. In *Nanotechnology and photocatalysis for environmental applications* (pp. 119–138). Elsevier. https://doi.org/10.1016/B978-0-12-821192-2.00008-5.

Taylor, G. I. (1950). The instability of liquid surfaces when accelerated in a direction perpendicular to their planes. II. *Proceedings of the Royal Society of London. Series A: Mathematical and Physical Sciences, 202,* 81–96. https://doi.org/10.1098/rspa.1950.0086.

Thareja, R., & Shukla, S. (2007). Synthesis and characterization of zinc oxide nanoparticles by laser ablation of zinc in liquid. *Applied Surface Science, 253,* 8889–8895.

Tsai, C.-J., Lin, J.-S., Deshpande, C. G., & Liu, L.-C. (2005). Electrostatic charge measurement and charge neutralization of fine aerosol particles during the generation process. *Particle and Particle Systems Characterization, 22,* 293–298. https://doi.org/10.1002/ppsc.200500961.

Tseng, K.-H., Ku, H.-C., Tien, D.-C., & Stobinski, L. (2019). Novel preparation of reduced graphene oxide–silver complex using an electrical spark discharge method. *Nanomaterials, 9,* 979.

Vander Wal, R. L., Berger, G., & Ticich, T. (2003). Carbon nanotube synthesis in a flame using laser ablation for in situ catalyst generation. *Applied Physics A: Materials Science & Processing, 77,* 885–889.

Vasiliev, A. A., Varfolomeev, A. E., Volkov, I. A., Simonenko, N. P., Arsenov, P. V., Vlasov, I. S., et al. (2018). Reducing humidity response of gas sensors for medical applications: Use of spark discharge synthesis of metal oxide nanoparticles. *Sensors, 18,* 2600.

Wang, W.-N., Purwanto, A., Lenggoro, I. W., Okuyama, K., Chang, H., & Jang, H. D. (2008). Investigation on the correlations between droplet and particle size distribution in ultrasonic spray pyrolysis. *Industrial and Engineering Chemistry Research, 47,* 1650–1659. https://doi.org/10.1021/ie070821d.

Widiyastuti, W., Wang, W.-N., Lenggoro, I. W., Iskandar, F., & Okuyama, K. (2007). Simulation and experimental study of spray pyrolysis of polydispersed droplets. *Journal of Materials Research, 22,* 1888–1898. https://doi.org/10.1557/jmr.2007.0235.

Wilkinson, N. J., Smith, M. A. A., Kay, R. W., & Harris, R. A. (2019). A review of aerosol jet printing—A non-traditional hybrid process for micro-manufacturing. *International Journal of Advanced Manufacturing Technology, 105,* 4599–4619. https://doi.org/10.1007/s00170-019-03438-2.

Wood, R. W., & Loomis, A. L. (1927). The physical and biological effects of high-frequency sound-waves of great intensity. *The London, Edinburgh, and Dublin Philosophical Magazine and Journal of Science, 4,* 417–436. https://doi.org/10.1080/14786440908564348.

Xie, J., Jiang, J., Davoodi, P., Srinivasan, M. P., & Wang, C.-H. (2015). Electrohydrodynamic atomization: A two-decade effort to produce and process micro – /nanoparticulate materials. *Chemical Engineering Science, 125,* 32–57. https://doi.org/10.1016/j.ces.2014.08.061.

Xue, M.-Z., & Fu, Z.-W. (2006). Electrochemical reaction of lithium with nanostructured thin film of antimony trioxide. *Electrochemistry Communications, 8,* 1250–1256.

Yu Vasilyev, A., Domrina, E. S., Kaufman, S. V., & Maiorova, A. I. (2019). Classification of atomization devices. *Journal of Physics Conference Series, 1359,* 012131. https://doi.org/10.1088/1742-6596/1359/1/012131.

Yudin, A., Shatrova, N., Khaydarov, B., Kuznetsov, D., Dzidziguri, E., & Issi, J.-P. (2016). Synthesis of hollow nanostructured nickel oxide microspheres by ultrasonic spray atomization. *Journal of Aerosol Science, 98,* 30–40.

Zeng, L., & Weber, A. P. (2014). Aerosol synthesis of nanoporous silica particles with controlled pore size distribution. *Journal of Aerosol Science, 76,* 1–12.

Zhang, D., Du, X., Shiab, L., & Gao, R. (2012). Shape-controlled synthesis and catalytic application of ceria nanomaterials. *Dalton Transactions, 41,* 14455–14475.

Zhang, H., Zhou, H., Wang, Y., Li, S., & Biswas, P. (2021). Mini review on gas-phase synthesis for energy nanomaterials. *Energy and Fuels, 35*, 63–85. https://doi.org/10.1021/acs.energyfuels.0c03264.

Zheng, J., Orcid, L. C., Valero-Vidal, C., Marega, C., Sekar, P., Shuang, S., et al. (2017). Ag-vanadates/GO nanocomposites by aerosol-assisted spray pyrolysis: Preparation and structural and electrochemical characterization of a versatile material. *ACS Omega, 2*, 2792–2802.

Zhu, L. Y. H., Lin, H., & Guo, H. (2007). Microwave absorbing property of Fe-filled carbon nanotubes synthesized by a practical route. *Materials Science and Engineering B, 138*, 101–104.

CHAPTER

13

Nebulizer-based detectors for liquid chromatography

Tadesse Haile Fereja[a,b] *and Guobao Xu*[a,c]

[a]State Key Laboratory of Electroanalytical Chemistry, Changchun Institute of Applied Chemistry, Chinese Academy of Sciences, Changchun, Jilin, People's Republic of China, [b]Department of Pharmacy, College of Medicine and Health Sciences, Ambo University, Ambo, Ethiopia, [c]University of Science and Technology of China, Hefei, Anhui, People's Republic of China

13.1 Introduction

In many research and industrial fields, including health-care scientific research areas, pharmaceutical manufacturing companies, and analytical chemistry, the applicability of nebulizers' ideology is increasing significantly. Generating the liquid sample as an aerosol as a way to prepare the sample for analysis is the principal advantage of nebulizers in chemical analysis. An ideal nebulizer should have an ability to generate an aerosol as fine, monodisperse, and reproducible as possible for analytical purposes. The nebulizer also has an efficient ability to convert a variety of liquids into aerosols, which should be suitable for the flow of gas. An ideal nebulizer need to be inert and have ability to work with organic and aqueous solutions even with high salt content solutions and slurries without clogging problems. The aerosols and their physicochemical properties are usually independent of the matrix solutions. In general, the nebulizer device should be portable, maintainable, and less susceptible to damage (Ari & Fink, 2020; Dolovich & Dhand, 2011).

In fact, there is no nebulizer that fulfills the ideal specifications. However, there is a wide variety of them in different shapes, sizes, and with varying operating principles whereby a huge number of applications in chemical analysis have been developed (Canals & Aguirre, 2015; Newman, 1991).

https://doi.org/10.1016/B978-0-323-91181-8.00012-6

Nebulizers are applicable in both molecular and atomic spectrometry. In atomic spectroscopy, nebulizer convert the liquid sample as an aerosol and then atomized (called atomization). Once atomize, the atoms get excited or ionized. In this area, nebulizers have evolved in order to expand the scope of applicability to the analysis of micro- and nanosamples. However, molecular spectrometry mainly utilizes nebulizers as a soft ionization technique (i.e., spray ionization techniques). Nowadays, the integration of microfluidic systems (i.e., LC and CE) to mass spectrometers has been widely studied and new nebulizers for soft ionization have emerged and received attention from the miniaturization community. A variety of spray ionization techniques have been developed as an interface to be used in LC- and CE-MS, including sonic spray ionization (SSI), thermospray ionization (TSI), electrospray ionization (ESI), and hybrid electrosonic spray ionization (ESSI) techniques (Acon, McLean, & Montaser, 2001; Canals & Aguirre, 2015; Copeland, Olson, & Skogerboe, 1972).

As shown in Fig. 13.1, there are two types of nebulizer nozzles. The first type of the nozzle is designed to allow the interaction of the liquid with the flowing gas internally prior to leaving the exit port. The second type of nozzle allows interaction of liquid with the gas externally after leaving the nozzles. Modifications on these designs are used by nebulizer manufacturers, without clear superiority of one approach over the other (Hess, 2000; Saeed, Elberry, Eldin, Rabea, & Abdelrahim, 2017).

In order to minimize the cohesive forces of the liquid such as surface tension and viscosity, energy should be applied on the liquid mass so as to generate aerosols. Based on the energy source, the most available nebulizers used in chemical analysis can be categorized as *Pneumatic, Hydraulic, Ultrasonic, and Electrostatic nebulizers* (Canals & Aguirre, 2015; Saeed et al., 2017).

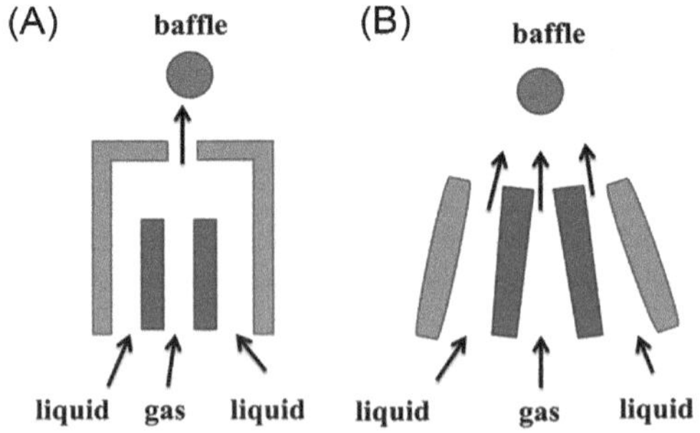

FIG. 13.1 Nebulizer designs for internal mixing (A) and external mixing (B) nebulizer designs. *Modified from Hess, D. R. (2000). Nebulizers: Principles and performance.* Respiratory Care, *45(6), 609–622.*

13.1.1 Pneumatic nebulizer

In this type of nebulizer, a relatively low-velocity liquid stream impinges on a high-velocity gas stream, and the kinetic energy of the pressurized gas is used to generate a liquid surface (i.e., the aerosol). These types of nebulizers are most applicable as a liquid sample introduction system in atomic spectrometry. In pneumatic concentric nebulizers (Fig. 13.2A), the liquid capillary is concentric to the gas capillary. Liquid-gas interaction occurs when high-velocity gas stream tangentially comes in contact with the liquid. The geometry of this interaction is annular. This is the most common nebulizer used in atomic spectrometry, as it is easy to handle, is relatively cheap, and has adequate robustness. According to its own operating principle, the sample can be introduced with and without (i.e., Venturi effect) external pumping, but the first mode is currently the most common one. In cross-flow nebulizers, the interaction between liquid and gas streams is perpendicular (Fig. 13.2B). The liquid inlet is positioned vertically and the gas inlet is positioned horizontally. The aerosol generated by the cross-flow nebulizer can be improved by optimizing different parameters including (i) distance between capillaries, which should be as small as possible (between 0.05 and 0.5 mm), (ii) relative position between capillaries, the optimal position being one in which the capillaries form an angle of 90°, and (iii) dimensions of the capillaries (i.e., inner diameters). The kinetic energy of the gas stream can be increased by reducing the inner diameter of the gas capillary. In the same way, clogging risks may be reduced by increasing the inner diameter of the liquid capillary (Almagro, Gañán-Calvo, Hidalgo, & Canals, 2006; Cresser, 1990; Park et al., 2021).

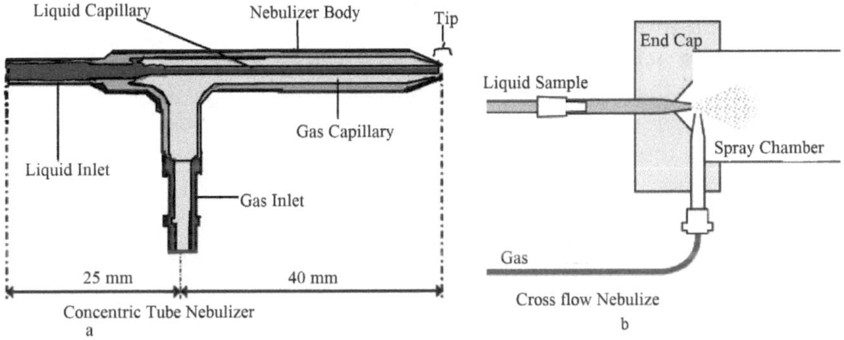

FIG. 13.2 General diagram of a pneumatic concentric (A) and cross-flow (B) nebulizer. *Reprinted with minor modifications from Canals, A., & Aguirre, M. Á. (2015). Roles of nebulizers in analytical chemistry,* Encyclopedia of analytical chemistry *(pp. 1–45).*

13.1.2 Thermospray nebulizer

The aerosol of a thermospray nebulizer is generated by heating a liquid stream followed by an adiabatic expansion of the solvent vapor generated. According to a conventional design, the solution is transported through a metal capillary heated by an electric current. When the heated metal capillary causes the liquid to evaporate, a partially vaporized mixture is produced consisting of a given fraction of vaporized liquid and an amount of remaining liquid. The evaporated solvent expands inside the tube and converts the remaining liquid into a heated aerosol. Therefore, the thermospray is a typical, concentric, pneumatic nebulizer in which the nebulizing gas is the evaporated solvent (Almagro et al., 2006; Park et al., 2021; Vestal & Fergusson, 1985).

In atomic spectrometry, the thermospray nebulizer has been extensively used as an aqueous and organic liquid sample introduction system in FAAS (Mora, Canals, & Hernandis, 1996), flame furnace atomic absorption spectrometry (FF-AAS) (Wen et al., 2012), ICP OES (Razpotnik, Budič, & Veber, 2002), and ICP-MS (Zhang & Koropchak, 2002). Thermospray nebulizers present better analyte transport efficiency and sensitivity than conventional pneumatic nebulizers because the primary aerosol created is finer and preheated. However, this model also poses serious problems: (i) it requires a high-pressure pump to introduce the liquid owing to the high pressures that, at a given temperature, depend on the inner diameter of the capillary, the nature of the liquid, and the liquid flow rate. However, a peristaltic pump has been successfully used by increasing the inner diameter of the capillary, although this adversely affects the accuracy and stability (Ding et al., 2000). (ii) The solvent load is very high and hence a desolvation system is mandatory, especially for organic solutions (i.e., the use of a desolvation system can produce memory effects). (iii) Potential clogging when nebulizing high salt content solutions. (iv) It is expensive and complex to handle. (v) Thermally labile molecules may decompose or certain compounds may present a different chemical reactivity.

However, in actual fact, the last problem could prove to be positive. It has been noted that the use of the thermospray nebulizer as a liquid sample introduction system in atomic spectrometry produces a valence discrimination effect (Zhang & Koropchak, 2002) (i.e., the sensitivity for certain elements is valence dependent). For example, the sensitivity for As(III), Sb(III), and Se(III) is significantly lower than that obtained for As(V), Sb(V), and Se(V). The mechanism identified as the cause of such results was that in the thermospray system, lower-valency elements are reduced to the nonvolatile and insoluble zero-valence metal form that is then trapped in the vaporizer. Reducing the sensitivity of the lower valence to a negligible level compared to that of the higher valence will make a nonchromatographic separation possible. This behavior has also been observed for chromium. Zhang and Koropchak (1999) have developed a nonchromatographic method for chromium speciation based on this phenomenon by properly controlling parameters such as pump speed of sample introduction, nebulizer temperature, and pH of the samples.

In all these works, the liquid sample was indirectly heated using an electrically heated metallic tube (i.e., the heat is transferred by convection and conduction). However, some studies have been performed in which the solution is directly heated by microwave radiation (Grindlay, Gras, Hernandis, & Mora, 2013). Grindlay, Gras, Montiel, Hernandis, and Mora (2010) developed a redesigned microwave-based thermal nebulizer (MWTN) based on the use of a TM010 cavity for ICP-OES. The new design overcomes the main experimental drawbacks shown by previous designs: (i) lack of robustness, (ii) poor short- and long-term stability, (iii) low heating efficiency, and (iv) high matrix effects (Ding et al., 2000). The experimental variables assessed were the incident microwave power, sample uptake rate, nebulizer dimensions, and matrix nature (i.e., HNO_3, HCl, NaCl, methanol, ethanol, and propan-2-ol) at different concentrations. Moreover, through the appropriate selection of the experimental conditions, the MWTN could be operated either as a conventional thermal nebulizer or as an online analyte preconcentration and nebulization device (Grindlay et al., 2013). The thermospray nebulizer has been extensively used in molecular spectrometry. It was first described by Blakley, Carmody, and Vestal (1980) as an ionization technique suitable for LC/MS. In a TSI interface, the sample is vaporized as it passes through a heated capillary tube (Setchell & Vestal, 1989).

13.1.2.1 *Pneumatic micronebulizers*

In recent years, the research on nebulizers in the field of analytical instrumentation has been mainly directed toward reducing the sample flow rate (i.e., analysis of micro- and nanoliter per minute). The reasons for this include (i) in certain applications (e.g., biological, forensic, and clinical analysis), the amount of sample available is limited and can be less than 1 mL, (ii) systems with low sample consumption have provided higher analyte transport efficiencies than those of conventional systems, (iii) different interferences caused by the solvent (e.g., polyatomic interferences, inorganic acids, and carbon-related interferences in plasma) can be reduced by working with low sample flow rates, and (iv) the toxic and radioactive wastes could be minimized (Maestre, Todolí, & Mermet, 2004).

Normally, when liquid samples are analyzed by ICP OES or ICP-MS, the nebulizer coupled to a spray chamber works at a sample flow rate between 0.5 and 2 $mL\,min^{-1}$. The time required to perform a signal registration is between 1 and 5 min depending on the detection system, the number of elements to be determined, and the number of replicates.

Therefore, the sample volume required varies from 0.5 to 10 mL. Furthermore, the transport efficiency of conventional pneumatic nebulizers ranges from 1% to 5%. In the case of USNs (Section 13.1.4), transport efficiency is approximately 20%, but this nebulizer requires the use of a desolvation system and the liquid flow rate is generally between 2 and 2.5 mL min^{-1}. As a result, the analysis of small volumes of sample with this nebulizer is very limited (Todolí & Mermet, 2006).

The main result of such previous research was the development of the so-called micronebulizers, being nebulizers that are able to work with a liquid flow rate of a few microliters or nanoliters per minute. Unlike conventional nebulizers, whose liquid flow rate is approximately 1 mL min^{-1}, the micronebulizer design has been modified in order to work at low flow rates (<100–200 μL min^{-1}). The most common micronebulizers are conventional pneumatic concentric nebulizers with reduced critical dimensions, primarily the gas and liquid outlet sections (Canals, Hernandis, & Browner, 1990). Such reduced dimensions enable them to work with liquid flow rates of few microliters per minute, obtaining a finer primary aerosol and better analytical performance than those rendered by conventional pneumatic nebulizers. This improvement is a direct result of the reduced dimensions of both outlets because this causes an increase in the amount of kinetic energy available from the gas stream for the generation of the aerosol and also a more efficient interaction between the liquid and the gas streams (Todoli, Hernandis, Canals, & Mermet, 1999).

13.1.3 Hydraulic nebulizer

The aerosol of the hydraulic nebulizer is generated when a high-velocity liquid jet, which emerges from a narrow orifice, impacts a solid surface placed in front of the orifice. Depending on the pressure applied through the nebulizer, two designs have been implemented in atomic spectrometry: the low-pressure hydraulic nebulizer (LPHN) and the high-pressure hydraulic nebulizer (HPHN).

13.1.3.1 Low-pressure hydraulic nebulizer

The first hydraulic nebulizer used in atomic spectrometry was the jet impact nebulizer (JIN) proposed by Doherty and Hieftje. The orifice diameter varied between 20 and 100 μm and the liquid pressure between 3 and 6 bar. The JIN generates a very stable aerosol obtaining limits of detection and transport efficiency values similar to those obtained with concentric, pneumatic, and cross-flow nebulizers. However, it is very prone to tip blockage with high salt content solutions and requires a very stable liquid flow rate (Doherty & Hieftje, 1984).

13.1.3.2 *High-pressure hydraulic nebulizer*

At a later date, the HHPN was developed by Berndt. The diameter of HHPN orifice is smaller (i.e., 10–30 μm) than that of the JIN, so the liquid pressure required is much higher (i.e., between 50 and 400 bar, according to the liquid flow rate, orifice diameter, and the solvent used). Consequently, the HHPN needs an HPLC pump to introduce the liquid in a stable and continuous manner. The primary aerosols generated are much finer than those generated by conventional concentric pneumatic nebulizers and therefore present a better analytical performance. In addition, memory effects are low, and acid solutions and high salt content solutions can be nebulized (Berndt, 1988).

13.1.4 Ultrasonic nebulizer

For this nebulizer, a liquid sample is pumped onto the surface of a rapidly vibrating solid surface (i.e., piezoelectric transducer) and a checkerboard-like wave pattern appears in the film that forms as the liquid spreads over the surface. As the amplitude of the surface vibration is increased, the wave crest height in the film also increases. Aerosol generation occurs when the amplitude of the vibrating surface increases to the point in which the wave crests in the film become unstable and collapse. Some interesting attributes of the USNs are their abilities to provide very fine and low-velocity aerosols. It has been mainly used in plasma-based elemental analysis. When the aerosol is generated, an auxiliary stream of argon transports the aerosol toward the plasma. Unlike the pneumatic nebulizers, aerosol generation in the USN is independent of the gas stream (Wilschefski & Baxter, 2019). The USN is an extremely popular nebulization system when high sensitivity is required. The analyte transport efficiency of the USN is higher than that of the pneumatic nebulizer (i.e., about 20% and 2%–3% for a USN and a conventional pneumatic nebulizer, respectively, using a liquid flow rate of 1 mL min^{-1}). Commonly, high analyte transport efficiency is associated with high solvent transport efficiency. Hence, it is essential to use a desolvation system to eliminate the high solvent loading in order to prevent plasma extinction. With simple matrices, the sensitivity obtained with the USN is higher and the LOD is considerably improved (i.e., one to two orders of magnitude compared to pneumatic nebulizers) (Masson, Vives, Orignac, & Prunet, 2000).

13.1.5 Electrospray

In electrospray (ES), the energy causing the liquid surface to disrupt comes from the mutual repulsion of like charges that have accumulated on the surface. When an electrically conductive liquid pumped through a capillary at a low flow rate (i.e., a hypodermic needle or a stainless steel capillary) is exposed to a very high voltage (kV), the shape of liquid starts to deform. Assuming positive potential, positive ions in the solution will accumulate at the surface; an electrical pressure is created that tends to expand the surface area. This pressure is opposed by surface tension forces, which tend to contract or minimize the surface area and the liquid takes the shape of a cone (Taylor cone). When the electrical pressure exceeds the surface tension forces, the surface becomes unstable and a liquid jet is ejected through its apex. At a certain distance from the nozzle, the liquid jet becomes unstable and charged droplet formation begins. Intact liquid jet length decreases with increasing electric field. As the voltage is increased, even more cone disappears and droplets are ejected from the capillary tip. If the electrical pressure is maintained above the critical value consistent with the liquid flow rate, then nebulization is continuous. A coaxial sheath gas (dry N_2) flow around the capillary results in better nebulization. The charged droplets diminish in size by solvent evaporation, assisted by the flow of nitrogen (drying gas).

The results of many experimental studies on electrostatic nebulization show that, in general, the size of droplets produced depends on the applied voltage, surface tension, electrode size and configuration, liquid flow rate, and electrical properties of the liquid, such as dielectric constant and electrical conductivity. The smallest droplets are produced by the highest voltage that can be used without encountering excessive corona losses (Canals & Aguirre, 2015).

13.2 Fundamentals of nebulizer-based LC-detection techniques

The term 'nebulizer-based detection' refers to the class of detection techniques that involve the conversion of liquid sample into an aerosol to generate a response to the specific physical properties exhibited by the sample residue left after desolvation (Fig. 13.3). Methods of nebulizer-based detection have evolved rapidly over the years, for example, AAS, ICP-MS, ELSD, condensation nucleation light scattering detector (CNLSD), and C-CAD. Among these, ELSD and C-CAD have attracted considerable attention because of their compatibility with LC, ease of use, low cost, and, most importantly, for their ability to provide detection response for a wide range of analytes independent of their optical absorbance and chemical properties (Khandagale, 2015).

High-performance chromatographic systems require a detector to provide responses for analytes that differ from that of the mobile phase, and these detectors can be classified as being either selective, if the response is only for a specific class of compounds, or as universal, if virtually any analyte can provide response. With most liquid chromatographic methods,

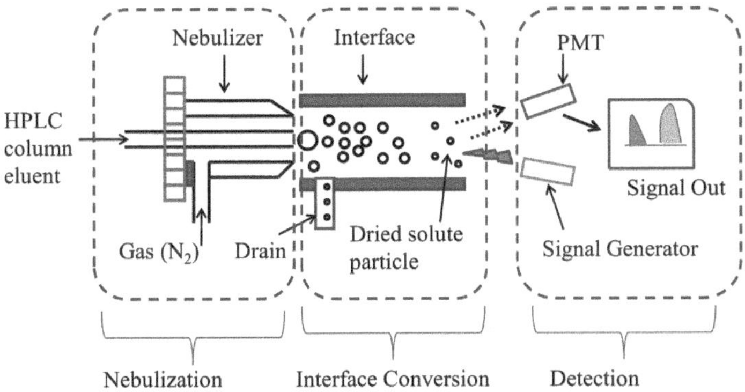

FIG. 13.3 Schematic illustration of the operation of aerosol-based detectors (Khandagale, 2015).

the mobile phase liquid is significantly more volatile than the analytes, and therefore with appropriate technology, this difference can be used to detect the presence of an analyte. Since aerosols have high surface area-to-volume ratios, they provide a particularly high efficiency means for selective evaporation of the mobile phase in the presence of an analyte. Aerosol nebulization methods convert the liquid chromatographic effluent into millions of droplets. After evaporation of the mobile phase, a virtually equivalent number of dry particles representing the less volatile species are present in the gas phase (Dixon & Peterson, 2002).

13.2.1 General principles of nebulizer-based LC detectors

With all aerosol-based detectors for liquid chromatography, the detection process begins with the conversion of the LC effluent to an aerosol, most simply by using a pneumatic nebulizer. As shown in Fig. 13.4, this solvated aerosol is converted to a solvent-dry aerosol using a desolvation system that may consist of a heated flow tube. The sizes of the dry particles (D_{dry}) reflect the sizes of the initial droplets (D_{wet}), the concentration of the nonvolatile analyte (C), and the density of the analyte (ρ), expressed by the following relationship:

$$D_{dry} = D_{wet}\left(\frac{C}{\rho}\right)^{1/3} \tag{13.1}$$

assuming that the analyte is the only nonvolatile species within the mobile phase. In practice, mobile phase components likely contain less volatile contaminants that will also remain behind in the "dry" particle phase, and these will contribute to the dry particle sizes. Typically, this background in the absence of analyte can introduce a background signal and noise level that will reduce analyte signal-to-noise and detection.

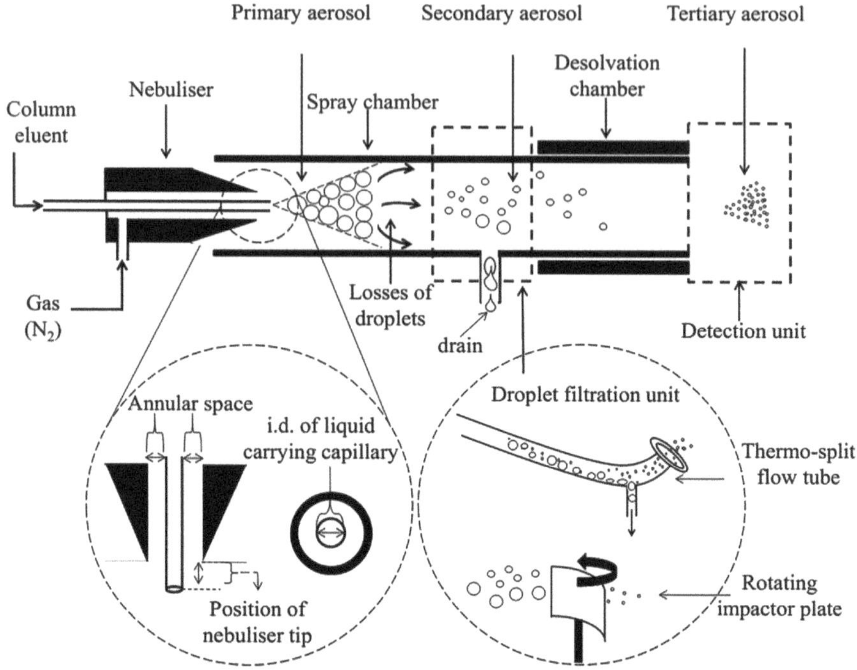

FIG. 13.4 Schematic representation of the terms related to the aerosol transport process, nebulizer design, and desolvation chamber, which are used to discuss the performance of aerosol-based detectors (Khandagale, 2015).

Methods that reduce this background, such as use of mobile phase components with low residue after evaporation, will generally improve detector response.

Once the aerosol is formed, the droplet size distribution is modified by many processes, including the impaction of larger droplets with surfaces within the flow system. These losses reduce the number of particles and mass of analyte that reaches the detection process, reducing signal levels. However, increased volatility of the mobile phase will increase the rate at which droplets decrease in size due to evaporation, reducing the level of these impaction losses, leading to increased signal (and perhaps background) levels.

13.2.2 Evaporative light scattering detection (ELSD)

The evolution of evaporative light scattering detector (ELSD) can be traced back to an invention at the Union Carbide Australian research laboratory in 1966. However, it was not until 1978 when the ELSD was reevaluated as a viable detector for modern HPLC systems. Starting in the early 1980s, commercially available detectors were introduced through companies such as the Varex Corporation (USA), S.E.D.E.R.E (France), and Applied Chromatography Systems Limited (UK) (Dixon & Peterson, 2002; Magnusson, Risley, & Koropchak, 2015).

13.2.2.1 ELSD operating principles

All commercially available ELSDs consist of three main stages: (1) nebulization, (2) evaporation, and (3) detection. These three processes are illustrated in Fig. 13.3 where evaporation is an interface conversion. The nebulization process typically uses an inert carrier gas, such as nitrogen, to transform the mobile phase into an aerosol of fine droplets that can be more easily evaporated. The uniformity and size of the aerosol droplets are important for detectability and reproducibility for sample injections, where larger droplets or particles result in a higher intensity of light scattered. Immediately following the nebulizer, the ELSD can be classified as either type A or type B designs. For type A design, the entire aerosol moves to the evaporation stage while in type B designs, the larger droplets do not make it to the evaporation stage and are diverted to a waste line. Type B design allows for lower evaporation temperatures, due to the aerosol splitting, and therefore better sensitivity for semivolatile or temperature-sensitive samples. However, during gradient elution (low to high organic composition), there are fewer larger droplets formed and therefore more sample (later eluting analytes) allowed into the evaporation process with type B design. Since there is no aerosol splitting in type A design, where the entire aerosol continuously enters the evaporation chamber throughout the gradient, this design is more suited for better accuracy in gradient elution, especially when standards are not employed. The gas flow to the nebulizer can be controlled and optimized affecting aerosol formation. Too low a gas flow can lead to spikes in the chromatogram, poor aerosol formation, or poor droplet evaporation of the mobile phase resulting in high background noise, while an optimum low gas flow results in larger droplets and better sensitivity. As the gas carries the droplets through a heated drift tube, the evaporation process removes the volatile mobile phase leaving particles of the nonvolatile analytes. The temperature of the drift tube can be controlled and it is preferable to use the lowest temperature possible. The optimum temperature helps maintain particle uniformity and particle formation. Too high a temperature can result in nonuniformity and poor particle formation as well as inhibition of semivolatile sample response, while too low a temperature can result in inefficient mobile phase evaporation and high background noise (Magnusson et al., 2015).

13.2.2.2 ELSD response

The nonvolatile analyte particles then proceed into an optical cell and pass through a light source where the amount of scattered light is detected. The size, shape, and number of particles affect the amount of light scattered. The ratio of D_{dry} to the wavelength of light emission can be used to classify three types of light scattering based on the size of the particles: Raleigh (ratio less than 0.1), Mie (ratio between 0.1 and 10), and refraction-reflection (ratio greater than 10). Unlike spectroscopic detectors, which follow Beer's Law, the ELSD response is described with a nonlinear equation:

$$A = amx$$

where A is the peak area response, m is the sample quantity, and a and x, respectively, are coefficients reflecting properties of the sample and the parameters of the chromatographic and detector systems.

A log-log plot of peak area response versus sample quantity will be linear, which allows for quantitative results as described by the equation:

$$\log A = x \log (m) + \log (a).$$

Quadratic equations can also be used. However, all nonlinear curves have a region of response that can be approximated as linear, and if calibration standards and samples are within this range, a linear equation can be applied. The ELSD is not influenced by the ultraviolet (UV) properties of solvents, buffers, or acids used for mobile phases; therefore, the ELSD is not subject to baseline drift from gradient elution. However, a limitation of the ELSD is that it requires the complete volatilization of all mobile phase components. The buffer, acid, and base additives can increase the baseline noise. Therefore, generally, these additives should be minimized for optimum detector performance (Magnusson et al., 2015; Webster, Jensen, & Diaz, 2004).

13.2.2.3 *Factors affecting ELSD response*

Light sources

Light sources for the different brands of ELS detectors can be tungsten-halogen, LED, or lasers. However, Gaudin et al. implemented a xenon arc lamp via a fiber-optic attachment into a Eurosep DDL 31 ELSD. These authors found the ELSD response was affected by higher intensity light source, showing a 16× increase in the sensitivity for ceramide (Gaudin, Baillet, & Chaminade, 2006).

Mobile phase characteristics

Megoulas and Koupparis used a Sedex 75 detector to study the effect of mobile phase volatility and ion-pairing agents on the response of neomycin. As the volatility of the mobile phase composition increased, the peak response also increased with acetone compositions leading to the best response, followed by methanol, acetonitrile, and then water. These authors also found that increasing the mass of the ion-pairing agents leads to an increase in response with heptafluorobutyric acid (HFBA) being the best and attributed to the larger mass formation of neomycin with HFBA (Gaudin et al., 2006). Optimization of ELSD temperature can also be highly important for semivolatile or temperature-labile compounds, such as the classic example of caffeine, where lower drift tube temperatures leads to greater caffeine response. With improvements in ELSD equipment and the combination of UPLC, HPLC, or monolithic columns, mobile phase flow rates (0.05–5.0 $\mathrm{mL\,min^{-1}}$) need to be matched with the appropriate nebulizer for optimum sensitivity.

Analyte chemical characteristics

ELSD is frequently described as a universal detector; however, Shock, Dennis, Guiochon, Dasgupta, and Shalliker (2011) showed that some diastereomers having the same mass and structures (but different spatial orientation). can have very different response factors (Shimadzu ELSD LTII), cautioning against the use of ELSD as a purely universal detector and recommending standards for accurate quantitation. The detector effects of increasing organic modifier composition during reversed phase gradient elution results in higher analyte response for later eluting analytes as well as compensate for the nonlinear behavior of the detector. Other factors leading to inaccuracies were related to the physicochemical properties of the analytes.

13.2.3 Condensation nucleation light-scattering detection (CNLSD)

Allen and Koropchak first introduced the condensation nucleation light-scattering detector (CNLSD) in 1993 (Allen & Koropchak, 1993). Multiple studies and applications of CNLSD have also been published since the last review in the areas of small- and large-molecule pharmaceutical analysis, analysis of dietary supplements, food industry applications (carbohydrates and lipids), environmental analysis (pesticides), and metals analysis, and are also subject to review herein.

13.2.3.1 CNLSD operating principles

Similar to ELSD and CAD, CNLSD utilizes the process of aerosol generation (typically via pneumatic nebulization for the case of liquid chromatography), producing a wet aerosol of mobile phase components and analyte (Fig. 13.3). The aerosol undergoes through desolvation (drying) to produce a dry (ideally) aerosol of nonvolatile analyte. Subsequently, the intermediate step of condensation nucleation (CN) is added. Here, supersaturated vapor, commonly n-butanol (as used in most CNLSD work at the research stage) or water (as in the currently commercially available NQAD), is condensed onto the dry particles, growing them in sizes smaller than what can be practically detected by direct light scattering to droplets that are easily detectable, the particle number concentration being used as the signal, typically using a laser diode and a photodetector. For n-butanol, supersaturation and condensation nucleation of the vapor are achieved by using a heated saturator and a cooled condenser (thermal diffusivity of the carrier gas being greater than the mass diffusivity of n-butanol), while for water, the condenser is held at a warmer temperature than the saturator (mass diffusivity of water being greater than the thermal diffusivity of the carrier gas) (Grindlay, Gras, Mora, & Hernandis, 2008).

13.2.3.2 Factors influencing CNLSD response

Condensing fluid

Decreasing the size threshold for particle detection would increase the sensitivity of CNLSD, assuming that background is not a limiting factor. However, for HPLC-CNLSD with pneumatic nebulization, the effect of lowering the size threshold would not be significant since the particles produced are typically large enough such that they would be detected with the commonly used CNLSD condensing fluids and since background is typically a limiting factor, usually necessitating the use of diffusion screens (Grindlay et al., 2010).

Analyte characteristics and condensing fluid

Water as a condensing fluid appears to be more efficient for hydrophilic/hygroscopic particles than hydrophobic ones, and detection of hydrophobic particles has been enhanced when a trace amount of hydrophilic species (sucrose) was also present, while n-butanol condensation appears to be less dependent on particle composition. Clearly, the effect of the condensing fluid chemical characteristics coupled with analyte characteristics may partially explain analyte-to-analyte differences in observed response with HPLC-CNLSD (Todolí & Mermet, 2006).

13.2.4 Charged aerosol detector (CAD)

The newest member of the class of aerosol-based detectors for liquid chromatography is the charged aerosol detector (CAD), as first described by Dixon and Peterson (2002).

13.2.4.1 Operating principles

The CAD approach involves the conversion of the liquid chromatographic effluent into aerosol droplets and desolvation (drying) to remove the majority of the mobile phase, leaving behind dry particles containing primarily the less volatile analyte species; up to this point, the process is similar to that of ELSD or CNLSD/NQAD (Fig. 13.3). Subsequently, with CAD, however, these dry particles interact with an ion source that results in the transfer of charge to the particles. The excess ions, which have higher mobility, are collected in a trap, and the charged particles associated with the analyte particles are collected and measured with an electrometer.

13.2.4.2 CAD response

The level of charging on particles is theoretically proportional to the particle size and surface area. In chromatographic practice, response for CAD typically follows a relationship:

$$A = aM^b$$

where A is the signal, a and b are response factors, and M is the analyte mass (Dixon & Peterson, 2002). Such a relationship can provide approximately linear calibration curves when plotted in the log-log fashion.

13.2.4.3 Factors affecting CAD response

Analyte chemical characteristics

Since the level of particle charging is theoretically only proportional to the particle surface area, the CAD response per unit mass analyte might be expected to be identical for virtually all substances that are nonvolatile under the operating conditions of CAD. Relative uniformity of response for a wide range of substances could simplify calibration and would be beneficial for quantifying unknowns or impurities.

Mobile phase characteristics

As indicated earlier, the initial dry particle size distribution is a modified reflection of the wet droplets produced by the nebulizer. Therefore, factors that change the droplet size distribution produced by the nebulizer may affect the dry particle size distribution and thus the relative response with CAD. Among these factors, liquid viscosity, density, and surface tension of the mobile phase are expected to influence droplet diameters for pneumatic nebulizers. However, the largest effect of changes in mobile phase composition (e.g., from a gradient) would likely result from increases in the volatility of the mobile phase, which would increase the rate of evaporation of the mobile phase and, in turn, decrease the probability of droplet losses due to impaction within the nebulizer spray chamber system (Khandagale, Hilder, Shellie, & Haddad, 2014).

13.2.5 HPLC-chip/MS technology

Microfluidics is the manipulation of minute amounts of liquid in miniaturized systems. In the life sciences, the most widely used microfluidics-based device today is known as the "lab-on-a-chip." Lab-on-a-chip technology allows various qualitative and quantitative biochemical analyses to be conducted quickly and efficiently, with minimum handling by the user (Fig. 13.5). Microfluidic devices are circuits of tiny closed channels and wells, etched onto a glass or plastic microchip. Pressure or electrokinetic forces push small volumes of fluids through selected pathways in a controlled manner. These mini-laboratories may include elements such as pumps, valves, mixing and reaction chambers, and separation channels. All mass spectrometers work in vacuum, which ranges from 10^{-3} to 10^{-10}mbar depending on the type of mass analyzer. The first step in measuring the mass spectrum of any sample is ionization. Thus, the sample has to be either transferred into vacuum before ionization or ionized at atmospheric pressure and the ions then transferred into vacuum.

The first devices used to couple LC to MS relied on ionization techniques in which ionization took place in vacuum or at significantly reduced pressure. The amount of eluent was reduced, for example, by splitting the liquid flow before ionization to decrease the gas load to the mass spectrometer. Of these techniques, thermospray was the most popular and allowed the broad breakthrough of LC-MS as an analytical technique. Microfabricated systems, such as lab-on-a-chip systems using microfluidics that can improve the control, the automation, and therefore the reproducibility, appear to be some of the best available candidates for improving this process. The chip allows the measurement of reproducible

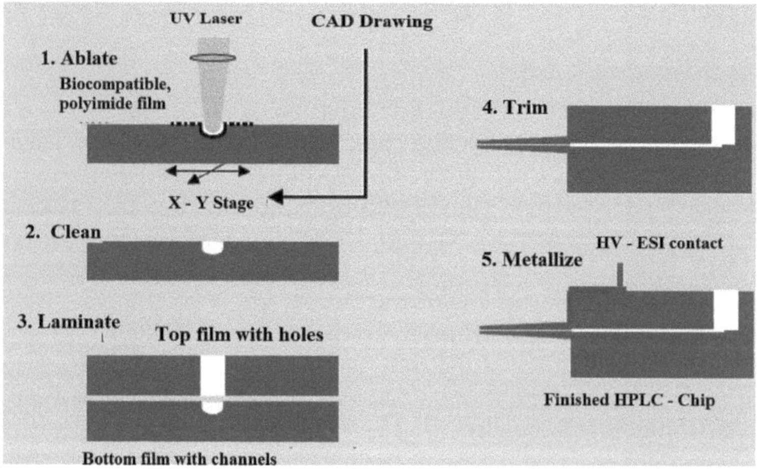

FIG. 13.5 HPLC-chip making process.

retention times that, in association with m/z ratios, was found useful for identifying peptide sequences without ambiguity. An increase in sensitivity is obtained compared to the results by conventional nano-LC/MS/MS on the same ion trap.

The function of the nebulizer in inductively coupled plasma-mass spectrometry (ICP-MS) is to convert a liquid sample into an aerosol for introduction into the plasma. The production of an aerosol is the easiest method to introduce a liquid sample into either an ICP or a flame. An aerosol facilitates a uniform sample introduction for both a reproducible signal output and the stable operation of the plasma (B'Hymer & Caruso, 2000). Heated nebulizer (HN) microchips are microfabricated devices that vaporize liquid and mix it with gas. They are used with low liquid flow rates, typically 0.5–10 μL min^{-1}, and have alternatively been utilized for atmospheric pressure chemical ionization (APCI) and atmospheric pressure photoionization (APPI). Conventional APCI and APPI sources are seldom feasible at such low flow rates. The high temperature of the HN microchips allows efficient vaporization of high-boiling analytes. With the chip, the sample flow rate could be reduced significantly without loss of performance and with greatly reduced contamination of the ion source and ion inlet of the MS. Microchip APPI provided efficient vaporization of high molecular-weight components in crude oil.

The HN microchips were also used to connect gas chromatography (GC) and capillary liquid chromatography (LC) to MS, using APPI for ionization. Microchip APPI allows efficient ionization of both polar and nonpolar compounds with capillary LC and other low flow-rate separation methods, whereas with electrospray ionization (ESI) only polar and ionic molecules are ionized efficiently. The combination of GC with MS with atmospheric pressure ionization showed that, with HN microchips, GCs can easily be used with MS instruments designed for LC-MS. The presented GC-μAPPI-MS method for polycyclic aromatic hydrocarbons (PAHs) and capillary LC-μAPPI-MS method for steroids showed good quantitative performance.

The same high-performance microfabrication methods used for the HN microchips were successfully utilized in the fabrication of the first integrated LC-HN microchip. In a single microdevice, there were structures for a packed LC column channel, micropillar frit, channel for optional optical detection, and heated vaporizer. Nonpolar and polar analytes were efficiently ionized by APPI, as demonstrated with PAHs and selective androgen receptor modulators. Ionization of nonpolar and polar analytes had not been possible with previously presented chips for LC-MS since they relied on ESI. With use of APPI-MS, preliminary quantitative performance of the new chip was evaluated in terms of limit of detection, linearity, and repeatability of signal response and retention time (Ehlert, Trojer, Vollmer, Van De Goor, & Tallarek, 2010; Haapala et al., 2010; Vollmer & van de Goor, 2009).

13.3 Application of nebulizer-based LC detectors

13.3.1 Pharmaceuticals

The development of pharmaceutical drug substances and drug products requires analytical methodologies for the determination of impurities, enantiomers, counter ions, potency and inactive ingredients, used in control strategies, and to ensure the safety, identity, strength, purity, and quality of the materials. Scheidl et al. used HPLC-ELSD for the separation and quantification of neomycin and bacitracin in sterile pharmaceutical formulations for content uniformity in quality control (Scheidl et al., 2009). They demonstrated that the ELSD was efficient without the need for derivatization or more expensive MS equipment.

Fekete, Ganzler, and Fekete (2010) developed and validated a reversed phase HPLC-CNLSD method for the analysis of PEG and polysorbate in a biopharmaceutical injection solution containing a PEGylated protein. The method was found to exhibit adequate performance, with an LOQ of 10 g mL^{-1} (polysorbate) and 2 g mL^{-1} (PEG). A linear range from LOQ to 150% of theoretical sample concentration (10–60 g mL^{-1} polysorbate 20 and 2–40 g mL^{-1} PEG) was obtained. Wang, Yang, Zhu, and Yin (2014) developed a reversed phase HPLC-CNLSD method for the determination of kanamycin and compared it to the Chinese pharmacopeia HPLC-ELSD method. Good linearity was achieved in the 195.2–1952 ng range. The detection limit was 19.4 ng, and an average recovery of 99.7% was achieved. The authors concluded that the method was more sensitive than the pharmacopeia method.

Phospholipid hydrolysis in a suspension of crystalline drug microparticles was monitored by quantifying free fatty acids using LC-CAD and LC-ELSD as a measure of the physical and chemical stability of the formulation (Nair & Werling, 2009). Log-log plotting of calibration data was used to linearize the CAD response, which provided lower LODs (0.26–0.75 g mL^{-1}) than did ELSD (0.96–3.4 g mL^{-1}). The LC-CAD method was also found preferable in terms of precision, accuracy, and recovery. A two-dimensional LC system coupled to UV, CAD, and MS detection was developed and applied to challenging pharmaceutical impurity analysis problems (Zhang, Li, Tsang, & Chetwyn, 2013). In this example, the CAD allowed analysis of a polar compound with no UV chromophore after a reverse phase by hydrophilic interaction LC separation.

13.3.2 Food products

Rodríguez-Alcalá and Fontecha validated an HPLC-ELSD method using a Sedex 85 ELSD for the routine quantitation of lipids and phospholipids in milk and dairy products (Rodríguez-Alcalá & Fontecha, 2010).

The LOD was determined to be as low as 0.3 g for phosphatidylinositol and phosphatidylserine. Rombaut, Dewettinck, and Van Camp (2007) used an Alltech 2000 ELSD for the analysis of polar lipids in 31 different dairy products from milk, cream, condensed milk, butter, fermented products, cheese, and whey.

A method for the quantitation of neutral and polar yeast lipids in *Saccharomyces cerevisiae* using LC-CAD was developed by Khoomrung et al. (2013) using a HILIC separation; the authors reported a nonlinear response and LODs in the 1.09–1.91 $g\,mL^{-1}$ range. Generally this method included microwave-assisted lipid extraction, and provided good recoveries (+90%) and highly reproducible results. Water-soluble vitamins (thiamine, folic acid, nicotinic acid, ascorbic acid, pantothenic acid, pyridoxine, and biotin) were determined in infant milk and dietary supplements using an LC-CAD method (Márquez-Sillero, Cárdenas, & Valcárcel, 2013). LODs in the range of 0.17–0.5 $g\,mL^{-1}$, and good precision (2.6%–4.6% RSD) and accuracy to declared values were reported.

13.3.3 Environmental analysis

The coupling of a high-power nitrogen (N_2) microwave-induced plasma (MIP) mass spectrometry-(MS) (1.3 kW) with high-performance liquid chromatography, connected with a concentric nebulizer (CN), an ultrasonic nebulizer (USN), and a hydride generation (HG) systems, for the optimization and determination of selenium compounds, has been carried out. The MIP-MS system fulfills the ideal requirement for an on-line real-time chromatographic detector for Se speciation analysis. Interchanging of MIP-MS system fabricated nebulizer (concentric) with an ultrasonic nebulizer increases about 3.4–12 (peak height) and 6.5–10 (peak area) times ion signals for the selenium compounds. The detection limits for selenate, selenite, trimethylselenonium ion (TmSe), selenomethionine (Semet), and selenoethionine (Seet) (in Milli-Q-water) obtained with the optimized HPLC-USN-N_2MIP-MS system are 0.11, 0.14, 0.09, 0.14, and 0.10 $\mu g L^{-1}$, respectively, about 12–48 times lower than the HPLC-CN-MIP-MS and 1.5–4.4 (peak height) times lower than the HPLC-CN-inductively coupled plasma (ICP)-MS coupling (Chatterjee, Shibata, Tao, Tanaka, & Morita, 2004).

The HPLC-USN-MIP-MS combination used is excellent, amplifying the ion signals by about 1.5–2 times for cationic and 1.3–2.8 times for the anionic arsenic compounds as compared to the HPLC-ICPMS coupling. The detection limits for As(V), MA, DMA, AB, TMAO, AC, and TMI (in Milli-Q-water) obtained with the optimized HPLC-USN-N2-MIP-MS system are 0.46, 0.36, 0.73, 0.21, 3.64, 0.39, and 0.32 $\mu g L^{-1}$, respectively, about 13–50 times lower than the HPLC-MIP-MS and about 3–11 times lower than the HPLC-ICPMS. The detection limits of As(V), MA, DMA, AB, TMAO, AC, and TMI, which spike in the urine, are deteriorated by 1.7–4.2 times compared with the detection limits of the seven different compounds (Chatterjee, Shibata, Yoshinaga, & Morita, 2000).

You and Koropchak (2003) developed a cation exchange method for the determination of the herbicide glyphosate and its major metabolite aminomethylphosaphonic acid (AMPA). The authors evaluated and compared two CNLSD detection systems: a laboratory-made CNLSD system and a prototype commercial CNLSD detector. Both systems were based on n-butanol and performed comparably. Good linearity was observed for glyphosate in the 41–3030 ng mL^{-1} range for the prototype system, and from 53 to 4320 ng mL^{-1} for the laboratory-built system. The linearity for AMPA was also good from 53 to 4320 ng mL^{-1} for the prototype system, and from 58 to 5040 ng mL^{-1} for the laboratory-built system. The average recovery for all samples was 100.0 ± 3.6%. The respective glyphosate and AMPA detection limits were 41 and 53 ng mL^{-1} on the prototype system, and 54 and 58 ng mL^{-1} on the laboratory-made system.

13.4 Conclusions and future trends

The response characteristics of nebulizer-based HPLC detectors depend on the measurement of optical or electrical properties exhibited by the plume of dried solute particles. Therefore, these detectors are commonly regarded as mass-sensitive detectors. In quantitative analysis, sensitivity is one of the important criteria in selecting an appropriate detection technique. The sensitivity of mass-sensitive detectors is expressed in terms of signal output per unit mass of the analyte entering the detector. The sensitivity of these techniques is commonly reported to be inferior to the conventional LC detectors such as UVD and electrospray ionization mass spectrometry (ESI-MS). Relatively low sensitivity compared to conventional LC-detectors is one of the major barriers to widespread acceptance of aerosol-based detection techniques. The mass flow and size distribution of the dried solute particle plume entering into the detection zone determine the sensitivity of these detectors and depend on several interrelated physical phenomena that occur at different stages of the detection process. Low sensitivity is attributed mainly to the poor nebulization efficiency and complex nature of the aerosol transport and, despite vigorous research from academic researchers as well as instrument manufacturers, it has remained a major challenge. The dependence of the detection response on physical properties of the solvents is another important factor that contributes to the response nonlinearity and sensitivity variation, thus complicating response prediction.

With the ability to provide nondiscriminating detection response, these techniques have a strong potential to meet universal detection requirements. Because of the inherently complex nature of these limitations, in recent years the focus of the research on aerosol-based detectors has shifted toward exploiting their existing capabilities to find alternate ways to improve their analytical performance and to extend the field of applicability.

Acknowledgments

We gratefully acknowledge Ambo University College of Medicines and Health Sciences pharmacy department (Ethiopia), Changchun Institute of Applied Chemistry, and the Chinese Academy of Sciences (CAS) President's International Fellowship Initiative (PIFI).

References

Acon, B. W., McLean, J. A., & Montaser, A. (2001). A direct injection high efficiency nebulizer interface for microbore high-performance liquid chromatography-inductively coupled plasma mass spectrometry. *Journal of Analytical Atomic Spectrometry, 16*(8), 852–857. https://doi.org/10.1039/b103085j.

Allen, L. B., & Koropchak, J. A. (1993). Condensation nucleation light scattering: A new approach to development of high-sensitivity, universal detectors for separations. *Analytical Chemistry, 65*(6), 841–844. https://doi.org/10.1021/ac00054a033.

Almagro, B., Gañán-Calvo, A. M., Hidalgo, M., & Canals, A. (2006). Flow focusing pneumatic nebulizer in comparison with several micronebulizers in inductively coupled plasma atomic emission spectrometry. *Journal of Analytical Atomic Spectrometry, 21*(8), 770–777. https://doi.org/10.1039/b518282d.

Ari, A., & Fink, J. B. (2020). Recent advances in aerosol devices for the delivery of inhaled medications. *Expert Opinion on Drug Delivery, 17*(2), 133–144. https://doi.org/10.1080/17425247.2020.1712356.

B'Hymer, C., & Caruso, J. A. (2000). Nebulizer sample introduction for elemental speciation. In *Vol. 33. Comprehensive analytical chemistry* (pp. 213–226). Elsevier. https://doi.org/10.1016/S0166-526X(00)80017-3.

Berndt, H. (1988). High pressure nebulization: A new way of sample introduction for atomic spectroscopy. *Fresenius' Zeitschrift Für Analytische Chemie, 331*(3–4), 321–323. https://doi.org/10.1007/BF00481903.

Blakley, C. R., Carmody, J. J., & Vestal, M. L. (1980). A new soft ionization technique for mass spectrometry of complex molecules. *Journal of the American Chemical Society, 102*(18), 5931–5933. https://doi.org/10.1021/ja00538a050.

Canals, A., & Aguirre, M.Á. (2015). Roles of nebulizers in analytical chemistry. In *Encyclopedia of analytical chemistry* (pp. 1–45). John Wiley & Sons, Ltd.

Canals, A., Hernandis, V., & Browner, R. F. (1990). Experimental evaluation of the Nukiyama-Tanasawa equation for pneumatic nebulisers used in plasma atomic emission spectrometry. *Journal of Analytical Atomic Spectrometry, 5*(1), 61–66. https://doi.org/10.1039/JA9900500061.

Chatterjee, A., Shibata, Y., Tao, H., Tanaka, A., & Morita, M. (2004). High-performance liquid chromatography-ultrasonic nebulizer high-power nitrogen microwave-induced plasma mass spectrometry, real-time on-line coupling for selenium speciation analysis. *Journal of Chromatography A, 1042*(1–2), 99–106. https://doi.org/10.1016/j.chroma.2004.05.001.

Chatterjee, A., Shibata, Y., Yoshinaga, J., & Morita, M. (2000). Determination of arsenic compounds by high-performance liquid chromatography-ultrasonic nebulizer-high power nitrogen-microwave-induced plasma mass spectrometry: An accepted coupling. *Analytical Chemistry, 72*(18), 4402–4412. https://doi.org/10.1021/ac000311s.

Copeland, T. R., Olson, K. W., & Skogerboe, R. K. (1972). Ultrasonic nebulization in atomic absorption spectrophotometry. *Analytical Chemistry, 44*(8), 1471–1473. https://doi.org/10.1021/ac60316a037.

Cresser, M. (1990). Pneumatic nebulization. In J. Sneddon (Ed.), *Sample introduction in atomic spectroscopy* Elsevier Science. https://doi.org/10.1016/B978-0-444-88229-5.50009-X.

Ding, L., Liang, F., Huan, Y., Cao, Y., Zhang, H., & Jin, Q. (2000). Low-powered microwave thermospray nebulizer for inductively coupled plasma atomic emission spectrometry. *Journal of Analytical Atomic Spectrometry, 15*(3), 293–296. https://doi.org/10.1039/a908855e.

Dixon, R. W., & Peterson, D. S. (2002). Development and testing of a detection method for liquid chromatography based on aerosol charging. *Analytical Chemistry, 74*(13), 2930–2937. https://doi.org/10.1021/ac011208l.

Doherty, M. P., & Hieftje, G. M. (1984). Jet-impact nebulization for sample introduction in inductively coupled plasma spectrometry. *Applied Spectroscopy, 38*(3), 405–412. https://doi.org/10.1366/0003702844555359.

Dolovich, M. B., & Dhand, R. (2011). Aerosol drug delivery: Developments in device design and clinical use. *The Lancet, 377*(9770), 1032–1045. https://doi.org/10.1016/S0140-6736(10)60926-9.

Ehlert, S., Trojer, L., Vollmer, M., Van De Goor, T., & Tallarek, U. (2010). Performance of HPLC/MS microchips in isocratic and gradient elutionmodes. *Journal of Mass Spectrometry, 45*(3), 313–320. https://doi.org/10.1002/jms.1719.

Fekete, S., Ganzler, K., & Fekete, J. (2010). Simultaneous determination of polysorbate 20 and unbound polyethylene-glycol in protein solutions using new core-shell reversed phase column and condensation nucleation light scattering detection. *Journal of Chromatography A, 1217*(40), 6258–6266. https://doi.org/10.1016/j.chroma.2010.08.028.

Gaudin, K., Baillet, A., & Chaminade, P. (2006). Application of a xenon arc lamp as a light source for evaporative light scattering detection. *Analytical and Bioanalytical Chemistry, 384*(6), 1302–1307. https://doi.org/10.1007/s00216-006-0297-5.

Grindlay, G., Gras, L., Hernandis, V., & Mora, J. (2013). On-line microwave-based preconcentration device for inductively coupled plasma atomic emission spectrometry: Application to the elemental analysis of spirit samples. *Talanta, 107*, 11–17. https://doi.org/10.1016/j.talanta.2012.12.049.

Grindlay, G., Gras, L., Montiel, J., Hernandis, V., & Mora, J. (2010). Design and evaluation of a new fully microwave-assisted liquid sample introduction device for inductively coupled plasma atomic emission spectrometry. *Journal of Analytical Atomic Spectrometry, 25*(4), 519–525. https://doi.org/10.1039/b922902g.

Grindlay, G., Gras, L., Mora, J., & Hernandis, V. (2008). Design and evaluation of an improved microwave-based thermal nebulizer for liquid sample introduction in inductively coupled plasma atomic emission spectrometry. *Journal of Analytical Atomic Spectrometry, 23*(1), 129–139. https://doi.org/10.1039/b707161b.

Haapala, M., Saarela, V., Pól, J., Kolari, K., Kotiaho, T., Franssila, S., et al. (2010). Integrated liquid chromatography-heated nebulizer microchip for mass spectrometry. *Analytica Chimica Acta, 662*(2), 163–169. https://doi.org/10.1016/j.aca.2010.01.005.

Hess, D. R. (2000). Nebulizers: Principles and performance. *Respiratory Care, 45*(6), 609–622.

Khandagale, M. (2015). *Investigation of the analytical performance of aerosol-based detectors in liquid chromatography*. Hobart: School of Physical Sciences, University of Tasmania.

Khandagale, M. M., Hilder, E. F., Shellie, R. A., & Haddad, P. R. (2014). Assessment of the complementarity of temperature and flow-rate for response normalisation of aerosol-based detectors. *Journal of Chromatography A, 1356*, 180–187. https://doi.org/10.1016/j.chroma.2014.06.055.

Khoomrung, S., Chumnanpuen, P., Jansa-Ard, S., Ståhlman, M., Nookaew, I., Borén, J., et al. (2013). Rapid quantification of yeast lipid using microwave-assisted Total lipid extraction and HPLC-CAD. *Analytical Chemistry, 85*(10), 4912–4919. https://doi.org/10.1021/ac3032405.

Maestre, S. E., Todolí, J. L., & Mermet, J. M. (2004). Evaluation of several pneumatic micronebulizers with different designs for use in ICP–AES and ICP–MS. Future directions for further improvement. *Analytical and Bioanalytical Chemistry, 379*(5–6), 888–899. https://doi.org/10.1007/s00216-004-2664-4.

Magnusson, L. E., Risley, D. S., & Koropchak, J. A. (2015). Aerosol-based detectors for liquid chromatography. *Journal of Chromatography A, 1421*, 68–81. https://doi.org/10.1016/j.chroma.2015.07.045.

Márquez-Sillero, I., Cárdenas, S., & Valcárcel, M. (2013). Determination of water-soluble vitamins in infant milk and dietary supplement using a liquid chromatography on-line coupled to a corona-charged aerosol detector. *Journal of Chromatography A, 1313*, 253–258. https://doi.org/10.1016/j.chroma.2013.05.015.

Masson, P., Vives, A., Orignac, D., & Prunet, T. (2000). Influence of aerosol desolvation from the ultrasonic nebulizer on the matrix effect in axial view inductively coupled plasma atomic emission spectrometry. *Journal of Analytical Atomic Spectrometry, 15*(5), 543–547. https://doi.org/10.1039/a909651e.

Mora, J., Canals, A., & Hernandis, V. (1996). Behaviour of the thermospray nebulizer as a system for the introduction of organic solutions in flame atomic absorption spectrometry. *Spectrochimica Acta - Part B Atomic Spectroscopy, 51*(12), 1535–1549. https://doi.org/10.1016/0584-8547(96)01518-2.

Nair, L. M., & Werling, J. O. (2009). Aerosol based detectors for the investigation of phospholipid hydrolysis in a pharmaceutical suspension formulation. *Journal of Pharmaceutical and Biomedical Analysis, 49*(1), 95–99. https://doi.org/10.1016/j.jpba.2008.10.027.

Newman, S. P. (1991). Aerosol generators and delivery systems. *Respiratory Care, 36*(9), 939–951.

Park, H. M., Chang, K. H., Moon, S. H., Park, B. J., Yoo, S. K., & Nam, K. C. (2021). In vitro delivery efficiencies of nebulizers for different breathing patterns. *Biomedical Engineering Online, 20*(1). https://doi.org/10.1186/s12938-021-00895-3.

Razpotnik, P., Budič, B., & Veber, M. (2002). Effects of matrix elements on the analyte emission signals in inductively coupled plasma optical emission spectrometry using a thermospray sample introduction system. *Applied Spectroscopy, 56*(8), 1000–1005. https://doi.org/10.1366/000370202760249729.

Rodríguez-Alcalá, L. M., & Fontecha, J. (2010). Major lipid classes separation of buttermilk, and cows, goats and ewes milk by high performance liquid chromatography with an evaporative light scattering detector focused on the phospholipid fraction. *Journal of Chromatography A, 1217*(18), 3063–3066. https://doi.org/10.1016/j.chroma.2010.02.073.

Rombaut, R., Dewettinck, K., & Van Camp, J. (2007). Phospho- and sphingolipid content of selected dairy products as determined by HPLC coupled to an evaporative light scattering detector (HPLC-ELSD). *Journal of Food Composition and Analysis, 20*(3–4), 308–312. https://doi.org/10.1016/j.jfca.2006.01.010.

Saeed, H., Elberry, A. A., Eldin, A. S., Rabea, H., & Abdelrahim, M. E. A. (2017). Effect of nebulizer designs on aerosol delivery during non-invasive mechanical ventilation: A modeling study of in vitro data. *Pulmonary Therapy, 3*(1), 233–241. https://doi.org/10.1007/s41030-017-0033-7.

Scheidl, C. O. R., Menzinger, F., Maier, E. J., Capek, E., Scheidl, O., & Huck, C. W. (2009). Simultaneous quantification of neomycin and bacitracin by LC-ELSD. *Chromatographia, 69*(11–12), 1181–1188. https://doi.org/10.1365/s10337-009-1100-5.

Setchell, K. D., & Vestal, C. H. (1989). Thermospray ionization liquid chromatography-mass spectrometry: A new and highly specific technique for the analysis of bile acids. *Journal of Lipid Research, 30*(9), 1459–1469. https://doi.org/10.1016/s0022-2275(20)38266-3.

Shock, D., Dennis, G. R., Guiochon, G., Dasgupta, P. K., & Shalliker, R. A. (2011). Anomalies in evaporative light scattering detection. *Analytica Chimica Acta, 703*(2), 245–249. https://doi.org/10.1016/j.aca.2011.06.060.

Todoli, J. L., Hernandis, V., Canals, A., & Mermet, J. M. (1999). Comparison of characteristics and limits of detection of pneumatic micronebulizers and a conventional nebulizer operating at low uptake rates in ICP-AES. *Journal of Analytical Atomic Spectrometry, 14*(9), 1289–1295. https://doi.org/10.1039/a900598f.

Todolí, J. L., & Mermet, J. M. (2006). Sample introduction systems for the analysis of liquid microsamples by ICP-AES and ICP-MS. *Spectrochimica Acta - Part B Atomic Spectroscopy, 61*(3), 239–283. https://doi.org/10.1016/j.sab.2005.12.010.

Vestal, M. L., & Fergusson, G. J. (1985). Thermospray liquid chromatograph/mass spectrometer interface with direct electrical heating of the capillary. *Analytical Chemistry, 57*(12), 2373–2378. https://doi.org/10.1021/ac00289a047.

Vollmer, M., & van de Goor, T. (2009). HPLC-Chip/MS technology in proteomic profiling. *Methods in Molecular Biology (Clifton, N.J.), 544*, 3–15. https://doi.org/10.1007/978-1-59745-483-4_1.

Wang, J. F., Yang, H. X., Zhu, L., & Yin, L. H. (2014). Determination of kanamycin sulfate injection by HPLC-NQAD and HPLC-ELSD. *Chinese Journal of Pharmaceutical Analysis, 34*, 644–648.

Webster, G. K., Jensen, J. S., & Diaz, A. R. (2004). An investigation into detector limitations using evaporative light-scattering detectors for pharmaceutical applications. *Journal of Chromatographic Science, 42*(9), 484–490. https://doi.org/10.1093/chromsci/42.9.484.

Wen, X., Zhao, Y., Deng, Q., Guo, J., Zhao, X., & Ji, S. (2012). The coupling of rapidly synergistic cloud point extraction with thermospray flame furnace atomic absorption spectrometry. *Microchimica Acta, 178*(1–2), 139–146. https://doi.org/10.1007/s00604-012-0826-1.

Wilschefski, S. C., & Baxter, M. R. (2019). Inductively coupled plasma mass spectrometry: Introduction to analytical aspects. *Clinical Biochemist Reviews, 40*(3), 115–133. https://doi.org/10.33176/AACB-19-00024.

You, J., & Koropchak, J. A. (2003). Condensation nucleation light scattering detection with ion chromatography for direct determination of glyphosate and its metabolite in water. *Journal of Chromatography A, 989*(2), 231–238. https://doi.org/10.1016/S0021-9673(03)00084-0.

Zhang, X., & Koropchak, J. A. (1999). Direct chromium speciation using thermospray: Preliminary studies with inductively coupled plasma mass spectrometry. *Microchemical Journal, 62*(1), 192–202. https://doi.org/10.1006/mchj.1999.1714.

Zhang, X., & Koropchak, J. A. (2002). Mechanistic studies of chromium speciation with thermospray. *Applied Spectroscopy, 56*(9), 1152–1160. https://doi.org/10.1366/000370202760295386.

Zhang, K., Li, Y., Tsang, M., & Chetwyn, N. P. (2013). Analysis of pharmaceutical impurities using multi-heartcutting 2D LC coupled with UV-charged aerosol MS detection. *Journal of Separation Science, 36*(18), 2986–2992. https://doi.org/10.1002/jssc.201300493.

Index

Note: Page numbers followed by *f* indicate figures and *t* indicate tables.

E

FSC
www.fsc.org
MIX
Papier | Fördert gute Waldnutzung
FSC® C083411

Kolibri 360 GmbH
Ferdinand-Jühlke-Straße 7
99095 Erfurt, Deutschland
produktsicherheit@kolibri360.de